Microcontrollers

Other Books of Interest

ANTOGNETTI • *Semiconductor Device Modeling with SPICE*

CLEMENTS • *68000 Sourcebook*

DEWAR, SMOSNA • *Microprocessors*

FORSYTHE, GOODALL • *Digital Control*

DI GIACOMO • *VLSI Handbook*

DI GIACOMO • *Digital Bus Handbook*

ELLIOTT • *Integrated Circuit Fabrication Technology*

PERRY • *VHDL*

SHERMAN • *CD-ROM Handbook*

SIEGEL • *Interconnection Networks for Large-Scale Parallel Processing*

TABAK • *Advanced Microprocessors*

TRONTELJ, TRONTELJ, SHENTON • *Analog/Digital ASIC Design*

TSUI • *LSI/VLSI Testability Design*

VAN ZANT • *Microchip Fabrication*

VIRK • *Digital Computer Control Systems*

Microcontrollers

Architecture, Implementation, and Programming

Kenneth J. Hintz

Daniel Tabak

*Department of Electrical and
Computer Engineering,
George Mason University,
Fairfax, VA 22030*

McGraw-Hill, Inc.

New York San Francisco Washington, D.C. Auckland Bogotá
Caracas Lisbon London Madrid Mexico City Milan
Montreal New Delhi San Juan Singapore
Sydney Tokyo Toronto

Library of Congress Cataloging-in-Publication Data

Hintz, Kenneth.
  Microcontrollers : architecture, implementation, and programming / Kenneth Hintz,
Daniel Tabak
      p.   cm.
  Includes bibliographical references and index.
  ISBN 0-07-028977-8 :
  1. Programmable controllers.   2. Microprocessors.   I. Tabak,
Daniel, date.   II. Title.
TJ223.P76H55   1992
629.8'9516—dc20   91-37333
                                                     CIP

  2 3 4 5 6 7 8 9 0   DOC/DOC   9 7 6 5 4 3 2

ISBN 0-07-028977-8

*The sponsoring editor for this book was Daniel A. Gonneau, the editing
supervisor was Fred Dahl, and the production supervisor was Suzanne
W. Babeuf. It was set in Century Schoolbook by Inkwell Publishing
Services.*

Printed and bound by R. R. Donnelley & Sons Company.

Contents

Preface   ix
Abbreviations   xi

**Chapter 1. Computer, Microprocessor,
and Microcontroller Architectures**   1

1.1   The Essential Elements of a Computer   4
    1.1.1   The Arithmetic Logic Unit Element   5
    1.1.2   The Input/Output Elements   8
    1.1.3   The Memory Element   13
    1.1.4   The Control Unit Element   16
1.2   Microprocessor: Most of a Computer on a Chip   27
    1.2.1   Microprocessor Buses   27
    1.2.2   Microprocessor ALU, I/O and Control Elements   29
1.3   Single-Chip Microcomputers   29
1.4   Microcontroller: I/O-Oriented Single-Chip Microcomputer   31
    1.4.1   Microcontroller Input/Output   32
    1.4.2   Interrupts   32
    1.4.3   ALU   34
    1.4.4   Timers   34
    1.4.5   Parallel and Serial I/O   35
    1.4.6   External Devices   35
    1.4.7   Configurations of Microprocessors and Microcontrollers   40

**Chapter 2. Computer, Microprocessor, and Microcontroller
Instruction Sets**   43

2.1   Computer Instruction Sets   43
    2.1.1   Desirable Characteristics of Instruction Sets   45
    2.1.2   Instruction Formats   45
    2.1.3   Addressing Modes   48
    2.1.4   SISC, RISC, and CISC   49
2.2   Task-Oriented Instructions   51
    2.2.1   Instructions for Business, Text Processing,
        and Data Manipulation   51

	2.2.2 Science-Oriented Instructions	52
	2.2.3 Control-Oriented Instructions	52
2.3	MCU Instruction Sets	53
	2.3.1 Capability Determined by Word Size	53
	2.3.2 A Comparison of Four MCU Instruction Sets	60
	2.3.3 I/O Instructions	61
	2.3.4 Arithmetic Instructions	65
	2.3.5 Bit Manipulation Instructions	65
	2.3.6 Program Flow Control Instructions	66
2.4	Programmable Controller Instructions	68
	2.4.1 Relay Logic	69
	2.4.2 Arithmetic, Data Manipulation, Data Transfer, and Program Control	71
2.5	Summary	71

Chapter 3. Controller Software Design   77

3.1	Finite State Machine Model	79
	3.1.1 The State Transition Function	80
	3.1.2 The Output Function	81
	3.1.3 FSM Table or FSM Diagram	81
	3.1.4 Control Flow and the FSM	84
	3.1.5 A Limitation of the FSM Model	86
3.2	Petri Nets	92
	3.2.1 Formal Definition of a Petri Net	93
	3.2.2 An FSM Coin Counter in Petri Net Notation	95
	3.2.3 The Coin Counter as a Colored Petri Net	96
	3.2.4 Interrupts, Flags, and Semaphores as Tokens	96
	3.2.5 Petri Tables as a Software Design Tool	98
3.3	Integrated Software Design Model	103
	3.3.1 Table Summarizing Design Steps for Petri Tables	104
	3.3.2 Example Problem: UAV Controller/Autopilot (UAV CAP)	106
	3.3.3 Petri Table for UAV Controller/Autopilot (UAV CAP)	109
	3.3.4 Example Problem: Infrared Sensor Target Tracker	115
3.4.	Summary	121

Chapter 4. Microcontroller Software Implementation   127

4.1	Software Development Process	127
	4.1.1 Software Development	128
	4.1.2 Include and Header Files	132
	4.1.3 HLL/Assembly Language Program Development	136
4.2	Real-Time Programming Requirements	137
	4.2.1 High-Level Languages	140
	4.2.2 Macro Expansion and Functions	143
	4.2.3 Assembly Language Programming	146
	4.2.4 C Language Programming	147

4.2.5   Comparison of C and Assembly Language   153
4.3   Conversion from Petri Table to Software   156
4.3.1   UAV Petri Table Implementation in C   156
4.3.2   Magnetometer Assembly Language Example   158
4.4   Interfacing C and Assembly Language   159
4.4.1   Example Linker Command File   163
4.4.2   Smart Compass Example of C and Assembly Language Linking   165
4.4.3   Interrupt Handling in the IR Tracker   165
4.4.4   Miscellaneous C/Assembly Language Interactions   167
4.5   Summary   169

Chapter 5. 4-bit and 8-bit Microcontrollers   173

5.1   4-bit Microcontrollers   174
5.2   Texas Instruments TMS1000 Family Members   175
5.2.1   TMS1000 Family Members   175
5.2.2   TMS1000 Architecture   176
5.2.3   Design and Application Tools   182
5.3   NEC μPD7500 Family of 4-Bit MCUs   182
5.3.1   μPD7500 Family Members   183
5.3.2   μPD7500 Architecture   184
5.3.3   Design and Application Tools   191
5.3.4   μPD75 Family Members   192
5.3.5   μPD75x Architecture   192
5.4   National Semiconductor COP400   198
5.4.1   National Semiconductor COP400 Family Members   198
5.4.2   National Semiconductor COP400 Architecture   198
5.4.3   Design and Application Tools   208
5.5   Other 4-Bit MCUs   208
5.6   8-Bit Microcontrollers   208
5.7   Motorola M6801 Family   209
5.7.1   Motorola M6801 Family Members   209
5.7.2   M6801 Architecture   210
5.7.3   Design and Application Tools for the M6801   231
5.8   Motorola M6805 Family   231
5.8.1   M6805 Family Members   231
5.8.2   M6805 Architecture   232
5.9   Motorola MC68HC11   235
5.9.1   Motorola MC68HC11 Family Members   237
5.9.2   Architecture   237
5.10   Intel MCS-51 Family   249
5.10.1   Intel MCS-51 Family Members   249
5.10.2   MCS-51 Architecture   251
5.11   Texas Instruments TMS370 Family   257
5.11.1   TMS370 Architecture   257
5.12   Summary   262

Chapter 6. 16-bit Microcontrollers   265

6.1   Intel MCS-96 Microcontroller Family   265
6.2   Motorola MC68332 Microcontroller   288
Appendix A: Design Example   297

Chapter 7. 32-Bit Microcontrollers   303

7.1   Intel 80960CA Superscalar Embedded Processor   303
   7.1.1   Superscalar Organization   305
   7.1.2   80960CA Architecture   313
   7.1.3   The 80960 Family and Applications   336
7.2   LSI Logic LR33000 Embedded Processor   353
7.3   AMD 29050 Embedded Processor   372
7.4   National Semiconductor Embedded Processors   385
7.5   Comparison and Evaluation of 32-bit Microcontrollers   400
Appendix A: 80960CA Pin Description   409
Appendix B: Design Examples   417

Chapter 8. Concluding Comments   453

Appendices   455

Appendix I.   Smart Compass   455
Appendix II.   UAV Code   464
Appendix III. Linker   468
Appendix IV.   IRTrack   470

Index   477

Preface

Microcontrollers are digital computers particularly designed to supervise, manage, monitor, and control various processes in industry, business, defense, aerospace, and other areas of application. With the advent of VLSI technology, microcontrollers are becoming essentially single-chip microcomputers. As the technology advances, microcontroller chips attain higher density, and more and more equivalent transistors can be placed on a single chip—over a million at the beginning of the nineties and over 50 million around 2000, as promised by some chip manufacturers. A microcontroller chip, being a self-sustained microcomputer, contains in addition to a central processing unit (CPU), cache, main memory (more memory is placed on a chip as technology develops), input/output (I/O) interfaces, direct memory access (DMA) controllers, interrupt handlers, timers, and other subsystems necessary to implement an efficient microcontroller.

Microcontrollers, being a particular case of digital computers, have to be programmed so that they can perform their assigned tasks. In the control of relatively simple processes, such as a single traffic intersection, one can probably get by with a relatively short and simple assembly language program. However, many processes are very complicated and require lengthy and sophisticated software. In such cases, assembly language programming is too time-consuming and impractical. Thus, high level language (HLL) programming (such as in C, ADA, or PASCAL) becomes an indispensable necessity. Naturally, one has to develop efficient approaches for generating useful and reliable microcontroller software, be it in assembly, HLL, or a combination of the above.

The goal of this book is to present to the reader a unified document containing a variety of aspects involved in the design and implementation of microcontrollers. The book covers microcontroller architecture, organization, structure, and assembly and HLL (with a stress on the C language) programming. In addition, the text presents a number of examples of actual microcontrollers of different capabilities, currently available on the market. Design examples based on existing commercial microcontrollers are also included.

The book is not intended for beginners in computers. It is assumed that the reader has had a basic course in computer organization and in programming of at least one assembly and one HLL computer programming language. The book can be used as a reference text by practicing engineers, computer scientists, and other professionals involved in the design and implementation of microcontrollers. It can also be used as a textbook in special senior level or first-year graduate courses on microcontrollers. Such a course (ECE 447) is indeed being taught at the authors' school, the George Mason University (GMU). It is a senior course, and a significant part of the material in the text was successfully presented in it. Parts of the book were also taught in other GMU courses, such as a senior course on computer design and a graduate course on advanced microprocessors. Since microcontrollers are applied in practically all industrial areas, the text can be used in any engineering curriculum. It can also be used by computer science students interested in practical applications.

The book is subdivided into two major parts. The first part (Chapters 1 to 4) presents a unified, generic coverage of the principles of microcontrollers, covering their architecture, organization, and software. The second part (Chapters 5 to 7) presents a selection of modern leading microcontrollers, including practical design examples involving some of them. The chapters covering the examples are subdivided according to the basic microcontroller word (or data bus) size: Chapter 5 deals with 4- and 8-bit microcontrollers, Chapter 6 with 16-bit, and Chapter 7 with 32-bit systems. Concluding comments are given in Chapter 8. Although every attempt has been made to include the most current manufacturers data, current detailed design information should be obtained from the individual manufacturers before designing a production microcontroller system.

Some of the examples of actual systems were reviewed by professionals within the particular manufacturing company. The authors would particularly like to thank Mr. Lindsay Wallace of Intel, Mr. Rob Tobias of LSI Logic, Mr. Mike Johnson of AMD (Advanced Micro Devices), Mr. Edward Goldberg of NEC Electronics, Inc., and Mr. Reuven Marko of National Semiconductor for their valuable comments and material made available to the authors. The authors would like to express particular appreciation to Prof. Jack Lipovski (University of Texas, Austin) for reviewing the text for McGraw-Hill and for many helpful suggestions. The authors acknowledge with thanks the editorship of Mr. Daniel Gonneau of McGraw-Hill. This work was partially supported by the Virginia Center for Innovative Technology (CIT) through the Center of Excellence in Command, Control, Communications and Intelligence (C^3I) at George Mason University. Last but not least, the authors would like to express their appreciation to their wives Sue Hintz and Pnina Tabak for their patience and moral support.

Kenneth J. Hintz
Daniel Tabak

Abbreviations

$xxxx:	Hexadecimal number	HSI:	High-speed input
%xxxx:	Binary number	HSIO:	High-speed I/O
A/D:	Analog-to-digital	HSO:	High-speed output
ALU:	Arithmetic logic unit	Hz:	Hertz, cycles/second
AMD:	Advanced Micro Devices	ICE:	In-circuit emulator
BIU:	Bus interface unit	ICU:	Interrupt control unit
CC:	Condition code	IMB:	Intermodule bus
CFG:	Configuration register	I/O:	Input/output
CFM:	Control flow machine	IOP:	I/O processor
CISC:	Complex instruction set computer	IR:	Instruction register
		IRQ:	Interrupt request
CMOS:	Complementary metal oxide silicon	ISR:	Interrupt service routine
		k:	1000
COP:	Computer operating properly, control-oriented processor	K:	1024
		LCD:	Liquid crystal dispaly
CPN:	Colored Petri net	LED:	Light emitting diode
CPU:	Central processing unit	LRC:	Local register cache
CSG:	Control signal generator	LRU:	Least recently used
DFM:	Data flow machine	LSB:	Least significant bit
D/A:	Digital-to-analog	MAR:	Memory address register
DMA:	Direct memory access	MCU:	Microcontroller unit, single-chip microcontroller
DOS:	Disc operating system		
DRAM:	Dynamic RAM	MDR:	Memory data register
DSP:	Digital signal processing	MIPS:	Microprocessor without interlocked pipeline stages
EBCDIC:	Extended Binary Coded Decimal Interchange Code		
		MMU:	Memory management unit
EBI:	External bus interface	MODEM:	Modulator/demodulator
EOC:	End-of-conversion (A/D)	MSB:	Most significant bit
FAM:	Facsimile accelerator module	N	Petri net with no specific initial marking
FAX:	Facsimile		
FIFO:	First-in, first-out	$(N,M_0):$	Petri net with a specific initial marking
FIP:	Fluorescent indicator panel		
FIR:	Finite impulse response (filter)	NC:	Normally closed (switch)
FP:	Frame pointer	NMI:	Nonmaskable interrupt
FPAL:	Floating-point arithmetic library	NO:	Normally open (switch)
		NRZ:	Nonreturn to zero (signaling)
FPU:	Floating-point unit	OEM:	Original equipment manufacturer
FSR:	Floating-point status register		
HCMOS:	High-density CMOS	Op-code:	Operation code
HLL:	High-level language	OS:	Operating system

PC:	Personal computer, program counter (register)	ROM:	Read only memory
		RPM:	Revolutions per minute
PC:	Programmable controller	RTOP:	Real time output ports
PCB:	Process control block	SA:	Successive approximation (A/D)
PFP:	Previous frame pointer		
PGA:	Pin grid array	SCI:	Serial communications interface
PLA:	Programmable logic array		
PLL:	Phase-lock loop	SFR:	Special function register
PMU:	Peripheral management units	S/H:	Sample and hold
PN:	Petri net	SIM:	System integration module
PW:	Pulse width	SIO:	Serial input/output
PWM:	Pulse width modulation	SP:	Stack pointer
PSR:	Processor status register	SPARC:	Scalable processor architecture
PSW:	Program status word		
PTS:	Peripheral transaction server	SPI:	Serial peripheral interface
QSM:	Queued serial module	SRAM:	Static RAM
RALU:	Register ALU	TDM:	Time division multiplexing
RAM:	Random access memory (read/write memory)	TLB:	Translation lookaside buffer
		TPU:	Time processing unit
RIP:	Return instruction pointer	UART:	Universal asynchronous receiver transmitter
RISC:	Reduced instruction set computer		
		WSR:	Window select register

Computer, Microprocessor and Microcontroller Architectures

There is a great distinction between computer architecture and organization; the following discussion focuses primarily on computer architecture. *Computer architecture* is a description (definition) of the attributes of a computing system as seen by a machine language programmer or a compiler writer.[1] Writable control stores for modifying microcode during computer operation are not considered available to the normal machine language programmer. *Computer organization* pertains to the various methods that can be used to implement a specific computer architecture. The demarcation between these two components of microcontroller implementation is becoming less distinct with the advent of reconfigurable designs. This will become more evident as this book progresses from the earliest to the most recent designs.

The categorization of computers was once easy because there was a clear difference in cost, amount of memory, type of peripherals, and physical operating speed. While there are still supercomputers, specialized array and pipelined processors, and expensive multiuser computers, the computational capabilities of the most powerful machines of a few years ago have trickled down to the level of personal computers (PCs). An inspection of computer specifications shows that the capabilities of computers in terms of complex arithmetic operations, memory addressing capabilities (virtual address space), as well as amount of connected memory (physical address

TABLE 1.1   MATLAB Benchmarks of Various Computers Showing the Favorable Performance of Microprocessor-Based Machines Relative to Larger Computers

Computer	Figure of Merit Relative to PC/XT @ 4.7 MHZ	KFlop/Second
MAC (8 MHz 68000)	0.2233	3
PC/AT (6 MHz/80286/EGA)	1.3570	15
PC/XT (4.7 MHz/8088/CGA)	1.0000	17
AT&T 6300 (8 MHz/8086)	1.8760	29
Mac II (68020/68881)	6.2358	85
Apollo DN3000 (16 MHz)	5.4287	72
Sun-3/50 (15 MHz with 68881)	5.1628	89
Apollo DN4000 (25 MHz)	9.0300	140
MicroVAX II (VMS/D_floating)	4.1810	140
Mac IIcx (68030/68882)	9.6776	168
803886/80387 (20 MHz, 386-MATLAB)	14.3466	232
Sun-386i (25 MHz)	10.7024	198
VAXStation 3100 (VMS/D_floating)	18.9047	365
Sun-3/260 (25 MHz with FPA)	19.4267	490
Sun-4/110	35.4397	730
Sun SparcStation	65.5196	1196
Ardent Titan	61.6884	3614

space), have inexorably migrated from the larger to the smaller machines. Even PCs that operate in a single-user mode have comparable computational capabilities to those of much larger machines as is shown by the comparison of MATLAB benchmarks in Table 1.1. The seven standard benchmarks that were run are

1. $N = 50$ real matrix multiply

2. $N = 50$ real matrix inverse

3. $N = 25$ real eigenvalues

4. 4096-point complex FFT

5. LINPACK benchmark

6. 1000 iteration FOR loop

7. $N = 25$ 3-D mesh plot

The geometric mean of the execution times was computed to determine a figure of merit relative to a PC/XT machine. Table 1.1 also lists the KFlop/s throughputs of the various machines. Initially microcontrollers were thought to be simpler implementations of computers with a limited instruction set and enhanced I/O capabilities that could be mass-produced inexpensively. With the advent of increasingly complex microcontrollers, including single-chip microcontrollers with complex computational capabilities and 32-bit architectures, microcontrollers have taken on a new set of problems such as digital signal processing, computer peripheral control, and real-time detection and modulation/demodulation of FAX signals as well as complex non-linear control problems. One can no longer look at just simple parameters such as number of instructions, quantity of virtual and physical memories, and processing speed to separate supercomputers from microcontrollers. The functional use of the computer determines whether it is a microcontroller, computer, or supercomputer. It is not too difficult to conceive of a massively parallel computer that uses all of its computational resources to maintain tracks of aircraft and direct the air traffic control (ATC) in a specific geographic area. There is a lot of computation performed in correlating radar returns, merging and splitting tracks, and predicting trajectories of aircraft for collision avoidance and routing. However, the computations are incidental to the primary function of the computer, which is air traffic control. There is, however, a clear distinction on closer inspection, when one studies the capabilities of an instruction set relative to a particular problem. Continuing with this example, if there are included in the instruction set single instruction operations for performing the discrete Fourier transform (DFT) as well as matrix operations, the computations necessary for target tracking can be performed more efficiently. If there are many I/O instructions that can interact directly with the contents of memory, then the computer can be interfaced more easily with radar inputs and control outputs. If there is a complex set of graphical primitive operators, then the computer can more effectively display the information to the human operators. When it comes to control, the application, more than the descriptive title of the computer or integrated circuit, must be used in selecting the correct hardware. However, microcontrollers will be considered to have all of the elements of a computer on a single integrated circuit (IC) and be able to operate with few or no external components.

Control and *microcontrollers* mean different things to different people. The ATC system is an example of control, but so is the control of motors in industrial settings, the control of microwave ovens, the control of videocassette recorders (VCRs), the control of drug administration to hospital patients, the replacement or bypassing of human nervous systems that have been damaged, the control of bank (automatic) teller machines (ATMs), the control of remotely piloted or autonomous vehicles for operations in hazardous environments and the control of automotive engines for pollution reduction and improved efficiency. It is not possible for one type of controller to meet all of the diverse requirements of these various applications; this prompts the impartial study of the various architectures and organizations of microcontrollers. No single computer or microcontroller is the best for all applications, and only a thorough understanding of the alternative implementations and their impact on design will allow one to select the correct hardware and software configuration for a particular application. This is the source of our egalitarian approach to computers, which implies that there is no hierarchy among computers, only the correct computer as determined by the particular application. There is no hierarchy between software and hardware, only the right application of each capability to solve the task at hand.

1.1 THE ESSENTIAL ELEMENTS OF A COMPUTER

There are five essential elements of a computer: the arithmetic logic unit (ALU), the input section, the output section, the control unit (CU), and the memory. These are shown in the block diagram of Fig. 1.1. The input and output sections are commonly referred to by their collective name of input/output (I/O), and depending on the convenience of the situation, they may be treated collectively or individually. All of these elements are necessary to form a complete computer. If there is no input, the computer cannot respond to its environment. If there is no output, the computer cannot effect changes in its environment. If there is no ALU, the computer cannot perform alterations of its input but only move it from one storage location to another and/or pass it from inputs to outputs. If there is no memory, the system is no more than a finite-state machine. And while most microcontroller applications can be represented by and behave as finite-state machines, computers' memory allows them to realize any finite-state machine as well as compute any computable number. For an expansion of

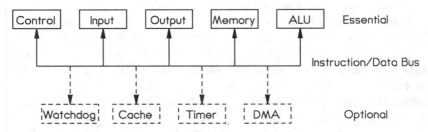

Figure 1.1 Required and typical optional elements of a single-chip computer.

the concept of computable numbers, refer to any of a number of books and articles on the Turing machine and the *Entscheidungsproblem*[2,3] (literally, the decision problem).

1.1.1   The Arithmetic Logic Unit Element

The arithmetic logic unit (ALU) is exactly that: the section of the central processing unit (CPU) that performs both the arithmetic and the Boolean logic operations on the data supplied to it. On some machines there may be separate internal implementations that separate the logic operations from the arithmetic operations and place the results in separate accumulators, but this is transparent to the programmer.[4] The ALU may also be used in modes hidden from the programmer to compute the effective address of the next instruction to be executed or the effective address of an operand in an instruction. But one must first look at the ALU when considering a new application since the presence or absence of particular instructions can make the realization of a particular control system either easy or difficult, as well as affect whether the application is able to perform the necessary control computations in a timely manner. The ALU typically operates on one or two values, called *operands*, and changes them in some way according to the operator that is specified. For example, in the instruction

$$C \Leftarrow A + B$$

A and B are the operands, C is the result, $+$ is the Boolean operation of OR, and $\Leftarrow$ is the replacement operator. This is shown diagrammatically in Fig. 1.2. The ALU can perform a multiplicity of operations, the particular one selected by the control lines. There are a maximum of 2^k distinct operations for k control lines. The ALU has no internal states and is nothing more than a combinational circuit that maps the combination of inputs A and B to output

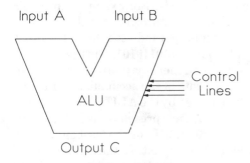

Figure 1.2   Basic ALU showing two input values, control signal lines to select the operation to be performed, and the output.

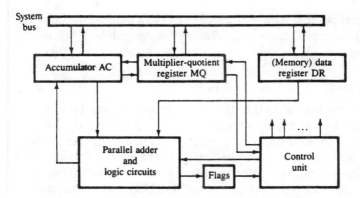

Figure 1.3   Integer-based ALU. *(John P. Hayes, 2nd ed., Fig. 3.59, p. 258, 1988).*

C. The ALU may have either a buffered input or output under the control of separate control signals, to interface it with the internal busses that route data among the internal sections of a computer. A more complete integer based ALU is shown in Fig. 1.3.

Logic operations.   Typical logic operations are the familiar ones of AND, OR, and INVERT since INVERT and either of the other two form a complete set of operations. These operations are typically bit-wise in that the lowest order bit of one operand is combined with the lowest order bit of the other operand and the result placed in the lowest order bit of the result. This procedure is replicated for each subsequent bit position. Of course, in actual operation all bits are operated on in parallel. For example, if the ALU is designed to work with 8-bit inputs (1 byte) to produce a 1-byte-wide output, and one operand represented in binary is 01001010 and the other operand 10000101, then the bit-wise Boolean OR of these is 11001111.

Not all ALU operations require two operands. For example, the *one's complement* is a typical operation that consists of taking the bit-wise complement of each individual bit of a single operand such that if the input were 01001010, its one's complement would be 10110101. Furthermore, if the computer is an accumulator-based design, there may be no operand at all but rather an implied operand of the contents of the accumulator. Logical rotate and shift operations are also performed by the ALU and allow one to move all of the bits of a single operand from their present location either to the left or right, with the most significant bit (MSB) or the least significant bit (LSB) moved or not moved through the carry bit. In the simplest case, this must be done one bit position at a time through the execution of a single instruction, or if a barrelshifter is implemented in the ALU, multiple bit positions at a time. A typical set of logic operations is exemplified by

TABLE 1.2 Logic Operations Available in the MC68HC11

EORA, EORB, ORAA, ORAB, ANDA, ANDB, BITA, BITB, COMA, COMB, COM, SEC, SEI, SEV, CLC, CLI, CLV, BSET, BCLR

those of the Motorola MC68HC11, whose mnemonics are shown in Table 1.2. It can be seen from this table that the 6811 includes the XOR, bit test, and bit manipulation instructions.

Arithmetic operations. Arithmetic operations take the same form as logic operations having none, one, or two operands and a result. Single-operand arithmetic operations can be as simple as the incrementing of a single operand or the slightly more complex operations of arithmetic shifts to the left and/or right a specified number of bits. The arithmetic shifts differ from

TABLE 1.3 Arithmetic Instructions of Intel 8096 and Motorola MC68HC11

	Intel 8096	Motorola 68HC11
Add (both signed and unsigned)		
Bytes	ADDB*	ADD<1>†
Accumulators A and B; result in A		ABA
Bytes plus carry	ADDCB	ADC<1>
Words	ADD*	ADDD
Words plus carry	ADDC	
Correction for BCD addition in accumulator A		DAA
Add B into X or Y (byte into word, unsigned)		ABX
		ABY
Subtract (both signed and unsigned)		
Bytes	SUBB*	SUB<1>
Accumulator B from A; result in A		SBA
Bytes minus borrow	SUBCB	SBC<1>
Words	SUB*	SUBD
Words minus borrow	SUBC	
Multiply		
Bytes, unsigned	MULUB*	MUL
Bytes, signed	MULB*	
Words, unsigned	MULU*	
Words, signed	MUL*	
Divide		
Word/byte, unsigned	DIVUB	
Word/byte, signed	DIVB	
Double-word/word, unsigned	DIVU	
Double-word/word, signed	DIV	
N-byte/word		IDIV, FDIV

SOURCE: Peatman, 1988.
*Includes both two-operand and three-operand versions.
†<1> represents accumulator A or B.

the logical shifts in that arithmetic shifts to the right maintain the sign bit and do not automatically shift a zero into the MSB position.

In addition to binary arithmetic operations, some computers that are intended to interface with the external world directly through the use of decimal numbers, and yet perform arithmetic, include decimal accumulator adjust (DAA) instructions which allow for the addition of binary coded decimal (BCD) numbers without first converting them into pure binary. If this instruction is available, it allows two BCD numbers to be added using the normal binary arithmetic, with the result then corrected to a BCD result using the DAA instruction. Typical arithmetic operations of the Intel 8096 and MC68HC11 are shown in Table 1.3.

More complex arithmetic operations are becoming increasingly typical. For example, in addition to the readily available integer multiply and divide operations, the National Semiconductor NS32FX16 32-bit FAX processor contains an internal floating point unit and allows for floating point vector manipulations at the assembly language instruction level.

1.1.2   The Input/Output Elements

Input/output is a topic of particular interest when microcontrollers are considered since their *raison d'etre* is the sensing and measurement of some aspect of the environment and the generation of one or more control signals that affect the environment in a desirable manner. These inputs may already be in the form of binary-valued signals that range from the simple, single-bit, binary-valued switch (which is either open or closed), to the output from a 21-bit digital shaft encoder that assigns to each angular position a unique binary-valued code. Other binary-valued input signals are those directly received from digital communications devices or from sensing devices that have already converted the analog (continuous) incoming signal to an equivalent binary value. These binary-valued signals can be further subdivided into those that are available to the microcontroller in parallel and those that arrive sequentially and are not all available at the same time. Of the serial signals, there are two methods for transmitting the data: synchronous and asynchronous. In addition to binary-valued signals, some analog signals may need to be input directly to the microcontroller without requiring high precision (in number of bits of quantization) or high speed (in A/D conversion time). To effect this, most current microcontrollers have on-chip A/D converters.

Another typical architectural consideration is whether I/O is considered part of memory or whether it has dedicated I/O addresses independent of the memory addresses. These two philosophies are best exemplified by Motorola's memory-mapped I/O, in which I/O devices occupy some of the memory addresses with the concomitant reduction in available memory, and by Intel's separate signal line that indicates whether the addresses and data are intended to be considered as memory references or I/O. The advantage of memory mapped I/O is that I/O ports can be directly operated on with many arithmetic and logic operations. The advantage of separate I/O is that the address decoding is simpler and there is no loss of available memory space due to its use by I/O devices.

Serial input/output.   Serial data transfer is by far the most common method of transferring binary-valued data over long distances because of the minimum number of wires required to transfer the data. The serial transmission of data can be at baseband, where the logic ones and zeros are represented as distinct voltage, or current levels as in the direct binary connection of microcontrollers to serial printers. Another familiar form of the serial transmission of signals over a telephone line from one computer to another uses audio frequency shift keying (AFSK) implemented in a modulator-demodulator (modem). For transmission, a modem converts each bit into one of two audio frequencies that can be transmitted within the audio passband of the telephone system. For reception of signals, two distinctly different audio tones are used so full duplex communications can be maintained between the transmitting and receiving sites. More recently, this technique of modulating digital signals as audio signals has been expanded to include facsimile (FAX) transmission of images (including printed pages) from one location to the other over telephone lines.

In serial transmission, one must know where information begins and ends or the data may be garbled. There are three methods for effecting this synchronization, two of which are commonly used in microcomputer communications and have widely varying data rates. The first (and slowest) is asynchronous transmission, in which the start of an individual byte of data is indicated by a transition from a marking to a spacing condition for the duration of one bit length. A typical asynchronous serial coding of the two letters GM with an arbitrary intercharacter time is shown in Fig. 1.4. The signaling rate is called the baud rate and has units of bits/second. The duration of one bit is 1/(baud rate). *Space* refers to the signal level (more positive than +3 volts in the RS-232 standard[5] that is transmitted when no data is sent and is interpreted as a binary zero. *Mark* refers to the signal level (more negative than −3 volts in the RS-232 standard) that represents a binary 1. The complete sequence that defines a single character or byte of data consists of one start bit, eight or nine data bits, and one stop bit resulting in a 10 or 11 unit code. There are two ways in which asynchronous serial communications are handled in microcontrollers. The least expensive method, but also the most software intensive, is for the program to generate the serial bitstream itself as well as measure the duration of each bit. One serial bit at a time is transferred to one bit of an output port. The program counts the duration of 1/baud, and then the next bit is transferred to the

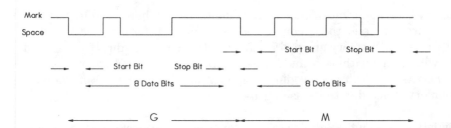

Figure 1.4 Asynchronous bit string of even parity ASCII characters GM showing start bit, data bits, and one stop bit.

output port. The sequence continues until all of the character(s) is sent. The advantage of this method is the minimum hardware required, and this may be a suitable alternative for large volume, cost sensitive applications that do not have stringent time constraints or large computational loads.

The second method for asynchronous data transmission is exemplified by the MC68HC11, in which there is an internal universal asynchronous receiver-transmitter (UART) that can have a complete data byte written to it by the program and then operate independently for the serial transmission of the data. This allows the program to continue with other operations without concern for the time consuming serial transmission of data. When the UART is finished transmitting a data byte, there is a variety of options for regaining the attention of the program, such as hardware interrupts.

The second and third serial transmission methods are the synchronous and the self-clocking methods. With synchronous methods, it is easy to achieve data rates of 1 million bits/s (Mbits/s) or more. Synchronous methods are typically used for local communications among peripherals and/or master and slave processors. With the introduction of wider band-width integrated services digital networks (ISDN), these methods may become more popular for distributed microcontroller communications. While synchronous signaling is faster than asynchronous signaling, in the first method discussed it requires an additional signal line, the clock. The master (transmitting device) sends two parallel signals to the slave (receiver). One of the signal lines contains the data itself and the other a clock signal that is used by the slave to synchronize its reception of the data. The protocol between the master and slave also includes a short period of synchronization at the initialization of the transfer of data, but unlike the asynchronous transmission, this occurs only once at the beginning of the transmission of a block of data rather than at the start of each character. So not only is the baud rate faster due to the wider bandwidth of the interconnection and the common clock between the master and the slave, but there is also less wasted channel capacity since there are many fewer start and stop bits. A sample synchronous character transmission for the MC68HC11 is shown in

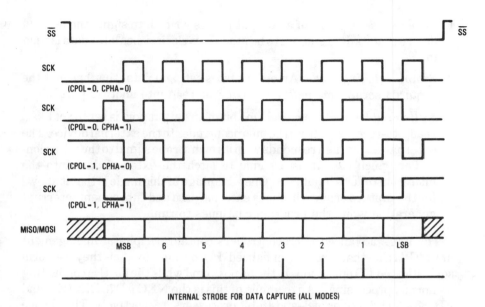

Figure 1.5 Synchronous data transmission showing clock signal and interconnecting data signals. *(Courtesy Motorola Corporation, Phoenix, Arizona)*

Fig. 1.5. The abbreviation *SCK* refers to the synchronous clock, *MOSI* refers to master-out-slave-in, *MISO* refers to master-in-slave-out, and *SS* is slave-select. The additional lines are required if there are multiple slaves and masters.

A second method for the serial synchronous transfer of data is self-clocking codes. Since these are not currently used in microcomputers, their details are not discussed here other than to say that they only need one signal line since the data is encoded in such a way that the received data can be manipulated to extract a clock. A local oscillator is slaved to this data-derived clock and used for demodulation of the signal. This method usually requires an initialization transmission to synchronize the clocks.

Parallel input/output. Parallel I/O pertains to the transfer of multiple simultaneous bits of information between the microcontroller and the environment. The number of bits that can be transferred is microcontroller dependent, as some have 8-bit parallel I/O ports and some have ports with a larger number of bits. The parallel output ports are usually little more than latches that are written to by the program and, because of their latching design, maintain the data available to the environment until the next time the port is written to. Input ports, on the other hand, particularly in microcontrollers, have a variety of methods for latching the data and making it available to the program. For example, multiple modes are present in the MC68HC11 and they are as follows:

1. *Strobe mode:* One bit of an output port is written to signal the desired device that its data is about to be read. Two clock cycles later data is read in.

2. *Simple latching mode:* An output bit is set, and data, that has met the required setup time before the edge, is read into another port.

3. *Full-input handshake mode:* One bit of an output port is set to act like a ready-to-receive bit from the microcontroller to the external device. The external device then responds with its own strobe signal to the microcontroller which uses it as a signal to latch the external data into the microcontroller's input port. One advantage of this mode is that it allows for the capture of asynchronous external data and subsequent interrupt generation while the program continues to run.

The more advanced microcontrollers contain peripheral management units (PMUs) that provide enhanced I/O options in that they perform sophisticated I/O operations in the background while the CPU is conducting its normal processing.[6] An example of this is the NEC μD782XX (K2) and μD783XX (K3) series of 8- and 16-bit processors, respectively. This device is capable of performing the following nine functions in the background: event counting, data comparison, bit shifting, bit logic, A/D converter buffering, block transfer, data differencing, data difference pointer, and data addition. The availability of these real-time output ports (RTOP) allows the improved control of devices that have time-critical requirements, such as the control of AC power sources or the generation of stepper motor control signals with separate characteristics for startup, run, and stop.

The hardware configuration is also flexible in microcontrollers since the use of particular I/O ports is determined by the contents of control registers that usually default to inputs but can be easily configured on a port or bit basis to be inputs or outputs. This dynamic reconfigurability is one of the strengths of microcontrollers since a single device can become well known to the designer and its configuration modified to suit the needs of a particular system design without the need to retrain on a different device. The only apparent limitation of this device is that there must be enough inputs and/or outputs to suit the application. Because the total number of I/O pins is limited on microcontrollers (which are typically implemented on a single integrated circuit, or IC), there is usually a provision for using some or all of the I/O pins as external data and address bus drivers. This flexibility allows for both easy expansion in the case of a particular need and the development of the design (with the addition of some temporary ICs) in an expanded mode that is easier to troubleshoot.

The I/O flexibility available through the simultaneous use of I/O pins both as input and output is also of interest. This is accomplished through the use of "weak" pull-ups that bias the output toward a 1 on output, with an active sink used to make the output a zero, while at the same time allow for reading the port as an input. Of course, the bit is read as a 1 if nothing is attached to it since it is passively pulled up. However, if an external device pulls down its device to a low level, that is also sensed and the data read in.

Analog input/output.   While not normally considered a defining part of a computer, analog I/O and analog-to-digital conversion (A/D) is considered here since it is such an integral part of single-chip microcontrollers and contributes to their power and widespread application. The analog I/O found most often in microcontrollers consists only of A/D input consisting of one or more channels. The A/D conversions are typically slow as measured by the fastest available external A/Ds and are implemented as ratiometric successive-approximation A/D converters. Typical conversion times are 16 μs (8 bits, MC68HC11 at 2 Mhz[7]), 22 μs (10 bits, Intel 8095BH at 12 Mhz[8]), and as fast as 13.8 μs (10-bit, Hitachi H-8/500 at 10 Mhz[9]) and accuracies of $\pm 1/2$ LSB. Multiple channels are handled by having one onchip sample-and-hold (S/H) circuit combined with a multichannel analog multiplexer. Some other models contain no onchip S/H, and the analog inputs must be held stable for the duration of the conversion. There is also a difference among manufacturers as to whether the reference voltage is generated onchip or whether it must be supplied externally.

Along with the convenience of having on-chip A/D conversion, there are a variety of operating modes for these converters including repetitive single-input measurements or cyclic measurements of multiple inputs. The multiple inputs can all be made and stored in registers followed by an interrupt generation, or they can continually cycle among the multiple inputs (MC68HC11). Other alternatives are to program an internal timer to initiate the conversion at some specified time with no CPU intervention (8095BH).

Usually digital-to-analog converters (D/A) are not found as part of single-chip microcontroller units (MCUs) since they are relatively easy to implement externally.[10] For example, a simple D/A converter can be made by using one of the programmable timers to act as a pulse-width modulator (PWM) and then integrating the result. This can be done with a simple resistor-capacitor (RC) circuit, or a more sophisticated op-amp circuit if more power-driving capability is required.

1.1.3   The Memory Element

The distinction between finite-state machines and computers is the presence of an (assumed) infinite quantity of memory. In reality, this is not realized, but if there is sufficient memory for the program to run, then practically it meets this requirement. Of course, even if there is not sufficient local memory, programs can be designed to swap the contents of memory to slower mass storage (magnetic or read/write optical disks) to produce a much larger virtual memory. The discussion of memory can be divided several ways. In terms of its ability to store data, there are two types, read/write memory (anachronistically called RAM for random-access-memory), and read-only memory (ROM). It is easiest to dispense with ROM first since it has few variations. In its most basic form, ROM is mask or electrically (EPROM) programmed to have a particular arrangement of bits that is either executable code—such as the bootstrap program, which is the first program executed by the computer—or some form of lookup table—such as the quick brown fox (QBF) ROMs, which contain a sentence comprised of all the letters of the alphabet and are usually used for communications channel testing. The simple ROM is nonvolatile in that it retains its contents even when power is removed from the system. It is usually not written to by the computer. Another variation on the ROM theme (EPROMS) allows for local programming by the engineer using special programming devices with the subsequent ability to erase the complete contents of the EPROM by exposure to intense ultraviolet light. This is particularly good during the development phase or when a small product run is anticipated that does not justify the expense of mask programming. Although it would seem that these EPROM devices would be the first of a new series that would hit the market, this is usually not the case for several reasons.[11] Typically, the required on-chip real estate for EPROM memory cells is larger than an equivalent number of masked memory locations. If imperfections resulting in faulty chips are assumed to be uniformly distributed across the wafer, this increases the probability of a faulty chip, reduces yield, and raises costs. The production costs associated with EPROM versions are also higher since the demand is low (because of their already higher cost and their additional testing costs to write/read/erase the EPROMs to verify their performance). There is yet an additional expense associated with the requirement for a quartz lid on the chip that must be transparent to the ultraviolet (UV) erasing light.

A second variation of alterable ROMs is EEPROM, which allows for the electrical modification of the ROMs while the computer is operating but without the volatile nature of RAMs. When the power is turned off, the contents of the EEPROM are maintained and are available for ready use the next time the computer is turned on. This type of alterable, nonvolatile memory is used for the storage of configuration or parameter information, which is necessary to reinitialize the computer when it is restarted. EEPROMS can also be partially or completely erased while the computer is operating; however, the erasure time is longer than for conventional RAMs and hence they are rarely modified. The emerging ROM technology is flash EEPROMs, which have a higher storage density than conventional EEPROMS.[10] The disadvantage of these newest EEPROMS is that they must be erased all at once, and the ability to erase by blocks has been lost.

RAMs also come in two flavors: static and dynamic. Static RAMs do not require continual clocking and hence tend to have shorter access times. They also tend to consume more power and require more chip space and are therefore used only when high speed is required. Dynamic RAMs, on the other hand, are made of technology that stores the data as charges on transistor gates. Since the charges slowly leak off, they must be continually refreshed to maintain data integrity. Power consumption tends to be lower and speed proportionately slower. The fabrication costs are lower and the storage density is much higher; therefore they tend to be the choice for large main storage or buffer memories that do not have stringent speed requirements.

Main memory. The maximum size of physical memory is theoretically limited only by the size of the virtual address space and the number of address lines that can leave the chip, but is practically limited for economic reasons. In some cases, the size of the physical memory can never equal that of the virtual address space because too few address lines are made available. Main memory may be comprised of a mixture of any or all of the types of memory mentioned previously, but typically it is mostly dynamic RAM with a small amount of ROM that contains the bootstrap loader to start the computer or contains the complete program to be executed. In unattended stand-alone systems, usually the complete program is in ROM so it can easily recover in case of power failure. Some systems, particularly those used to control communications networks, need only a small bootstrap program in ROM since they can be readily downloaded with the complete operating program in the event of a power failure.

Up to the limit of the physical address space of the machine, the main memory can take on any size required by the application. In some cases in which the complete operating program is contained in ROM, the RAM may be small and used only for temporary storage of a few variables. In the case of a communications network node, the amount of RAM may be large to account for all of the buffers necessary to store incoming data streams before they are passed on to another channel or converted to a different communications protocol. In a few cases where program execution speed must be maximized, the entire program may be stored in ROM and then transferred to RAM on startup since RAM typically has a shorter access time than ROM. A typical example of this is the common practice of moving the basic input/output system (BIOS) in a PC from ROM to RAM, in essence converting it from firmware to slushware. This not only allows for faster execution, but also allows for configuration modifications to be embedded in the BIOS without having to jump to a nonvolatile EAROM to get the information that is needed on a regular basis.

Cache memory. One technique for speeding up program execution that is becoming more common even in small computers is cache memory. Since fast dynamic RAM is expensive, an alternative to completely populating the physical address space of the computer with fast dynamic RAM is to use a high-speed, local memory called a *cache*. This memory is either closely connected to the CPU or onchip, since it does not have to be large to increase the execution speed of a program. Cache takes advantage of the locality-of-memory effect that results from instructions being executed from sequential locations in memory and data tending to be located in isolated regions of memory. Cache memory is not just another fast memory but requires a cache controller since the cache's contents duplicate the contents of sections of main memory. Two methods are available for the implementation of a cache controller, the difference being important only if multiple processors or a direct-memory-access (DMA) device are accessing the same main memory. The first method is *read-back* and does not continually update main memory with the contents of cache until some other device requests a memory access from that same location or the cache is full and must restore some of its contents to main memory to make room for new data. The second method is *read-through* and is implemented by the cache automatically updating the main memory on every access by the central processing unit (CPU), although not at the same rate. This method of caching in effect provides a high-speed buffer between the CPU and main memory. The operation of a cache and cache controller is transparent to the programmer. In the case of multiple devices that bypass the cache memory there are two effects that must be handled effectively by the cache controller. The first has already been alluded to in that if a device such as a DMA controller wants to transfer data from main memory, it must first assert a signal line to the cache controller to indicate this fact and wait for a response, since the cache controller may need to transfer the contents of cache back to main memory before the DMA transfer takes place to ensure valid data. This is known as the *stale data problem*. The other difficulty arises when the DMA transfers data to a section of main memory and there is already a copy of that memory in cache. In this case, the cache controller must set a "dirty bit," which indicates that the data in at least a section of cache is corrupted and a read to main memory must be executed if the CPU tries again to access that data.[12,13]

1.1.4   The Control Unit Element

To understand the workings of a control unit, it is helpful to consider a simple bus-oriented central processor. The simple machine is shown in Fig. 1.6 and consists of a single databus for transferring data among the various elements that have been previously described, a program counter (PC) register for storing the address of the next instruction, an ALU for performing both data and effective address computations, an instruction register (IR)

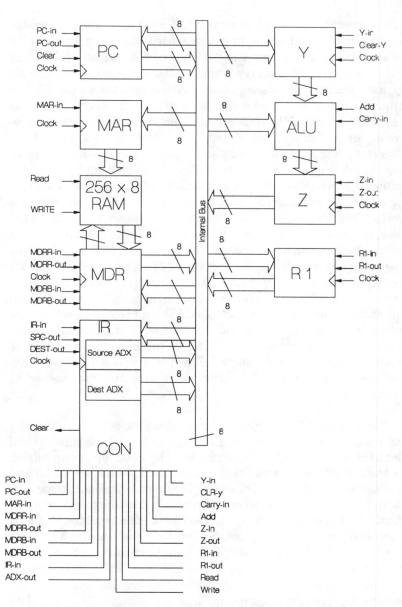

Figure 1.6   Block diagram of internal organization of simplified computer.

to hold the current instruction being executed, a memory address register (MAR) for storing the address of the main memory location currently being accessed for either reading or writing, a memory data register (MDR) for storing the data that has been read from or is being written to the main memory, and a general-purpose register called the accumulator (ACC). The control unit is connected to the IR and can be of different organizations, as discussed in the following. The bus-oriented central processor is not representative of any particular machine but is used to describe in a simplified way how a computer works internally and to provide a plausible argument that even more complex control schemes can be implemented.

The internal data bus is nothing more than a set of parallel lines that allow any single internal register to put data on the bus and any or all other registers to take it off. The contents of the PC can be transferred to the ALU on the bus, incremented, and then returned to the PC to point to the next instruction. The function of the control unit is to effect this as well as any other operation available in the instruction set. This represents organization as opposed to architecture, since this architecture could remain the same (in terms of available registers and instructions) and be implemented with two or more busses and/or additional hidden registers that are not directly usable by the programmer but can be used by the controller in the execution of each macroinstruction. The term *macroinstruction* is used to distinguish a single assembly-language-level instruction from *microinstructions* that are used internally by a microcoded controller to execute the macroinstruction.

Hardwired controller.   The hardwired controller is a good example of a finite-state machine used to effect the operation of the computer and implement the execution of its instructions. The need for synchronizing clocks is basic to this operation; the simplest kind is shown in Fig. 1.7 and consists of a seven-phase clock. The seven phases comprise one macrocycle, and it may take one or more complete macrocycles to execute a single instruction. More complex clocking schemes can be implemented that may include termination of a macrocycle before it is complete and/or a variable number of clock cycles. A computer is initialized such that the program counter is set to a preset value, typically 0000H or FFFEH. Assume here that it is initialized to 0000H and that it is desired to fetch the first instruction for execution. During each clock phase, the bus is utilized to transfer data from one register to another as shown in Fig. 1.7. This is done in the hardwired controller through a combinational circuit that generates control signals dependent on the particular phase of the clock, the macrocycle, and the contents of the instruction register.[13]

Returning to the fetching of the instruction, it can be seen that when the first phase of the clocks is active, the PC puts its contents onto the data bus, the MAR receives the address, the Y register is cleared, the ALU is

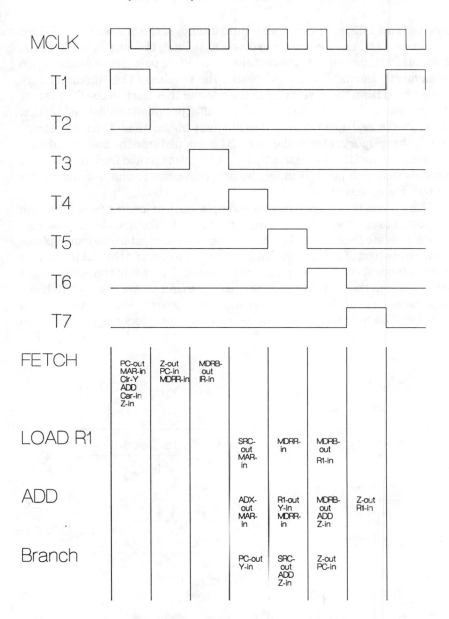

Figure 1.7 Seven-phase clock used in the simple computer example.

directed to ADD (with carry set), the contents of the bus with the cleared Y-register, and the Z-register is directed to latch the output of the ALU. This, in effect, increments the contents of the PC to point to the next location in memory. During the second phase, the results of the summation are transferred from the Z-register to the bus and then back to the PC to store the address of the next instruction to execute (provided the instruction that is being executed does not alter it as in a jump instruction). During the third phase, the memory returns the data that is enabled onto the bus. This data is latched into the IR. The instruction has now been fetched from the memory location that was pointed to by the program counter and is available for decoding and execution.

If the instruction to be executed is simple, such as loading register 1 with the contents of a memory location, then the controller decodes the instruction to activate the appropriate control signals such that in the fourth phase the address part of the instruction is put onto the bus, and the MAR receives it and latches it on the falling edge of the clock. The rest of the clock phases actually fetch the data from memory and transfer it into register R-1. An implementation of this hardwired controller is shown diagrammatically in Fig. 1.8. The first stage of the decoding of the instruction is just a 4-to-16

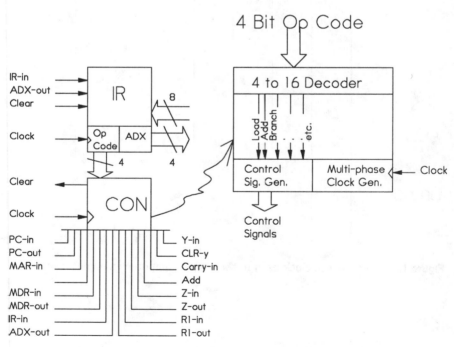

Figure 1.8 Block diagram of hardwired control unit for simple computer.

decoder that converts the four encoded bits of the operation code (opcode) of the instruction into single control lines. The combinational circuit combine these control lines logically with the phases of the clock to generate the register and bus control signals. The detailed implementation of the control signal generator to the gate level for several instructions is shown in Fig. 1.9. It can be seen that several of the control signals, such as PC-out, take place on each occurrence of the first phase since this implementation assumes a single macrocycle in which to execute all instructions. Were the instruction set more complex, requiring multiple macrocycles to execute some instructions, additional logic would be required to inhibit this PC incrementing and instruction fetching except during the first macrocycle of a multi-macrocycle instruction.

Hardwired control units were the preferred method because of their high speed of operation. The difficulty was in the development and troubleshooting of these units since if there were an error in design, it might not be detected until the masks for the IC were completed and the first ICs made. In this event, the complete design cycle had to be performed again, including laborious determination of whether the new circuit would meet all of the timing constraints. Clearly, another method was needed that would improve the efficiency of the design and development phase while at the same time not be detrimental to the execution speed of instructions. That alternative method is the use of a microcoded control signal generator.

Microcoded control signal generator unit. A microcoded control unit produces exactly the same signals as the hardwired unit, however, it generates them in an entirely different way. This is another example of an organizational difference that has no effect on the architecture of the computer. For this reason, the block diagram of Fig. 1.6 need not be repeated, and only a new design for a control signal generator (CSG) must be considered. A simplified, microcoded CSG is shown in Fig. 1.10. The design of this and the other microcoded controllers presented here is in the fashion of White.[14] In this first implementation, the combinational circuit that generates the control signals has been replaced with a read-only memory containing the sequence of microcode that will cause the identical sequential generation of control signals. Two methods are available for this encoding of control signals within the ROM,[15] which is also referred to as the control store. The first method, and perhaps the easiest to understand because of its one-to-one mapping of the control signals to data bits, is *horizontal microcode*. In horizontal microcode, an example of which is shown in Fig. 1.11, each data bit of the control store is itself a control signal. The value of each control signal at a given phase of the clock is determined by the contents of a memory location in the control store. The operation of the control signal generator of Fig. 1.10 starts with the opcode of the instruction being loaded in the instruction register as a result of the fetch part of the control cycle. The

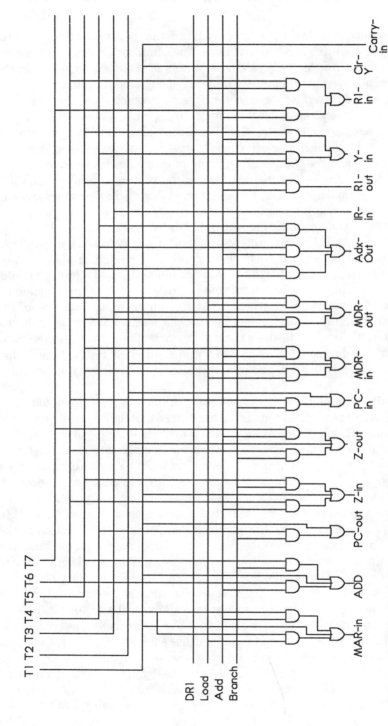

Figure 1.9 Gate-level diagram of hardwired controller implementing several instructions.

22

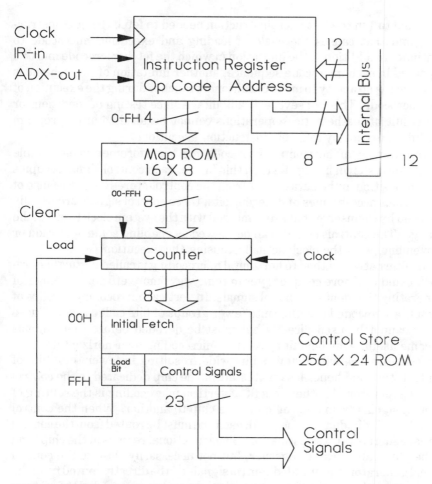

Figure 1.10 Microcoded control signal generator with horizontal microcode and address mapping ROM.

opcode is then used as an address to access the mapping ROM. Although this slows down the CSG, it allows for the unrestricted location of starting addresses of microcode. In this manner, the complete mapping ROM can be used since there are no unused portions between sequences of microinstructions, as might happen if a paging scheme for addressing of microcode were implemented. Since there is no ability for this CSG to branch or jump to other segments of microcode, several microcode instructions must be duplicated in each sequence since the last thing that an instruction must do is fetch the next instruction. Because the most significant bit (MSB) of the microcode is interpreted as a load bit, the number of microinstructions that need to be executed can be anything less than the maximum. The load bit is set equal to 1 in the last microinstruction needed to fetch the next instruction, and that causes the cycle of loading and executing microcode to continue. In addition to the limitation in which the fetch microcode must be repeated in each microcode sequence, another limitation of this scheme is that there is no ability to make conditional branches during the execution of the microcode. This is a severe disadvantage since testing of conditions or the result of mathematical operations cannot be used to alter program execution except by multiple instruction executions.

An example of horizontal microcode for the aforementioned simple computer is shown in Fig. 1.11. In this figure, each control signal occupies one bit position in the data output from the control-store ROM. Because of the various access times of different bits, these control signals are usually not used by themselves but are combined with the system clock to effect the change. The controls signals can be viewed as enabling some operation or movement, with the clock actually causing the execution.

An alternative method to horizontal microcoding is called *vertical microcoding* and is a more complex, yet in some ways more efficient, manner of generating the identical control signals. In vertical microcoding, groups of signals are encoded together into fewer groups of bits called *fields*. These fields cannot be used directly but must be decoded before their signals become useful. One benefit of vertical microcoding over horizontal is that the control store will be fewer bits wide, resulting in fewer total bits of control store and hence less real estate on the chip dedicated to the control signal generator. Another benefit of vertical microcoding is the routing of control signals from one region of the chip to another. When the control signals are individually generated, each one must be routed from the control signal generator to its destination. Since functional regions of the chip tend to be physically located together, but not necessarily close to the control signal generator, the encoded control signals can be directly routed from the vertical coded control store to the region where they are needed before they are decoded into the individual control bits.

The particular implementation of vertical microcoding in the CSG of Fig. 1.12 is not complex in that it is limited to the encoding of a multiplexer address for conditional jump selection, yet it does show how this can be done in principle. The additional feature of this implementation is a control bit (ADX SEL) that selects the source of the address for the microcode that is to be executed next. This allows for conditional jumps within the microcode and allows a single sequence of microcode to be used to fetch the next instruction since each sequence can end with a jump to the fetch sequence. The COND bits control a multiplexer that results in one of four conditions to the counter that stores the address of the next microinstruction to be executed. Conditions 0 and 3 are never branch or always branch, respectively. This allows for the normal sequential execution of instructions or the unconditional jump within microcode. The other two inputs, 1 and 2, are

Figure 1.11 An example of the horizontal microcode implementation for several simple instructions in the microcoded control signal generator of Fig. 1.10.

Clock Phase	Control Store Address	Cond Bit 1	Cond Bit 0	Adx Sel	Br Adx 7	Br Adx 6	Br Adx 5	Br Adx 4	Br Adx 3	Br Adx 2	Br Adx 1	Br Adx 0	PC-in	PC-out	MAR-in	IR-in	ADX-out	Y-in	Clear-Y	Carry-in	Add	Z-in	Z-out	SP-in	SP-out	Read	Write	MDRR-in	MDRR-out	MDRB-in	MDRB-out	SUB	R1-in	R1-Out	Macro Instr.
T1	00H													1	1				1	1	1	1													Fetch
T2	01												1										1			1		1							
T3	02	1	1	1												1															1				
T1	10H															1										1	1									RTS
T2	11																			1	1		1						1							
T3	12	1	1	0	0	0	0	0	0	0	0	0	1										1	1							1	1			
T1	30H																		1	1	1	1													JSR #
T2	31														1								1	1	1										
T3	32													1									1							1					
T4	33	1	1	0	0	0	0	0	0	0	0	0	1				1										1		1						

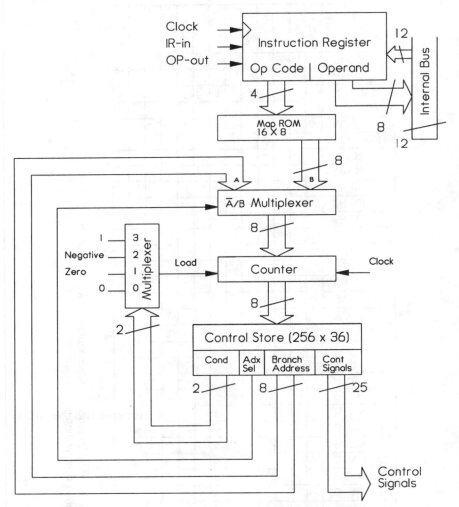

Figure 1.12 Microcoded CSG with vertical microcode and conditional jump capability.

connected to the status bits that are generated by the ALU, indicating that the result of the last operation was a zero ($Z = 1$) or a negative number ($N = 1$). The N-bit is nothing more than the MSB of the result of the operation. Although the width of the microinstruction has been narrowed through the use of the condition code for selecting the channel of the multiplexer, the overall width has been increased to allow for the incorporation of the branch address within the microcode itself.

Even faster execution of microcode can be realized through the pipelining of instructions. This is possible because not all data paths are used simultaneously and the next instruction can be fetched while the current one is being executed. This is subject to some limitations, such as those caused by the execution of a conditional jump in which the next sequential instruction is not the one to be executed but rather the one pointed to by the jump. In this case, the pipeline must be "flushed" and several clock cycles lost. A detailed discussion of pipelining is outside the scope of this book, but some details of its use in specific microcontrollers are found in later chapters.

1.2 MICROPROCESSOR: MOST OF A COMPUTER ON A CHIP

The first commercial microprocessor was the Intel 4004, introduced in 1971.[16] Since then a number of significant improvements have been made by various manufacturers, expanding the microprocessors from 4 bits to 8, 16, and now 32 bits. Additionally, many of the capabilities of full scale computers have been incorporated in these chips, including supervisory modes, floating point computations, and memory management units. Initially, microprocessors evolved as a result of attempts to incorporate more of the elements of a computer on a single chip and even originally existed as chip sets that were designed to interface and work together. There are several motivations behind this move including decreasing cost and reducing interconnections among ICs, the latter being a significant source of system failures. Now most microprocessors are single-chip implementations of all of the elements except memory. Once the architecture is established around the register and instruction format, the only variables remaining are the quantity of connected physical memory and the number and type of I/O. Because the applications of microprocessors are so diverse and the cost in terms of space and heat on an IC is so high, this is certainly the rational thing to do since it allows the microcomputer system designer the greatest flexibility consistent with minimizing production costs.

1.2.1 Microprocessor Busses

There are two distinct sets of busses that can be identified in computers based on their location and whether the hardware designer has access to them. The internal busses are part of the organization of the computer and are of great concern to the chip designer but are of little concern to the microprocessor applications designer. These internal busses have no effect on the external interfacing of the chip and only affect the chip's effective operating speed.

The second difference among microprocessor busses is their external bus structure. The alternatives can be differentiated by several easily quantifiable features such as the number of address or data bits that are simultaneously present (the width of the bus), the type of interaction on the bus between the CPU and external devices, and the dedication of the bus to one or more uses.

Wider data paths.   Because many of the elements of a computer have been concentrated on a single chip, the ability of the microprocessor to communicate with the environment becomes limited. At the least, since there is no on-chip memory, there must be a bus that transfers addresses out, data in and out, and power lines and control signals. On one extreme, the bus could be 1 bit wide and consist of the serial transmission of addresses that are then decoded and presented in parallel to the memory, the data being handled the same way. The other extreme is for all of the address, data, and control lines to be available continually to the environment as separate lines. The 1-bit-wide serial implementation is too restrictive on speed of operation since all of the program is external to the microprocessor and needs to be fetched from memory. The other extreme of all lines being available continually ignores the fact that address lines are unidirectional and are only needed to point to memory or I/O locations during part of the instruction execution cycle. Between these extremes is the usual compromise of time division multiplexing (TDM) some of the signals on the same physical lines. This same technique is used in microcontrollers and is illustrated by examples presented in later chapters.

Multiple busses: Harvard Architecture.   Although not often used in microprocessors, Harvard architecture is an alternative architecture used in some special applications. In this architecture, separate memories are maintained with their independent address and data busses. The two memories are the program memory and the data memory. This allows for two parallel streams of data and program to be maintained without their having to TDM the single bus to main memory. While this seems little more than the application of hardware parallelism to improve the performance of a computer, it allows a different internal organization of a computer such that instructions can be prefetched and decoded while multiple data are being fetched by the CPU and being operated on.

Memory handshaking.   The idea of memory handshaking is strictly an external bus problem but can significantly affect the operating speed of a microprocessor, particularly if it is operating with slow external devices and a cache memory has not been implemented. *Slow* means that the memories do not respond to the microprocessor within a single bus cycle. There are actually three handshaking options: no handshaking, synchronous handshaking, and asynchronous handshaking. No handshaking is an implementation of an open-loop control system in which there is no feedback from the devices that are being communicated with. In other words, the designer has chosen RAM, ROM, and I/O devices that can respond to the microprocessor's control signals and both capture information from the data bus and put data on the data bus fast enough that the microprocessor can read it off of the bus at the end of the bus cycle. Synchronous handshaking, which is exemplified by the Intel line of microprocessors, involves the addition of an integer number of "wait states" to complete the memory access time. When a memory is accessed, a return line from the memory to the microprocessor can be asserted that indicates that it has not yet finished either the read or write cycle. If this line is asserted, the microprocessor's CPU cycles in its wait state until this signal line tests as not asserted. It then continues with the execution of the instruction. The third alternative, asynchronous handshaking, is exemplified in the Motorola line of microprocessors. While the handshaking is similar to that of the synchronous wait-state method, the CPU does not wait a complete machine macrocycle, but rather responds on the next phase of the clock. So while it is synchronous handshaking in the sense of being slaved to a clock, it is asynchronous in the sense that it is not tied to a macrocycle of the CPU and therefore can respond more quickly.

1.2.2   Microprocessor ALU, I/O, and Control Elements

Since the major physical difference between microprocessors and microcontrollers is in the location of the memory (a microprocessor must access it offchip, and a microcontroller has at least some onchip), the other four elements of a computer are discussed in the following section. It is seen that the ALU, the I/O, and the control elements of a microcontroller all differ from those of a microprocessor in apparently minor yet significant ways when the application is considered. However, their basic operation in microcontrollers is the same.

1.3   SINGLE-CHIP MICROCOMPUTERS

It is now easy to define a single-chip microcomputer, of which microcontrollers are a subset, as a single integrated circuit that contains the five essential elements of a computer: input, output, memory, ALU, and a control unit. While it is necessary to have all of these elements, it is not necessary that they exist to an extreme degree. An example is the first single-chip microcomputer, the Intel 8048 which was introduced in 1976.[17] The internal structure of the Intel 8048 is shown in Fig. 1.13. This family of single-chip microcomputers may alternatively be referred to by the name MCS-48 since there is a variety of models that differ only in amount of memory and I/O capability. This computer, which operates from a single 5-volt supply, has an identical architecture across a variety of implementations. The primary difference among the various models is the type of on-chip program storage, the two options being no program ROM (8035), 1024 (8048), 2048 (8049), or 4096 (8050) bytes of mask-programmed ROM, or 1024 bytes of EPROM (8748). The amount of RAM is limited to 64 (8048), 128 (8049), or 256 (8050) bytes, but this is adequate for a number of applications. The on-chip I/O is

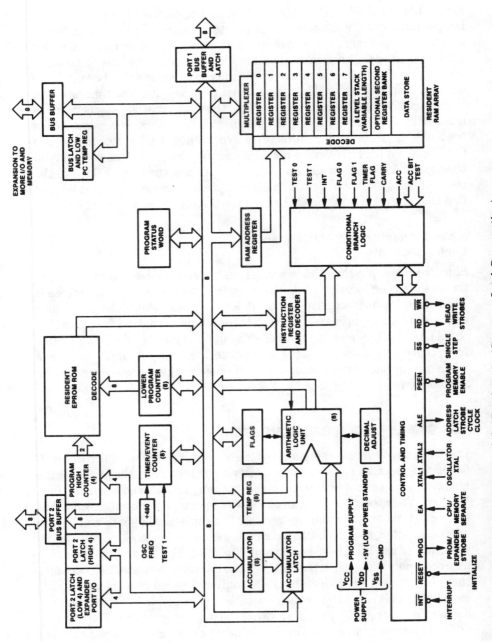

Figure 1.13 Internal structure of the Intel 8048. *(Courtesy Intel Corporation)*

30

implemented as 27 programmable (either input or output) lines consisting of three ports of eight bits each and three test inputs that can be tested by conditional jump instructions.

Although not considered an essential element of a computer, timers are useful in control environments and even in noncontrol environments for generating periodic interrupts. The MCS-48 has one 8-bit binary counter that is presetable and readable. The counter generates an interrupt on overflow and sets an overflow flag flip-flop if interrupts are not desired. The timer can also be used as an event counter and timer since it can respond to external inputs. If more capabilities are needed than can be provided by the single chip, additional interface circuits are available that can expand memory addressing up to the 12-bit limit of its program counter as well as supply additional I/O capability. This chip itself is typically able to support one TTL load, so either low-power ICs are required for more than the simplest of systems.

1.4 MICROCONTROLLER: I/O-ORIENTED SINGLE-CHIP MICROCOMPUTER

Most of the discussion thus far can apply as well to any computer. Henceforth, the discussion is restricted to single-chip microcontrollers (MCU), although reference may occasionally be made to more complex machines for the sake of explanation or comparison.

The distinction between MCUs and other single-chip microcomputers lies primarily in three areas: the I/O capability of MCUs, the interrupt handling capability, and the instruction set. Because the orientation of MCUs is toward their interaction with the environment, both in the sensing and control of the environment, much more real estate (in the form of I/O pins and chip are) is dedicated to I/O. For example, it is typical to find both synchronous and asynchronous communications devices, A/D and D/A converters, and timers on the single chip. Along with this expanded I/O capability is the increased ability to prioritize levels of interrupts and mask individual interrupts within the CPU rather than relying on external hardware to do this. While there are various single-chip microcomputers that have some of these capabilities, MCUs have most or all of these capabilities. The second major difference is in the instruction set. Many of the instructions are oriented toward bit manipulation, reflecting the need to mask certain bits that are of no importance to a particular operation or perform logic operations on I/O ports on a single-bit basis.

The following discussion of specific capabilities of MCUs is comparative in nature. A more exhaustive treatment of unique capabilities of typical MCUs is found in later chapters.

1.4.1   Microcontroller Input/Output

A major architectural difference among microcontrollers and computers in general is the method of accessing input/output ports. The three different approaches are memory-mapped I/O, separate I/O, and I/O processors. The first two are exemplified by Motorola and Intel, respectively. The third method, that of downloading I/O programs to separate I/O processors, is primarily limited to large machines and is exemplified by mainframe IBM machines.[18] Memory-mapped I/O means that all I/O is performed as if the I/O devices were memory addresses. The read/writes are from/to memory locations whose addresses are dedicated to I/O ports. While this complicates I/O addressing somewhat, particularly in applications where it is necessary to populate the virtual address space with physical memory, it offers other computational advantages. For example, all mathematical and logical operations that can be performed on the contents of memory can now be performed on I/O ports. The contents of an I/O port can be read, incremented, and written back to the I/O port with a single instruction. Data can be moved from one port to another with a single move instruction, and block transfers of data from a section of memory to a single I/O can be done with a limited amount of software overhead. Typically, all I/O port addresses are placed in the same block of memory so only one major address needs to be decoded identifying I/O with the lower-order bits used to access specific I/O ports. Separate I/O has, in addition to a read/$\overline{\text{write}}$ bar control line, an (I/O)/$\overline{\text{memory}}$ bar control line. This method can use identical address decoders for both I/O and memory with the final determination of which of these is accessed determined by the status of the (I/O)/$\overline{\text{memory}}$ bar control line.

1.4.2   Interrupts

Thus far, the discussion of handshaking has been limited to the direct interconnection of memory and I/O to the CPU through the address and data busses. There is an additional level of handshaking that is important for control and real-time operations, although it is being supplanted by the introduction of peripheral management units (PMUs) in the NEC K2 and K3 microcontrollers.[16] This other type of handshaking, what one might call *operational handshaking*, is through the form of interrupts. Interrupts, when enabled, allow for the asynchronous interruption of the order of the program execution, diverting its resources to handle a specific subtask that must be attended to in a timely manner. This immediate handling of an interrupt is at the expense of the currently executing code which is halted. In response to an interrupt, the state of the machine is saved and execution is resumed from where it left off after the interrupt-service-routine (ISR) is completed.

The earlier discussion of memory was limited to primary storage that was directly connected to the CPU's busses. Additional memory, with less expensive, longer access times, can also be connected to the microcontroller. This secondary memory exists in the form of magnetic or optical disks. Because of its slow access (relative to primary memory) and the block structured nature of the data, accesses to secondary memory tend to be for the transfer of blocks of data rather than single bytes or words. This transfer is typically handled through disk controllers that access the disk and convert the data into a form readable by the CPU. Rather than have the CPU wait for the disk access, the CPU may request data, continue to execute instructions, and expect to be interrupted by the disk controller when it is finished so the CPU can read the data from its buffer memory. An alternative to this is for the disk controller to use direct-memory-access (DMA) techniques to transfer the data from the disk to primary memory, interrupting the CPU only when the block transfer is complete. This saves the CPU the overhead of having to fetch the data from the disk controller's buffer and transfer it to primary memory. The disk controller can use cycle-stealing techniques to perform the DMA transfer. In cycle stealing techniques, the DMA controller only uses the bus for data transfers from one peripheral or memory location to another when the CPU is not accessing the bus.

Interrupts can be decomposed into two types: maskable and nonmaskable. The ability of maskable interrupts to interrupt the CPU can be controlled by the programmer either by enabling or disabling blocks or levels of interrupts, or by enabling or disabling individual interrupt lines. Nonmaskable interrupts will always generate an interrupt and cannot be inhibited. These are usually reserved for events that threaten the integrity of the microcontroller system, such as power failure or stack overflow. Other interrupts of various levels of priority can be enabled or disabled as the needs of the program dictate. For example, once an alarm has generated an interrupt to a microcontroller, it is more convenient to disable the interrupt within the CPU rather than have the additional hardware that would be required for the microprocessor to reply to the alarm and disable it at a remote point. From another point of view, interrupts can be divided into hardware and software interrupts, although software interrupts are usually called *exceptions* or *traps*. The types of software interrupts are (1) those that cause the execution of the software equivalent of an interrupt service routine in response to a system integrity threatening situation such as a divide-by-zero operation, and (2) those that are used to interrupt ongoing processes without affecting the state of the machine. The execution of a software interrupt (SWI) as a line of code has the same effect as a normal hardware interrupt in terms of automatically saving the machine's state before executing the SWI specific instructions. An SWI can be executed from any place in a program without giving any consideration to which registers must be saved and restored. The use of an SWI rather than a subroutine or function call is somewhat less efficient than a simple subroutine call, since the complete state of the machine is saved prior to execution of the code and the state is also restored before the calling process resumes execution of its own instructions.

An SWI can also be used to execute, in response to a software instruction, the same sequence of instructions that would be executed in response to an actual hardware interrupt. It is often used in the debugging of software through the use of *breakpoints*. Breakpoints are used to halt the execution of instructions in a program to inspect the value of variables at that point, alter variable values, or determine if certain branches have been taken in the execution of the program. A breakpoint can be implemented by replacing the instruction at the breakpoint with an SWI that returns the program to a monitor program that is being used to perform the debugging.

The control of interrupts and the execution of interrupt service routines that respond to interrupts are two important matters that are dealt with in more detail in the chapter on real-time programming and the integration of assembly and higher level languages.

1.4.3 ALU

The significant difference in the ALUs of microcontrollers and those of general purpose computers is the invariable inclusion of bit manipulation instructions. As with all computers, general logic instructions could be used to perform the same functions, but in significantly less convenient fashion. For example, if it is desired to test whether a single bit of an input port representing the state of a switch is asserted in a non-memory-mapped machine, it is first necessary to read the contents into a register. After the port is read in, a mask consisting of all zeros except for a single 1 in the bit position being tested is ANDed with the register. The register is then tested for a zero condition and the appropriate branch instruction is executed. Microcontrollers typically have instructions that allow branching based on testing particular bits of an input port as single instructions. In some cases, separate logic units are included in parallel with the internal arithmetic units to speed execution and allow for parallel execution of instructions. At least one, the Intel 8051, has a separate Boolean processor (BP) that allows individual bits to be set, cleared, complemented, moved, tested, and used in logic computations.[4]

1.4.4 Timers

When it comes to control, timers, and plenty of them, seem to be a necessity rather than a nicety. It is typical to load and have several timers operating simultaneously, programmed so that they generate interrupts at regular or deterministic, irregular intervals. Most MCUs have versatile timers that can be configured in real time by writing appropriate codes to control registers. These timers can be reprogrammed as to the duration that they will measure; the type of interrupt, if any, that they will generate; and whether they will reload and continue to time the next period automatically. They can also usually be reconfigured, again in real time, to act as event counters or to count the duration between transitions on the inputs. There is a great deal of flexibility in these onchip timers, and as such they present a significant challenge in initialization and proper use.

A second type of timer is also usually present and necessitated by the use of microcontrollers in unattended environments. This timer is called a *watchdog timer* and is used to restart the microcontroller if it fails to be reset at regular intervals. The interval between clearing the watchdog timer is usually programmable and is chosen to suit the application. The assumption in using a watchdog timer is that as long as the system is operating properly, the code that is executing will clear the timer on a regular basis. If it does not do this, it is assumed that the microcontroller has malfunctioned and the program is not running in the manner that was intended. In this case, the only thing that can be done is to restart the microcontroller under the assumption that the malfunction was temporary and restarting it will clear the fault and allow it to resume operation. This hardware backup is not usually required in microprocessor systems and so is rarely found as part of the microprocessor architecture. If needed for a microprocessor system, it can easily be incorporated in external hardware.

1.4.5 Parallel and Serial I/O

The complete range of binary I/O, both parallel and serial and synchronous and asynchronous, is also available on most MCUs to a greater or lesser degree. The goal is to make these devices not primarily message handlers, yet allow effective communications with other similar devices in a distributed system of multiple processors and/or communicate easily with the operator. These communications interfaces are designed to be as autonomous as the timers, performing their function and only interrupting the CPU when they have received a complete byte or block of data or when they are ready to transmit the next byte or block of data.

1.4.6 External Devices

While a computer's primary use is the computation and manipulation of data, necessitating only connection to data I/O devices, a microcontroller's primary purpose is interacting with the environment. This interaction includes much more than prepackaged units of binary data and therefore requires much more extensive capabilities to interface with the external world. As with all engineering disciplines, the requirements for interacting extensively with the external world and the limited physical size of the MCU require tradeoffs to optimize the design and minimize the design and production costs. The following sections examine some of these conflicting considerations.

Pinout considerations.   At this stage of development, there are several pins that are required on all MCUs, including the power and ground pins, external oscillator pins (or external crystal pins), and a hardware reset pin.[19] After this, the choice is up to the chip designer regarding how to allocate I/O pins, external memory access pins, and interrupt pins. Hardware reliability and ease of fabrication dictate the minimum number of pins, while flexibility and usefulness demand the maximum number of pins. Without getting into the detailed design tradeoffs that are required and that are inevitability based on the intended market for the MCU, there are some general observations concerning available tradeoffs. The first of these is the use of pins for multiple functions. The easiest to understand is TDM use, of which there are two varieties identified by the duration of their operation. In the most limited case, a pin can be used to set an initial value of an internal register or mode of operation during the first few cycles of MCU operation after it is reset. If these pins are not used properly during startup, no later programming can change the operation of the MCU. This is a reasonable use for these pins, which otherwise would have to be set to a fixed potential for the entire operation of the system with no need to change them. In fact, changing them might result in unpredictable operation of the MCU since the mode is not intended to change. For example, in the Motorola M68HC11 MCU, two mode pins select at startup whether the MCU is to be operating in normal single chip mode, normal expanded mode, special bootstrap mode, or special test mode.[20] Once the mode is read in from these two pins, it cannot be changed. After the two mode pins are read, they then serve the alternate function of indicating the status of internal operations of the MCU.

A more typical TDM operation is the use of common pins for address and data. Since the address lines are output only, the address (or part of the address) can be output on normal port lines during the first clock cycle of an instruction execution and latched externally for use in accessing memory or I/O. These same pins can now be used for I/O since the address data that they contained is no longer valid or necessary and the address is latched in an external latch.

External memory.   Some MCUs make no provision for external memory and are designed to operate autonomously in a confined environment. Others allow for the possibility of adding external memory through the previously discussed method of TDM of the address and data lines through the I/O ports. A disadvantage of this approach is that the I/O ports are lost, and this may not be acceptable. Another need for external memory is during the development phase, in which it is inconvenient continually to erase and reprogram on-chip EPROM program memory. In this case, external memory can be attached if the I/O requirements can suffer the losses. An alternative to this is provided by some manufacturers in the form of external chips that are intended to be used during the development phase to emulate the register operation of the single-chip mode yet allow the use of external memory. This approach provides a second benefit in that it makes the normally internal address and data busses available to the development engineer for hardware troubleshooting and testing.

ROM: program development. MCUs are usually specified as families since there are a wide variety of I/O and memory configurations attached to the same basic architecture. While mask programmable ROM is the cost effective alternative for large production runs, program development requires either the use of an in-circuit emulator (ICE) or EPROM program ROM on chip. The decision regarding whether to use only EPROM units for a production run or convert to a mask programmed version is strictly economic. The cost in lost time when (not if) the first masked version is not correct is normally not considered in these decisions. Typical crossover points are around 500 to 2500 units depending on the number of mask revisions required.[11] An alternative to this is the Hitachi ZTAT (zero turnaround time) approach in which MCUs are made with one-time programmable PROMS.[19]

RAM: data/code. If a sufficient amount of RAM is available onchip or externally connected, then an MCU can be loaded with slushware that is an operating program residing in RAM. This may be useful in network node controllers in which it is necessary to change their operating program depending on the type of protocol that is being passed through or reformatted. Of course, this presupposes that the MCU is connected to the communications network, has a small bootstrap loader in ROM, and can have its operating program downloaded or modified by a central network controller. The disadvantage of this approach is that if the node goes down, it cannot operate as a node controller until the central network controller can download the operating program to it.

Math coprocessors. Math coprocessors are not typical units associated with MCUs. If the control process needs that type of computational power, a more comprehensive device is used such as the newer 32-bit superscalar processors, which are discussed in a later chapter. Math coprocessors are also expensive due to their low yield and demand and hence do not pose a cost effective alternative to the use of the more sophisticated 32-bit controllers.

DMA controller. DMA controllers also make sense in some low-end applications where the MCU performs a supervisory function for the transfer of data to and from a peripheral and controls its operation in such a tightly coupled manner that it does not have time to transfer the data using software. But, as was mentioned earlier, newer MCUs with PMUs can provide this same type of data throughput and handling without disturbing the MCU with time consuming interrupts.

TABLE 1.4 Capabilities of Representative 8-bit Microcontrollers from Two Families

Chip	RAM	ROM/EPROM EAROM/EEPROM	Clock μs	I/O Ports	A/D	Timers
M68HC11A0	256	--	0.476	4x8 1x6	4/8	9
M68HC11A1	256	512 EEPROM	0.476	4x8 1x6	4/8	9
M68HC11A2		2048 EEPROM	0.476	4x8 1x6	4/8	9
M68HC11A8	256	8k ROM, 512 EEPROM	0.476	4x8 1x6	4/8	9
M68HC11E0	512	--	0.476	4x8 1x6	8	9
M68HC11E1	512	512 EEPROM	0.476	4x8 1x6	8	9
M68HC11E2	256	2048 (EE)	0.476	4x8 1x6	4/8	9
M68HC11E9	512	12k (ROM), 512(EE)	0.476	4x8 1x6	8	9

M68HC11D3	192	4096 (ROM)	0.476	4x8 1x6	8	9
M68HC11F1	1024	512 (EEPROM)	0.476	4x8 1x6	8	9
Intel 8021	64	1024 (ROM)	2.5	2x8 1x4	-	2
Intel 8022	64	2048 (ROM)	2.5	3x8	-	2
Intel 8035	64	--	2.5	3x8	-	2
Intel 8039	128	--	1.4	3x8	-	2
Intel 8041	64	1024 (ROM)	2.5	3x8	-	2
Intel 8048	64	1024 (ROM)	2.5	3x8	-	2
Intel 8049	64	2048 (ROM)	1.4	3x8	-	2
Intel 8748	64	1024 (EPROM)	2.5	3x8	-	2
Intel 8031	128	--	1	4x8	-	2
Intel 8051	128	4096 (ROM)	1	4x8	-	2
Intel 8751	128	4096 (EPROM)	1	4x8	-	2

1.4.7 Configurations of Microprocessors and Microcontrollers

Table 1.4 lists some of the capabilities of various microcontrollers. It is not meant to be exhaustive or complete, but only to illustrate the variety of capabilities currently available from two major manufacturers.

REFERENCES

1. G. J. Myers, *Advances in Computer Architecture*, 2nd ed. New York: Wiley, 1982.
2. A. M. Turing, "On Computable Numbers, with an Application to the Entscheidungsproblem," *Proceedings of the London Mathematical Society*, Series 2, 42:230–65 and 43:43:544–46, 1937.
3. H. R. Lewis and C. H. Papadimitriou, *Elements of the Theory of Computation*. Englewood Cliffs, NJ: Prentice-Hall, 1981,
4. H. Bates, "Next Generation 8051," *WESCON '82 Conference Record*, Anaheim, CA, session 5C/3, pp. 1–5, September 14–16, 1982.
5. Electronic Industries Association, *EIA Standard RS-232-C*, August 1969.
6. M. Birnkrant and I. Olsen, "Boosting System-level Throughput in a Microcontroller," *High Performance Systems*, 10(11):53–55, November 1989.
7. *HCMOS Single-Chip Microcontroller (MC68HC11A8)*. Motorola Semiconductor Technical Data Publication #ADI1207R2, Phoenix, AZ, 1988.
8. *Microcontroller Handbook*. Intel Publications #210918-004, 1986.
9. S. Baba, K. Matsubara, and K. Noguchi, "High-performance Single-chip Microcontroller H-8 Series," *Hitachi Review*, 38(1):1–10, 1989.
10. J. J. Vaglica and P.S. Gilmour, "How to Select a Microcontroller," *IEEE Spectrum*, pp. 106–9, November 1990.
11. G. B. Nelson, "The Use of EPROM-based Microcomputers," *WESCON '82 Conference Record*, Anaheim, CA, session 5c/2, pp. 1–4, 14–16, September 1982.
12. V. C. Hamacher, Z. G. Vranesic, and S. G. Zaky, *Computer Organization*, 3rd ed. New York: McGraw-Hill, 1990.
13. J. P. Hayes, *Computer Architecture and Organization*, 2nd ed. New York: McGraw-Hill, 1988.
14. D. E. White, *Bit-Slice Design: Conrollers and ALUs*. New York: Garland STPM Press, 1981.
15. M. Andrews, *Principles of Firmware Engineering in Microprogram Control*. Potomac, MD: Computer Science Press, 1980.
16. D. P. Siewiorek, C. G. Bell, and A. Newell, *Computer Structures: Principles and Examples*. New York: McGraw-Hill, 1982.
17. *MCS-51 Family of Single Chip Microcomputers*. Intel publication #121517-001, July 1981.
18. W. Stallings, *Computer Organization and Architecture*. New York: Macmillan, 1987.
19. G. Ramachandran, "Evolution of a High Performance Microcomputer Architecture," *WESCON '82 Conference Record*, Anaheim, CA, session 5c/4, pp. 1–8, 14–16, September 1982.
20. *M68HC11RM/AD*. Englewood Cliffs, NJ: Prentice-Hall, 1989.

PROBLEMS

1. List the essential elements of a computer and differentiate between microprocessors and microcomputers.

2. Is there any computation that can be done in software that cannot be done in hardware, and vice versa? What is the essential difference between computer hardware and software?

3. Show, through a sequence of diagrams for each phase of the seven phase clock, the actual connections that are made through the internal bus of Fig. 1.6 while executing the instruction LOAD R1 as defined in Fig. 1.7.

4. Explain the advantage of synchronous serial communications over asynchronous serial communications.

5. Discuss the relative merits of horizontal and vertical microcode as it pertains to the use of memory space on a single-chip microcontroller.

6. Discuss the relative merits of masked ROM, one-time PROM, EPROM, and EAROM as alternatives for program store in MCUs.

7. Is an MCU a computer? Why?

8. As a designer of a heart-monitoring system for an intensive care unit at a hospital, you are particularly concerned about the possibility of power failure causing your unit to malfunction. What capabilities of an MCU might you use, and how would you use them to minimize the effect of a power outage? Assume no battery backup is available except for a small alarm. Draw a general block diagram indicating the general capabilities you would use and any external hardware required.

9. Your company's product is in a cost-competitive market that necessitates close scrutiny of all aspects of your design. You are particularly concerned with whether you should design your system using EPROM units only or whether you should use mask programmed ROMs for the production run using EPROM units only for the initial development. Determine the break-even number of EPROM units if each EPROM MCU cost $14/ decreasing to $12/unit after the first 500 units. The nonrecurring engineering cost for setting up the mask is $650 with a unit price per PROM MCU of $8. Perform the analysis assuming one, two, and three different masks before the device is correctly programmed. Ignore the cost of the EPROM units in evaluating the PROM route.

10. Assume that you are required to monitor four analog channels using an MCU that can perform a single A/D conversion in 2 μs with 8-bit resolution. If the channels are sampled sequentially, what is the maximum latency between a signal occurring and your detecting it, assuming no overhead other than the A/D conversion process? If the dynamic range of the A/D converter is 10 volts, what is the minimum voltage change that can be detected assuming 1/2 LSB accuracy?

11. Draw a block diagram similar to that of Fig. 1.6 but with a Harvard architecture.

12. What is the difference between a hardwired controller and a microcoded controller?

13. Explain the difference between daisy chain interrupt acknowledge and polled interrupt.

14. What is the first order limit on the size of memory (number of words) that a computer can have connected to it?

15. Why are register/register arithmetic operations faster than register/memory operations?

16. How can conditional branching be implemented in a hardwired controller? Draw a suggested circuit.

17. What is DMA, and why does it effectively increase the operation speed of a computer?

18. What is the function of a mapping ROM in a microcoded controller?

19. Discuss the differences between a vertical and horizontally coded microcoded controller and list the advantages of each.

20. What could be gained by having two or more complete and identical sets of registers in a computer?

21. Can a hardwired controller be replaced by an equivalent microcoded controller? Given that the same technology is used in both, what is the unavoidable difference?

Computer, Microprocessor and Microcontroller Instruction Sets

2.1 COMPUTER INSTRUCTION SETS

Hardware alternatives, which are a significant part of computer architecture, were discussed in Chap. 1. Another aspect of computer architecture is instructions and their implementation, which make that hardware architecture accessible to the engineer. This chapter discusses the structure of instructions, the modes of addressing operands, and the functions of instructions. While this discussion is presented without explicitly trading hardware for software, it is implicit that software constitutes hardware that can be reconfigured without resorting to reconnecting it physically. While the two can be viewed as separate issues, the effective implementation of a microcontroller system recognizes that there are inevitable design tradeoffs between hardware and software implementations of particular functions. This is becoming more evident with the introduction of reduced-instruction-set computers (RISC),[1] and it must be recognized that there is a continuum from the single instruction computer to the direct implementation of high-level language computer.[2,3,4] The actual selection of the instruction set for an MCU as well as a computer is a complex interaction of five fundamental issues:[5]

1. *Operation repertoire:* The number and complexity of the operations selected;

2. *Data types:* Which types of numerical, character, or logical data types are desired as operands;

3. *Instruction format:* The structure of the instruction including the number of opcodes, number of addresses, and whether it is a fixed or variable size;

4. *Registers:* The amount of on-chip, directly addressable storage; and,

5. *Addressing:* The modes of addressing that are allowed.

In its most general sense, a computer instruction is a mapping from a set of inputs consisting of n operands to an output:[6]

$$X_1 \leftarrow f(X_1, X2, \ldots, X_n). \tag{2.1}$$

Usually, only some of these n operands are specified explicitly, and the others are implicit. If there are m explicitly specified operands associated with an instruction, then it is called an *m-address* machine. The value of m is usually either 0, 1, 2, or 3. To implement the stored program approach of von Neumann, the instruction must be encoded in such a way that it can be decoded and executed by the CPU. Without considering the efficiency of encoding it, an instruction can be represented as a binary string as shown in Fig. 2.1. The leftmost part of the instruction is the OPCODE, which is a binary code indicating which operation is to be performed on the operands. The rest of the instruction is occupied with information in a variety of forms that yields the *effective address* of one or more operands. If the instruction has only implicit operands, the entire length of the instruction can be used to encode the OPCODE. In some machines, the number of explicit operands is a function of the OPCODE, and in others, such as the zero-operand or stack machines, the number of operands is fixed. For example, in the MCS-51,[7] the DEC (decrement) instruction can have either an implicit opcode address of the accumulator with an instruction bit pattern of 00010100 (14 H) or specify explicitly one of seven registers by replacing bit 3 (counting 0 to 7 from the right) with a 1 and using bits 0 to 2 as the 3-bit address of the register.

Operand data types must also be considered in instruction set design. Although all machines have numerical operands to facilitate counting, indexing, and looping, they may not be stored directly as binary words but may be encoded in some form such as binary coded decimal (BCD). For example, numerical data types include binary (logical) data, signed and

Opcode	Operand 1 Address	Operand 2 Address	...	Operand m Address

Figure 2.1 Binary representation of an *m*-operand instruction.

unsigned binary integer, floating point, BCD, and character, including ASCII or EBCDIC. Decimal and floating point values are not inherent in the binary representation of numbers in computers, and hence there are many variations, from the IBM-370's packed decimal format of 1 to 16 bytes including sign[8] to the excess-16 form of floating-point representation in the PDP-11.[9]

2.1.1 Desirable Characteristics of Instruction Sets

While any instruction set with the property of completeness (in the sense that it is able to compute any desirable function), meets the functional requirement of a computer, there are several other desirable properties of computer instruction sets,[6,10,11] including the following:

1. *Completeness:* Any computable function can be implemented by a reasonable number of machine language instructions.

2. *Efficiency:* Often used instructions can be executed rapidly and not occupy much memory storage, and appropriate instructions should be available to allow for the easy translation of higher-level languages into executable programs.

3. *Regularity* (consistency): Expected opcodes should be included. If all registers can be decremented, then one would expect to be able to increment all registers; there should be very few exceptions to a consistent use or form of instructions.

4. *Orthogonality:* All addressing modes should be available for all instructions.

5. *Condition codes:* Sufficient conditions should be set by the results of the execution of all instructions, and they should all be testable to allow for effective conditional branching.

6. *Appropriateness:* The instructions should be tailored to the particular task the computer is being asked to do. *Floating-point* instructions are not usually necessary in the control of a microwave oven, nor are single-bit I/O instructions usually required for a large-scale scientific computer.

In all cases, any particular instruction set is at best a compromise of these desired properties and the requirements for implementation.

2.1.2 Instruction Formats

Although the basic form of an instruction is shown in Fig. 2.1, this section covers in more detail the various options available as determined by the number of operands m. In its most basic form, an instruction can contain the opcode, two source operand addresses, the destination address, and the address of the next instruction to be executed.[9] Except for completeness, this form is never practically implemented since it wastes memory. An easily implemented alternative is to have a program counter (PC) that automatically increments at the fetch of each instruction so it always points to the next instruction, provided that there is no branch or jump as a result of the execution of the current instruction. The first useful instruction is the three-operand instruction as shown in Fig. 2.2, which includes the OPCODE, the two source operand addresses, and the destination address. An example of this type of instruction is the Control Data CDC 6600 scientific computer.[6] For example, the integer multiply operation might be

MUL Src1, Src2, Dest   (2.2)

which is interpreted as "take the product of the operands specified by the address Src1 and Src2 and put it in the location specified by address Dest."

An even more parsimonious use of instruction length is to consider one of the sources also to be the destination of the operand, leading to a two-operand instruction. For example, in the IBM 360,[12]

ADD Dest, Src1   (2.3)

means "add the operand at address Src1 to the operand at address Dest and store it at address Dest." Nothing is lost with this two-operand instruction except a modest amount of flexibility in programming, but much is gained in saved storage since the address of one of the operands is no longer needed for each binary-operand instruction.

Opcode	Source1 Address	Source2 Address	Destination Address	3–Address Instruction

Opcode	Destination Address	Source Address	2–Address Instruction

Opcode	Source1 Address	1–Address Instruction

Opcode	0–Address Instruction

Figure 2.2 Format of instructions with various number of operands m.

If one must reduce the number of operands even further, or the hardware architecture of the machine dictates a single-operand instruction can be implemented. In this case, since most operators are binary operators (i.e., they require two operands), one of the operands must be implicit. If the machine is an accumulator-based machine in which most, if not all, operations involve a unique register called the *accumulator*, then one can have a single-operand instruction such as the Intel 8048 instruction[13]

ADD A, Src   (2.4)

which initially looks as if it were a two-operand instruction, but the A is just a reminder that the ADD is performed with the accumulator since all ADD instructions in the 8048 have the accumulator as one of the operands.

Although it would appear at first that a *zero-address* machine is not possible, it is, if one allows a MOV instruction to have an operand that permits data to be moved from memory to a special set of registers called a *stack*, and from the top of stack to memory. In this way, operations can be performed on the two top operands stored on the stack. These operations can be performed with no operand address included since the two operands are implicit—they are always considered to be the two operands on the top of the stack. These operands are removed from the stack, and the result of the operation is placed back on the top of the stack. For example, the PDP-11/40 contained an optional floating instruction set (FIS) that had four zero-address instructions for operating on the contents of a specified stack.[9]

The best utilization of any available number of bits is to have one opcode per unique bit pattern such that if there are k bits for the opcode, then there would be 2^k instructions. In practice this is not realized for several reasons. First, although there could be an arbitrary assignment of bit patterns to opcodes and a mapping ROM used to reorganize the bit pattern into a more useful, orthogonal structure, this would result in an additional delay in decoding the instruction which would slow down the execution of all instructions. Second, if the opcodes are themselves divided into fields such that individual bit locations indicate certain properties, then the decoding of the instruction is faster. For example, one particular bit position in the opcode could be used to determine whether the instruction was a single-address or zero-address instruction. If the opcode indicates a zero-address instruction, then the remaining bits in the instruction that would have been the address of an opcode could be used to indicate additional instructions. An extension of this is discussed in the following sections, in which the addressing modes of operands are controlled by certain bit locations in the instruction.

2.1.3   Addressing Modes

While most addressing could be performed by simple immediate addressing (the address field contains the data itself) and direct addressing (the address field contains the absolute physical address of the operand in memory), this is overly restrictive in that it does not directly reflect typical usage of memory references in complex, pointer-based data structures. More efficient use can be made of the computer's resources by adding addressing modes. Most of the additional addressing modes are the results of attempts to implement, in hardware and machine-level instructions, address operations that are needed at a higher level to implement commonly used algorithms. This "native mode" addressing simplifies the compiler's job of converting from high-level language constructs to assembly language and machine language implementations. For example, the available addressing modes are as follows:

1. *Immediate:* The address field contains the data itself.

2. *Direct (absolute):* The address field contains the address of the operand.

3. *Indirect:* The address field contains the address of the address of the operand.

These are shown in Fig. 2.3. These are not necessarily the actual physical addresses in memory because these address fields can be interpreted in several different ways. If they are interpreted as *absolute* addresses, then the address fields are indeed the actual binary addresses of the operands in physical memory. A fourth method of determining the *effective address* (EA) of the operands of an instruction is in the following:

4. *Relative:* The effective address is the contents of the address field of the instruction interpreted as a two's complement binary number and added to the contents of an offset register.

Relative addressing can be subdivided further into program counter relative (PC relative) addressing and addressing relative to some other register, usually called an *index register*. *PC-relative* addressing is particularly important since it provides the ability to compile or assemble the program once and execute it at any physical location in memory. If this addressing mode is not used, then each time the program is to be run, the effective addresses of all operands and jumps must be computed by a relocating linker before the program can be loaded and executed. Indexed addressing is also of great value since it can be used to change the base address for indexing into a database.

Of course, there are additional variations that can be applied to addressing modes by making various combinations or adding autoincrement or autodecrement. For example, there could be an indexed address field with

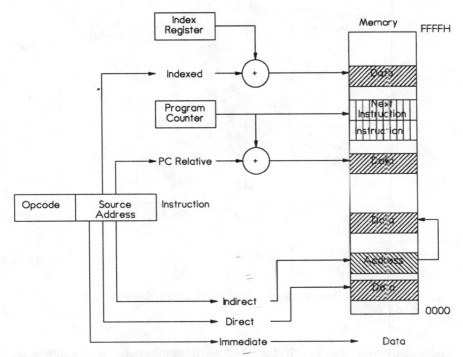

Figure 2.3 Diagram showing interpretation of three modes of address interpretation in address field of instructions.

PC-relative addressing and pre-auto-increment. The effective address would then be computed as shown in Fig. 2.4.

2.1.4 SISC, RISC, and CISC

The tradeoffs in execution speed among various reduced instruction-set computers (RISC)[1] and complex-instruction-set computers (CISC) are of current interest. Although it is theoretically possible to have a single-instruction-set computer (SISC),[1,14] it is not very practical, and the discussion of size of the instruction set is directly tied to the total speed of program execution. Although there are a variety of ways to measure the speed and performance of a computer,[15] including various benchmarks, the execution time of any particular program is

$$\text{Program execution time} = \sum_{i=1}^{n} t_i \qquad (2.5)$$

the sum of the execution time of all of the instructions. Ignoring the reordering of instructions by an optimizing compiler, there are two ways to

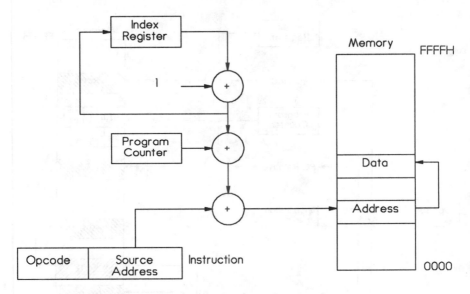

Figure 2.4 Computation of effective address of preautoincrement, indexed, PC relative addressing.

reduce this total time. The first is to reduce the number of instructions n that must be executed. This is the CISC solution since the availability of complex instruction allows for fewer instructions to be used to implement each high-level language construct. The second way to reduce the program execution time is to reduce the individual instruction execution times. This is the RISC alternative. Reducing the number of instructions in the computer program has the net effect of decreasing the execution time for each of the instructions since there is a temporal overhead associated with the decoding of the opcode before the particular instruction can be executed. If there are fewer alternative instructions to decode, the decoding is faster. A second reason for the decreased execution time is that fewer bits are required to represent all of the opcodes, which may result in a reduction in the length of the instruction. The longer instruction, if it is multiple bus widths long, may require multiple memory fetches before it can be completely decoded and execution begun. There are alternatives that have been implemented to speed up the execution of CISC computers, such as instruction caches and pipeline instruction execution, but these are fixes that do not completely obviate the advantages of a RISC computer. There is another consideration that is not explicitly expressed in Eq. (2.5): the types of instructions necessary to execute the program. If the program under study consists primarily of difficult operations that are easily expressed in native CISC instructions, then a RISC computer may need many more instructions to implement these CISC instructions, perhaps somewhat inefficiently. Hence, there will be no net gain in program execution speed. It is considerations such as these that make the RISC/CISC tradeoffs difficult to assess since they are data (program) dependent, although the current trend in computer design is toward RISC implementations. Essentially, a RISC will perform better for a program with fewer instructions, but with multiple reusable data.[1]

2.2   TASK-ORIENTED INSTRUCTIONS

Taking into account the desired properties of instruction sets, the choice of which particular instructions to implement is not overdetermined, and there is a wide latitude in which to make instructions fit the problem at hand. Although all computers can implement all computable functions, some will be able to compute certain computable functions faster than others, and vice versa. Thus, another component of instruction set design is the choice of instructions such that they are the most efficient at implementing those functions for which the device is intended. Decimal arithmetic is important in a cash register or a hand-held calculator, and it may be more efficient to implement BCD arithmetic instructions directly rather than go through the time-consuming conversion from decimal to binary and back. Likewise, there is no need to work in decimal in a control system that does not interact with a human operator, such as in an aircraft autopilot. All of the mathematical operations implemented can be in two's-complement notation. In an even more complex control example, there may need to be significant signal processing, such as Fourier transforms, to isolate and cancel a particular vibration component. In this case, a single-instruction fast Fourier transform (FFT) may be appropriate, but this instruction would be useless if the computer's intended use were in business calculations.

2.2.1   Instructions for Business, Text Processing, and Data Manipulation

Special data types, such as packed decimal, have already been introduced, but there must be instructions to manipulate these native data types. Business machines are characterized by character oriented and I/O oriented instructions. Data representation reflects this orientation, and in addition to packed decimal, data is stored in the IBM 360[12] in extended binary coded decimal interchange code (EBCDIC) and 2-, 4-, or 8-byte fixed-point integers. There are also a significant number of data conversion instructions such as PACK and UNPACK, convert to binary (CVB), and convert to decimal (CVD) to facilitate character and decimal manipulation. The I/O orientation of the machine is also evident in its load/store instructions such as store multiple (STM), which stores multiple registers with a single command, the move characters (MVC) instruction, which transfers up to

256 bytes of storage from one location to another, and GET and PUT macros, which control external I/O processors. Rather than the CPU being bound by manipulating all of the data from memory to peripherals either for input or output, the IBM 360 is designed to have I/O channels that are controlled by the CPU but operate autonomously until their task is finished. This offloading of I/O responsibility enhances the computational speed of the computer.

Another aspect of the IBM 360 and 370 is that there is no indirect addressing.[8,12] There is also no autoincrement or autodecrement capability as there is in the PDP-11 and VAX computers, which leads to difficulties in implementing a stack.[16] The large word size of the 360/370, 32 bits, also allows for large programs and a multiuser computing environment.

2.2.2 Science-Oriented Instructions

One would like to think that there is a fundamental difference in instructions for a scientific computer as opposed to a business computer; however, this does not appear to be the case. There are two possible areas in which a difference might be expected: first, the availability of complex mathematical operations as machine language instructions; and second, floating-point number representation as a native data type. Inspection of the instruction sets of various scientific computers shows only an occasional "unusual" scientific instruction such as the polynomial instruction on the VAX computer. The primary distinguishing feature of scientific computers is their speed and word length.[17] Even floating-point data types are native to the IBM 370, so this cannot be used to distinguish scientific computer instruction sets. It is the implementation of the computer's floating-point operations through vector or pipeline hardware that distinguishes scientific computers, rather than their instruction set.

2.2.3 Control-Oriented Instructions

There are two levels at which one can discuss control oriented instructions. The first comes from the field of automatic control, in which the value of a control variable is calculated to be output as a digital or digital/analog value to be used as the control input to a plant. This is typically done as a discrete proportional-integral-derivative (PID) controller approximation, but it can be extended for more difficult control problems to include plant identification, state estimation, and nonlinear control of the plant. These computations either require scaled integer arithmetic to meet control loop bandwidth requirements, or floating-point calculations performed offchip by a math coprocessor. In essence, from the previous analysis of the instruction sets of scientific computers, there is little missing in microprocessors or MCUs except for real-number native data types, and this can be compensated for by the addition of math coprocessors.

The second type of control oriented instruction is the one that allows direct, single bit manipulation of I/O port values in machines with separate I/O or the same type of single-bit manipulation on the contents of memory for memory-mapped I/O machines. These single-bit manipulations are re-quired to turn on devices that have binary controls, or step through a series of single-valued decisions such as ramping up the speed of a high power motor through the control of current through separate windings. They also make it easier to read and test individual input bits such as fluid level sensors or limit switches. Although both of these operations of reading and clearing or setting single bits on I/O devices are important, they can be done on any machine that has logic functions through the use of AND and OR functions, but much less efficiently. Despite the presence of this single-bit manipulat-ing capability in most computers, it seems to be a prime component of MCU instruction sets.

2.3   MCU INSTRUCTION SETS

The preceding discussion showed that MCU instructions sets are basically the same as all other computers, with the primary difference being the lack of native floating-point data types (except in the most recent 32-bit MCUs) and the addition of more extensive single-bit manipulation capability. The instruction sets have been optimized to word size, with the smallest-word-size MCUs having the most restricted instruction sets and the 32-bit MCUs having extensive instruction sets designed around dedicated applications. Since many MCUs are designed with specific applications in mind, the RISC/CISC debate seems to be reversed from the current trend towards RISC, emphasizing the decoding and execution of complex instructions that will be executed often, thereby yielding a net savings in program execution time. For example, the Hitachi H8 Series of MCUs[18] fetches the address portion of each instruction before it fetches the opcode portion. This allows for simultaneous decoding of the opcode and pipelining of the effective address computation. Another reason for this net improvement is the design of the MCU toward a specific application while at the same time retaining a general instruction set for the computational support that may be required.

2.3.1   Capability Determined By Word Size

The word size of an MCU determines several computing capabilities, including the complexity of the instruction set (CISC versus RISC), the size of the stored program (and hence the unit cost of the resulting MCU system), the on-chip and off-chip implementation complexity (the size of the parallel bus and address decoding), and the speed of execution (whether there is a single or multiple memory fetches and instruction decoding time). With respect to MCUs, the restrictions imposed by unit cost necessitate looking primarily at the ability of the MCU to meet the functional require-ments of the system while at the same time keeping the recurring costs to a

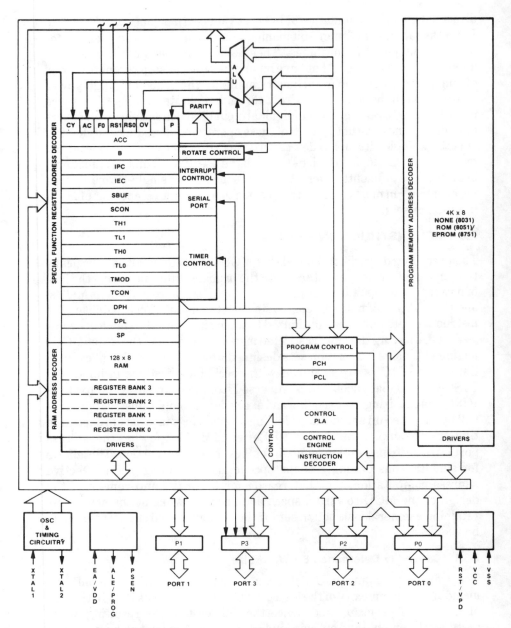

Figure 2.5 Block diagram of MCS-51 MCU. *(Courtesy Intel Corporation)*

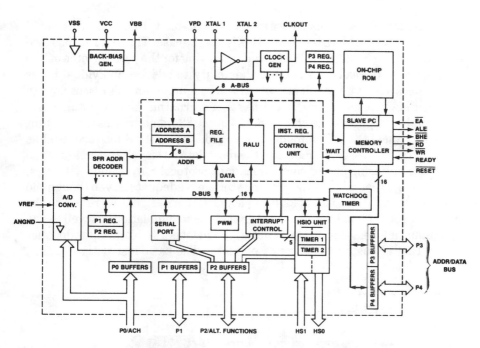

Figure 2.6   Block diagram of MCS-96 MCU. *(Courtesy Intel Corporation)*

minimum. These performance requirements have brought about the recent introduction of 32-bit MCUs tailored to specific applications. Some of the applications for which 32-bit MCUs are required are office imaging peripherals,[19] computation intensive embedded-control applications,[20] and facsimile, modems, voice mail, and laser printers.[21] The speed improvement is due to both the wider bus width and the associated ability to fetch floating-point operands as native data types as well as the ability to encode a greater quantity of instructions in the larger word. The use of the larger word size also allows for a more general set of operands. That is, more operations can be performed directly on memory rather than using several instructions to fetch operands into the CPU, perform the operation, and then restore the result.

Since MCUs are not intended for general computational use, they do not need a large physical memory. To minimize recurring costs in volume production of systems with embedded controllers, the external buses are narrower than one would expect in a 32-bit computer. For example, the National Semiconductor NS32000 series of MCUs have a virtual address space of 2^{32} or 4 Gaddresses. The term *Gaddresses* is used to point out that although there are 2^{32} unique addresses, in this machine they are used uniquely to address bytes of data rather than 32-bit words. While this would imply an external bus 8 bits wide, this is not the case. The external bus (see Fig. 2.7) of the NS32000 is 16 bits wide, allowing for the parallel fetch of two bytes simultaneously. Of course, the ability to address individual bytes introduces odd/even problems, which will be ignored for this discussion. In another effort to minimize the width of the external bus (and simultaneously the number of pins on the chip), only 24 of the 32-bit addresses are brought off chip. Since the I/O is memory mapped, the upper 8 bits are used as peripheral addresses. For any but the most data intensive systems (such as message buffering and/or communications protocol conversion), 2^{24} bytes, 16 Mbytes, of combined program and storage is adequate. Even though the NS32000 family consists of 32-bit machines, this designation refers to the width of the internal data bus and arithmetic manipulations as well as the maximum address that can be generated. That it is a 32-bit machine does not imply that all instructions are 32 bits long. The basic instruction format,

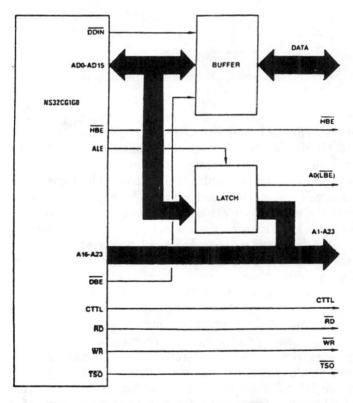

TL/EE/1075

Figure 2.7 National Semiconductor N32000 external bus structure. *(Courtesy National Semiconductor Corporation)*

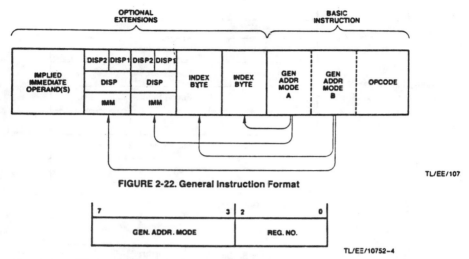

FIGURE 2-22. General Instruction Format

TL/EE/107

TL/EE/10752-4

Figure 2.8 National Semiconductor N32000 32-bit MCU instruction format. *(Courtesy National Semiconductor Corporation)*

shown in Fig. 2.8, is one to three bytes long and contains the OPCODE and up to two 5-bit, general-addressing mode fields. This basic instruction may have appended to it optional extensions. The combination of the restricted-width external bus and the variable-length instruction means that not all instructions can be fetched in a single memory access; however, because of the instruction prefetch capability of the NS32000, this does not reduce its performance significantly. For completeness, and for later reference to the tables with the N32 instructions, the register structure of the N32 is shown in Fig. 2.9

On smaller machines, such as the 8-bit MC68HC11, approximately the same number of pins (60 on the N32, 52/48 on the MC6811) have been used differently. Since the virtual address space of the MC6811 is 2^{16}, only 16 pins need to be used for addressing external memory. There is no need to address external memory if the capabilities of the on-chip RAM and I/O are sufficient. In this case, these same pins are used for I/O rather than addressing external memory. Clearly, this chip is oriented toward applications that do not need large amounts of buffer storage and can easily get by with the maximum 1024 bytes of on-chip RAM, and require programs that can be contained in the maximum 12K bytes of ROM. The block diagram of Fig. 2.10 shows the number of pins that are used for I/O and that the external address lines replace I/O ports if the chip is operated in an expanded mode. The self-contained eight-channel A/D converter is also shown as being part of Port E. This port can alternatively be used for binary inputs.

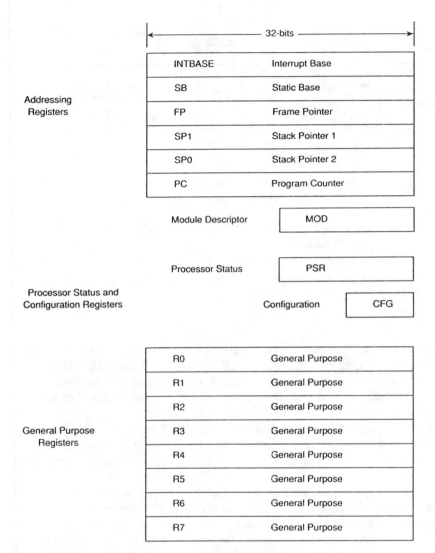

Figure 2.9 Register structure of National Semiconductor N32000 family of MCUs. *(Courtesy National Semiconductor Corporation)*

The register structure of the MC6811, shown in Fig. 2.11, exhibits the primarily 8-bit concatenated accumulator structure along with the 16-bit register structure. The difference in the size of the registers is that the 8-bit accumulators are used for small value arithmetic, although they can be used as a single 16-bit accumulator if necessary, and the registers are intended

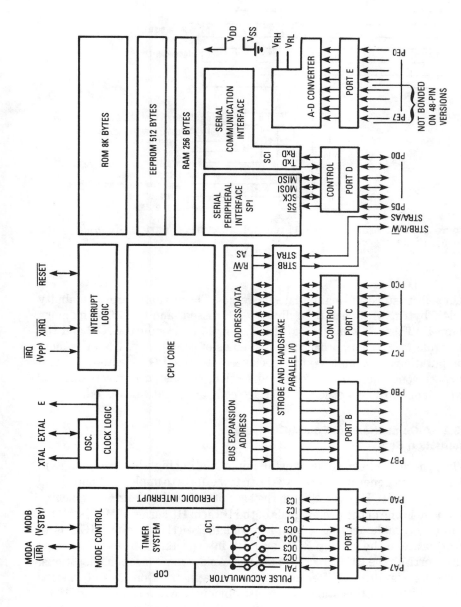

Figure 2.10 Motorola MC68HC11 block diagram. (*Courtesy Motorola Corporation*)

59

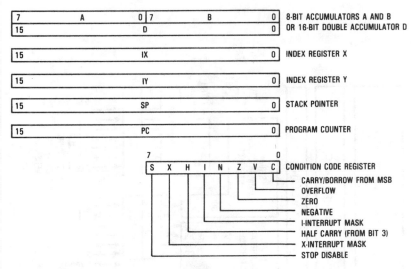

Figure 2.11 Motorola MC68HC11 register structure. *(Courtesy Motorola Corporation)*

primarily to store and manipulate addresses. The virtual address capability is 64K bytes, with each byte directly addressed, and the I/O is memory mapped. These two machines, the National Semiconductor N32 and the Motorola MC6811, approach the extremes of complexity, from the most computationally complex and specialized, to the I/O oriented, general purpose controller. In the following section, the instruction sets of these two, along with two other MCUs, are compared and contrasted.

2.3.2 A Comparison of Four MCU Instruction Sets

While there has been some effort in implementing high-level languages in computers in general,[4] there have been few efforts to implement high-level languages in MCUs even though the ease of converting from C to machine language is one of the design goals of the Hitachi H8 series of MCUs.[18] One method of implementing a high-level language (HLL) in an MCU is to incorporate in the on-chip ROM an HLL interpreter. This has been done with Forth on the RCA CDP1804 MCU.[22] Aside from these considerations, there are significant differences among the instruction sets of various MCUs, although this stands out most obviously when comparing MCUs with different inherent word sizes. To illustrate this point, the instruction sets of four representative MCUs have been listed in Tables 2.1 to 2.3. Table 2.1 loosely groups instructions that affect the control flow and sequence of execution of instructions within the computer. This includes conditional and unconditional jumps, branches, and subroutine calls. Also included are instructions that test either single bits or the contents of registers or memory locations. Table 2.2 contains data movement instructions. For accumulator-based machines, this means movement instructions to transfer data from/to memory and from/to the accumulator(s). For register- and memory- based machines, this includes instructions that transfer data between registers or between memory locations or any combination of them. Table 2.2 also includes shifts and rotates as data movement instructions even though they can be used to implement computations such as modulo 2 multiplies or divides. Table 2.3 lists all of the computational instructions. These range from the simplest of one's complement to the more difficult floating-point divides through the rather unusual mathematical operations such as polynomial step and multidimensional array indexing.

The four MCUs selected are the Intel MCS-51 of Fig. 2.5, the Motorola 6811, the Intel MCS-96, and the National Semiconductor NS32FX16. The MCS-51 and the MC6811 address two alternative implementations of 8-bit MCUs, the former with no internal A/D and the latter with a multiplexed 8-channel 8-bit A/D. The MCS-96 is a 16-bit MCU that has 10-bit A/D capability, a 16-bit CPU, a pulse-width modulation (PWM) output, high-speed output, and a watchdog timer. As an example of the latest generation of computationally capable 32-bit MCUs, the National Semiconductor NS32FX16 is included. This is one of a set of MCUs built around a common core of instructions with the addition of specific instructions customized to the specific purpose of the MCU. In this case, the NS32FX16 (referred to henceforth simply as the NS32000) is used as a facsimile controller. This range of MCUs allows for a comparison of the capabilities of instruction sets as a function of CPU processing width.

2.3.3 I/O Instructions

An investigation of Tables 2.1 to 2.3 is surprising in that there are no explicit input or output instructions. Intel microprocessors have been characterized by their use of I/O instructions distinct from memory references. Both the MCS-51 and MCS-96 have dedicated regions of low-address memory that access internal RAM and I/O registers at dedicated addresses generally referred to as the *special function registers*. Some of these low order memory locations are also dedicated to internal hardware control registers and internal hardware data registers, as can be seen in the memory map of the MCS-96 of Fig. 2.12. This memory-mapped I/O approach is reasonable when considered in the context of an MCU's use, which is I/O intensive and rarely involves a large memory. This also minimizes the number of pins required to bring signals off chip if the MCU is operated in an enhanced mode with additional I/O that is not already one of the on-chip ports. Memory-mapped I/O also streamlines the instruction set in that operations, and in

TABLE 2.1   Control Operations of Representative 8- to 32-bit MCUs

Action	MCS-51	MC6811	MCS-96	NC32FX16*
Jump, Unconditional; Subroutine	AJMP, LJMP, SJMP,JMP; ACALL, LCALL,RET	BRA,BRN,JMP; JSR,BSR,RTS	SJMP,LJMP,BR, SCALL,LCALL, RET,	JUMP,BR,JSR,BSR, RET,CXP,CXPD, SVC,FLAG,BPT, ENTER,EXIT, RXP,RETT
Jumps, Bit or Condition Code Directed	JC,JNC,JB,JNB, JBC,CJNE*, JZ,JNZ	BCC,BCS,BEQ, BGE,BGT,BHI, BHS,BLE,BLO, BLS,BLT,BMI, BNE,BPL,BRCL BCLR*,BRSET*, BVC,BVS	JC,JNC,JE,JNE, JGE,JLT,JGT, JLE,JH,JNH, JV,JNV,JVT, JNVT,JST, JNST,JBS,JBC	Bcond,CASEi,
Jump, With Computation	DJNZ*		DJNZ	ACBi,
Test, Bit				TBITS,TBITi, FFSi
Test, Data		BITA,BITB, CBA,CMPA, CMPB,CPD, CPX,CPY, TST*,TSTA, TSTB	CMP,CMPB	CMPi,CMPOi, CMPMi,CMPf
Test and Alter				SBITi,SBITIi, CBITi,CBITIi, IBITi
Interrupt	RETI	CLI,RTI,RTS, SWI,WAI	DI,EI	RETI,WAI

*Memory-to-memory operations are also allowed; i,f: Integer operand or floating point operand.

some case direct single-bit operations, can be performed directly on I/O rather than having to use an I/O instruction to bring the data into an accumulator, operate on the data, and then restore the data to the I/O port.

The other two MCUs considered also have memory-mapped I/O. Externally, there are no dedicated I/O ports on the NS32000, but within a dedicated section of memory a hardware facsimile accelerator module is located. This device is treated as a separate I/O device with its own dedicated registers and arithmetic unit for vector operations on complex variables. The addition of this facsimile accelerator module (FAM) allows for easy implementation of finite impulse response filters (FIR) and other digital signal processing (DSP) primitives. It is treated as a separate I/O device, even though it is physically located on the same chip, because it has its own address and data registers, which allows it to have a local store as well as fetch its own operands. This maximizes the usage of the internal data bus by not requiring all data to pass through the CPU core.

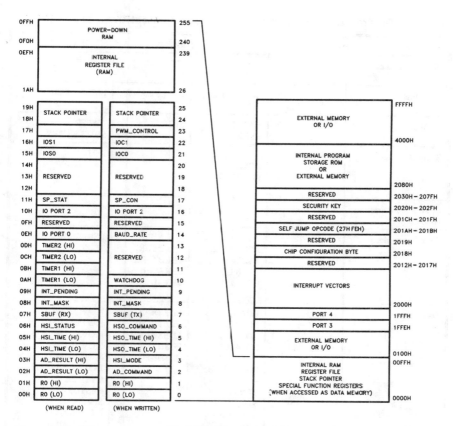

Figure 2.12 Memory space of Intel MCS-96 8-bit MCU with 16-bit CPU. *(Courtesy Intel Corporation)*

All other I/O associated with the NS32000 is treated as memory-mapped, external I/O since there are no dedicated I/O ports on the chip. Since all I/O is memory-mapped and the MOV instructions of the NS32000 can transfer data from memory locations to memory locations without passing them through the accumulator, I/O data can be transferred from one I/O address to another I/O address without the intervening steps of loading data into the accumulator. This feature is unavailable in either the MCS-96 or the MC6811, which have only load and store accumulator instructions. The MCS-51, on the other hand, does have a MOV instruction that can transfer data from one memory location to another, and hence, from one I/O port to another.

Table 2.2 also includes shift instructions and bit-clearing and -setting instructions. The MCS-51 has only rotate instructions that shift the data left or right, moving the LSBit (MSBit) to the MSBit (LSBit) through the carry bit. The MCS-96 has only shift instructions, including both logical and

TABLE 2.2 Data manipulation instructions from four representative 8-32 bit MCUs

Action	MCS-51	MC6811	MCS-96	NC32FX16*
Clear Accumulator	CLR	CLR*,CLRA, CLRB	CLR,CLRB	
Clear Bits	CLR	CLC,CLI,CLV, BCLR*	CLRC,CLRVT,	BICi,BICPSRi
Set Bits	SETB	SEC,SEI,SEV, BSET*	SETC,	BISPSRi,SETCFG, SBITS,SBITPS
Rotate Accumulator	RL,RLC, RR,RRC	ROL*,ROLA, ROLB,ROR*, RORA,RORB		ROTi
Swap Accumulator Nibbles	SWAP			
Shift Accumulator		ASL*,ASLA, ASLB,ASLD, ASR*,ASRA, ASRB,LSL*, LSLA,LSLB, LSRD	SHL,SHLB, SHLL,SHR, SHRB,SHRL, SHRA,SHRAB, SHRAL	LSHi,ASHi
Data Transfer	MOV*,MOVC, MOVX,PUSH, POP,XCH, XCHD	LDAA,LDAB, LDD,LDS, LDX,LDY, PSHA,PSHB, PSHX,PSHY, PULA,PULB, PULX,PULY, STAA,STAB, STD,STS,STX, STY,TAB,TAP, TBA,TPA,TSX, TSY,TXS,TYX, XGDX,XGDY	LD,LDB, ST,STB, LDBSE,LDBZE, PUSH,POP, PUSHF,POPF	MOVi,MOVQi, MOVMi,MOVZBW, MOVZiD,MOVXBW, MOVXiD,ADDR, SCONDi,EXTi, INSi,EXTSi,INSSI, CVTP,SAVE, RESTORE,LPRi, SPRi,ADJSPi, MOVf,MOVLF, MOVFL,MOVif, BBOR,BBAND, BBFOR,BBXOR BBSTOD,BITWT, EXTBLT,MOVMPi,

*Memory-to-memory operations are also allowed; i,f: Integer operand or floating point operand.

arithmetic, the difference being whether the value of the sign bit is maintained in the right arithmetic shift. Another difference is that the MCS-96 allows byte, word, and double-word shifts by a specified number of bits as opposed to the MCS-51, which only rotates one bit per instruction execution. If one is doing software serial I/O, the MCS-51 single-bit shift is adequate, but if multiple bits need to be shifted, the single-bit shift becomes time consuming since it must be implemented either as a subroutine call or a sequence of inline rotates equal to the number of bits by which it is desired to rotate the register. While the MCS-51 is limited to rotating the accumulator, the MCS-96 can rotate any of the internal registers, which implies that I/O ports can be directly shifted.

The MC6811 has both rotate and shift instructions and can directly shift memory locations, hence I/O data, but with the same limitation as the MCS-51 in that the shifts and rotates can only be performed one bit per instruction, necessitating multiple instruction fetches for multiple bits of shift. The NS32000 appears to have fewer rotate and shift instructions than the other three MCUs until one realizes that the operands of the NS32000 are all general operands, which includes all registers and memory locations. In addition to rotate, the NS32000 has both arithmetic and logical shifts, making for a complete set of shift instructions.

2.3.4   Arithmetic Instructions

The computational instructions of the four MCUs are shown in Table 2.3. There are no particular surprises in that they all have the same basic set of arithmetic and logic operations. The arithmetic operations include addition, subtraction, increment, decrement, as well as integer multiply and divide. The logic operations include AND, OR, and XOR. The only significant difference is in the range of registers and memory locations over which these operations can be performed, with the MCS-51 and NS32000 allowing logic operations directly on memory locations.

The operation differences are due to the MCS-51 and MC6811 having decimal adjust operations for correcting binary operations on BCD coded numbers and the additional mathematical operations unique to the NS32000. These NS32000 operations include vector and array operations as well as native floating point mathematical operations: multiply, divide, polynomial, dot product, scaling, and others unique to floating-point calculations. These are to be expected in an MCU such as the NS32FX that is tailored to a specific application.

2.3.5   Bit Manipulation Instructions

In control applications, the ability to manipulate individual bits is important, as can be seen by the inclusion of extensive bit manipulation capability in all four of the MCUs. While there is no unique instruction explicitly to clear one or more of the general registers in the NS32000, it is easy enough to do with other operations. The other MCUs have direct accumulator clear instructions. All four MCUs except the MCS-96 have the capability to set and clear individual bits, with the only significant difference being the range of addresses over which the operation can be performed. For example, the MCS-51 is limited to clearing individual bits in the first 256 bits of internal RAM memory, which equates to all of the general registers as well as the I/O ports. The MCS-96 would be required to clear or set a bit by ANDing or ORing an operand with a memory location. Of course, it still has the capability of directly setting the carry bit and setting or clearing the carry and overflow bits. The MCS-51 has one feature not shared by the other

TABLE 2.3 Computational instructions of four representative 8-32 bit MCUs

Action	MCS-51	MC6811	MCS-96	NC32FX16*
Add to Accumulator(s), Add Packed	ADD	ABA,ABX, ABY,ADDA, ADDB,ADDD	ADD, ADDB	ADDi,ADDQi, ADDPi,ADDf
Add with Carry to Accumulator	ADDC	ADCA,ADCB	ADDC, ADDCB	ADDCi
Subtract from Accumulator, Subtract Packed			SUB, SUBB	SUBi,SUBPi, SUBf
Subtract from Accumulator with borrow	SUBB	SUBA,SUBB, SUBD,SBA, SBCA,SBCB	SUBC, SUBCB	SUBCi
Increment/Decrement Accumulator(s) or registers	INC,DEC	INCA,INCB,INS, INX,INY,DECA, DECB,DES,DEX, DEY,INC*,DEC*	INC,INCB, DEC,DECB	INDEXi
Multiply Registers	MUL	MUL	MUL,MULU, MULB,MULUB	MULi,MELi,MULf
Divide, Divide with Rounding	DIV	IDIV,FDIV	DIV,DIVB, DIVU,DIVUB	DIVi,QUOi, DELi,DIVf
Other Arithmetic, Modulo,Remainder, Absolute Value, Rounding,Floor, Polynomial, Dot Product,Scale			EXT,EXTB, Norml	MODi,REMi,ABSi,R OUNDfi, TRUNCfi, FLOORfi,ABSf, POLYf,DOTf, SCALBf,LOGBf
Decimal Adjust Accumulator(s)	DA	DAA		
Logical with Accumulator (AND,OR,XOR, 1's Complement, 2's Complement)	ANL*, ORL*, XRL*, CPL	ANDA,ANDB, EORA,EORB,ORAA, ORAB,COMA, COMB,COM*, NEG*,NEGA,NEGB	NEG,NEGB, NOT,NOTB, AND,ANDB, OR,ORB, XOR,XORB	ANDi,ORi, XORi,COMi, NEGi,NOTi, NEGf
Array Operations				CHECKi,INDEXi
String Operations				MOVSi,MOVST, CMPSi,CMPST, SKPSi,SKPST

*Memory-to-memory operations are also allowed; i,f: Integer operand or floating point operand.

MCUs: the SWAP instruction. SWAP interchanges the two 4-bit nibbles of the accumulator. This is particularly useful when doing decimal arithmetic, shifting out BCD numbers, or when 4-bit rotates are required.

2.3.6 Program Flow Control Instructions

There appears to be no consistent interpretation of the flow control instructions. The usage here follows that of Hennessy and Patterson[15] in that *jump* is used to indicate unconditional change in control and *branch* to indicate when the change is conditional. This is in contrast to another usage that interprets branches as program-relative transfer of control, independent of cause, and jumps as transfers of program control to absolute addresses not relative to the PC. Using the *cause* definitions is difficult here since the nomenclature among the four example MCUs is not consistent, as can be seen in Table 2.1. The MCS-51 uses JMP and its variations to refer to both conditional and unconditional changes in program flow. The MC6811, which has many more conditions that it can branch on, makes the distinction between *branch* and *jump* based on whether the address of the next instruction to be executed is relative to the program counter or whether it is specified by a complete 16-bit address. It has both conditional and unconditional PC relative branches and subroutine calls as well as unconditional relative and absolute jumps. The MC6811 does lack one type of instruction that the other three incorporate: one that performs an arithmetic operation and then branches on the result. In the MCS-51 and MCS-96 there is a decrement and jump on nonzero. In the NS32000, the ACBi instruction adds a 4-bit constant to any register or memory and branches if the result is nonzero. This type of instruction is particularly useful when indexing though loops or arrays, as the indexing and testing operation are combined in a single instruction.

As with the MC6811, the MCS-96 has a great variety of conditions on which it can execute conditional branches (referred to by the MCS-96 as *jumps*). The single branch instruction BR is an unconditional jump to a 16-bit address as retrieved from any location in memory. The SJMP and LJMP instructions refer to PC-relative unconditional jumps with the offset being 11 bits ($+1024$, -1023) or 16 bits, respectively. The 16-bit offset allows for a jump to any place in memory relative to the PC. In addition to the jumps and branches, the MCS-96 has the normal jumps to subroutines.

The NS32000 adheres to the standard of a branch being PC relative and jump being a PC change to an absolute value specified by an operand. In addition to the usual conditional branches (Bcond), the NS32000 also has a multiway branch (CASEi) as well as a breakpoint trap (BPT), flag trap (FLAG), supervisor calls (SVC), and calls to external procedures (CXP, CXPD). The CASEi instruction can directly implement such high-level instructions as C's "switch" statement.

Testing without acting on the result of the test is not available in the MCS-51 and very limited in the MCS-96 (CMP, CMPB). Both the MC6811 and the NS32000 allow for the testing of accumulators and memory and the setting of the associated flag or status word bits while only the NS32000 allows for the testing of individual bits and/or the testing of bits immediately followed by the setting or clearing of that bit.

The handling of interrupts is also distinctly different among the four MCUs studied. The enabling and disabling of interrupts in the MCS-51, MC6811, and NS32000 is controlled by the state of a bit in a register, where that bit is read/written by a normal data movement instruction. In the MCS-96, the interrupts are enabled or disabled by separate EI and DI instructions. In addition to the normal maskable and nonmaskable interrupts, the MC6811 has a software interrupt (SWI) that behaves exactly as a hardware interrupt, although under program control. That is, the execution of an SWI instruction causes the state of the machine to be saved on the stack and restored on the return from interrupt. The traps and breakpoints of the NS32000 are an extension of this.

2.4 PROGRAMMABLE CONTROLLER INSTRUCTIONS

For the rest of this chapter, the abbreviation PC refers to programmable controllers. The context should make this apparent, but if there is any ambiguity, the term is spelled out. Process control (also abbreviated PC) and programmable controllers are introduced here since process control is a natural application of MCUs. The original programmable controllers were relays controlled by a series or parallel connection of normally open (NO) or

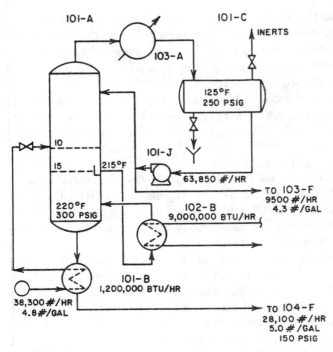

Figure 2.13 A typical process which can be controlled with a programmable controller. (*Béla G. Lipták and Kriszta Venczel (eds)*, Instrument Engineers' Handbook, *rev. ed., Radnor Penn.: Chilton Book Company, 1982*)

normally closed (NC) switches. A process is a system that should be controlled, such as a pulp paper mill for making paper or a water desalination plant. An example of a process is shown in Fig. 2.13, which indicates motors, valves, and heating elements, all of which can be under the control of a PC. Program control (implemented in a PC) is a "control system in which the set point varies with time according to a predetermined program."[23] Program control has been implemented with hardwired relay logic, but it is increasingly being done with MCUs and other digital computers because of the versatility and relative ease of altering a software control program rather than physically adding series or parallel switches or relays. Some process control involves servo actions in which it is desired to control the positioning of a milling machine or ship that is stationkeeping over an oil well several hundred feet below the surface while the well is being drilled. Some of these process controls, which are more appropriately called *automatic control*, are beyond the scope of the simpler MCUs but may not be beyond the capability of the newest generation of 32-bit MCUs.

Ladder logic and *ladder diagrams* are a common form of documenting and designing process control since they are visual representations of the system as well as wiring diagrams for the switches and actuators. On the one hand, relating ladder logic to the instructions available in an MCU will help someone with a process control background and a familiarity with that paradigm to relate that experience to the programming of MCUs. From the other extreme, the MCU applications engineer may be asked to convert an existing switch and relay PC to one that is under the control of an MCU or more capable computer. In either event, it is instructive to relate the two symbolic methods since engineers with both backgrounds will need to understand the capabilities of the MCU.

2.4.1   Relay Logic

The format of a single rung of a ladder diagram is shown in Fig. 2.14 with a source of current and condition symbols on the left and output symbols and a sink for the current on the right, with each element given a unique identifying number. An output is actuated if there is a continuous line from the source to the sink through the output. The conditions represent conditional switches, which can be either NO or NC and are actuated by the value of some variable such as time or temperature. The outputs are represented by a coil symbol which is indicative of their origin as the coil windings of a relay which, when energized, allows current to flow through its contacts to the device being controlled. This, in effect, allows a low actuating current to control a large current device. There are six categories of PC instructions:[24] relay type, timer/counter, arithmetic, data manipulation, data transfer, and program control. The first of these, relay logic and timer/counters, are related and defined in terms of the programming language for computers and are shown in Table 2.4.

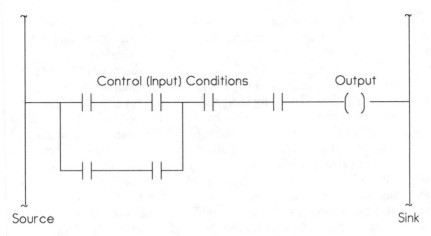

Figure 2.14 Basic ladder diagram showing conditions on the left and the controlled output on the right.

An abbreviated section of a ladder diagram is shown in Fig. 2.15. The simplicity of this ladder diagram should not lull one into a false sense of security since these diagrams can get as convoluted and difficult to interpret as computer wiring diagrams. The symbol under *CR13* is a NO switch, *PB13* refers to a pushbutton, NO switch, below *Alarm* is an indicator light; and below that, a motor. This ladder rung is read as "the motor and the alarm will be energized when the NO switch controlled by output CR13 and the NO switch controlled by output CR2 are closed, and when pushbutton PB3 is pushed." All elements of a ladder diagram are labeled both symbolically as to the type of condition necessary to make the element true as well as to the alphanumeric cross-reference to the input or output that activates that condition. The element below the motor, a clear circle, refers to a relay labeled CR13 which, when actuated, causes the NO switch of the previous ladder rung to close and allow current to flow. This second rung of the ladder also shows that there can be parallel paths which can enable an output, in this case the relay labeled *CR13*. To summarize, when enough of the conditions on the lefthand side of the ladder diagram are true so that there is a complete path for current to flow from the source to the sink through the output on the right, then that output will be enabled.

These two rungs of the ladder can be converted into individual subroutines that can be called sequentially by an executive routine that supervises the PC. Although the process of conversion to assembly language is straightforward, an example is not given here since the process of programming MCUs is introduced in Chap. 3, where it will be seen that Petri nets are a more complete way of describing ladder diagrams. Other ways of documenting the operation of an MCU or PC are also in Chap. 3. In the context of Petri nets, the conditions equate directly to the tokens that enable transitions.

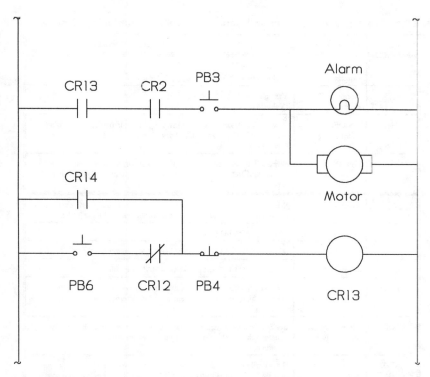

Figure 2.15 Sample ladder diagram showing interaction of control relays between different rungs. *(From* Instrument Engineers' Handbook *by Béla G. Lipták and Kriszta Venczel (eds). Copyright 1982 by the author. Reprinted with the permission of the publisher, CHILTON Book Company, Radnor, PA)*

2.4.2 Arithmetic, Data Manipulation, Data Transfer, and Program Control

A complete listing of ladder logic[24] as it applies to PCs would include conditions and outputs other than the relay logic and timer/counters of Table 2.4. The additional conditions and outputs would be arithmetic, data manipulation, data transfer, and program control instructions. It seems as if these were added to the basic relay and timing/counter symbols to make the ladder diagrams compatible with the improved capabilities of the computers that were used to implement them. These additional "computer-like" conditions and outputs are not included here since they are not necessary to understanding ladder diagrams and they are directly duplicable in the assembly language instructions of MCUs.

2.5 SUMMARY

This chapter introduced MCU instruction sets by first considering elements that are desirable in an instruction set, including completeness and orthogonality of instructions (particularly when considering the addressing modes

TABLE 2.4 Basic Relay and Timer and Counter Instructions for Ladder Diagrams

	Symbols	Meaning	MCU Equivalent Instruction Sequence
C o n d i t i o n s	--] [--	Normally-Open Contact	Variable with positive logic, (Value ≠ 0) ≡ True
	--]/[--	Normally-Closed Contact	Variable with negative logic, (Value = 0) ≡ True
	--]↓[--	ON-OFF Transition	When a ↓ is detected on one MM I/O port, write 010 sequence to a second MM I/O port, maintain 1 level for predefined length of time (one-shot)
	--]↑[--	OFF-ON Transition	When a ↑ is detected on one MM I/O port, write 010 sequence to a second MM I/O port, maintain 1 level for predefined length of time (one-shot)
O u t p u t s	--( )--	Energize Coil	Write 1 to memory mapped (MM) I/O port, continually test actuating conditions and de-energize coil when conditions no longer TRUE.
	--(/)--	De-energize Coil	Write 0 to MM I/O port
	--(L)--	Latch Coil	Write 1 to MM I/O port and shadow RAM variable
	--(U)--	Unlatch Coil	Write 0 to MM I/O port and shadow RAM variable
	--(TON)--	Time Delay Energize (ON)	Set timer compare register to value equal to preset time plus current counter value, enable timer interrupt, when interrupt occurs, enable condition
	--(TOF)--	Time Delay De-energize (OFF)	Set timer compare register to value equal to preset time plus current counter value, enable timer interrupt, when interrupt occurs, disable condition
	--(RTO)--	Retentive On-Delay Timer (Non-volatile)	Periodically copy contents of timer and timer compare registers to non-volatile EEPROM (shadow EEPROM)
	--(RTR)--	Retentive Timer Reset	Zero the appropriate shadow EEPROM variable value
	--(CTU)--	Up-counter (increment)	Counter registers can be configured to accumulate external values or an internal variable can be incremented
	--(CTD)--	Down-counter (decrement)	Counter registers can be configured to count external values (preload register with 2's complement) or an internal variable can be decremented
	--(CTR)--	Counter Reset	Clear the counter register or the counter variable

available to the programmer). Instruction formats were then discussed, from the simplest single-word, no-operand instructions through the multibyte, complex instructions that include all of the necessary information for a binary operator (i.e., the addresses of both sources and the address of the destination). The general addressing modes were discussed before describing the differences in SISC, RISC, and CISC approaches to MCU design.

Instruction sets were then discussed in terms of their intended use, whether for business, science, or control applications. It was shown that the basic differences between machines with different computational orientations were not found in the complement of instructions available, but rather in the hardware used to implement them. The chapter ended with a comparison of the specific instruction sets of four representative 8- to 32-bit MCUs and a brief description of relay logic and ladder diagrams, and how the conditions and outputs relate to the programming steps in an MCU.

REFERENCES

1. D. Tabak, *RISC Systems*. United Kingdom: RSP Ltd; and New York: Wiley, 1990.
2. D. R. Ditzel, "Reflections on the High-level Language Symbol Computer System," *Computer*, 14(7)55–66, July 1981.
3. Y. Chu, ed., *High-Level Language Architecture*. New York: Academic Press, 1975.
4. S. R. Vegdahl, "A Survey of Proposed Architectures for the Execution of Functional Languages," *IEEE Transactions on Computers*, C-23(12):1050–1071, December 1984, as reprinted in *Software-oriented Computer Architecture*, E. B. Fernandez and T. Lang, eds. Washington, DC: IEEE Computer Society Press, 1989.
5. W. Stallings, *Computer Organization and Architecture*. New York: Macmillan, 1987.
6. J. P. Hayes, *Computer Architecture and Organization*, 2nd ed. New York: McGraw-Hill, 1988.
7. *MCS-51 Family of Single Chip Microcomputers*. Intel publication #121517-001, July 1981.
8. H. Katzan, Jr., *Computer Organization and the System/370*. New York: Van Nostrand Reinhold, 1971.
9. R. H. Eckhouse, Jr., and L. R. Morris, *Minicomputer Systems, Organization, Programming, and Applications (PDP-11)*, 2nd ed. Englewood Cliffs, NJ: Prentice-Hall, 1979.
10. G. W. Gorsline, *Computer Organization*. Englewood Cliffs, NJ: Prentice-Hall, 1986.
11. N. Marovac, "A Systematic Approach to the Design and Implementation of a Computer Instruction Set," *Computer Architecture News*, 11(1):19–24, 1983.
12. G. K. Kapur, *IBM 360 Assembler Language Programming*. New York: Wiley, 1970.
13. *MCS-48 and UPI-41 Assembly Language Manual*. Intel publication #9800255-06.
14. W. L. Van der Poel, "The Essential Types of Operations in an Automatic Computer," *Nachrichtentechnische Fachberichte*, 4:144–145, 1956.
15. D. A. Patterson and J. L. Hennessy, *Computer Architecture: A Quantitative Approach*. San Mateo, CA: Morgan Kaufmann, 1990.
16. V. C. Hamacher, Z. G. Vranesic, and S. G. Zaky, *Computer Organization*, 3rd ed. New York: McGraw-Hill, 1990.
17. D. P. Siewiorek, C. G. Bell, and A. Newell, *Computer Structures: Principles and Examples*. New York: McGraw-Hill, 1982.
18. S. Baba, K. Matsubara, and K. Noguchi, "High-Performance Single-Chip Microcontroller H-8 Series," *Hitachi Review*, 38(1):1–10, 1989.
19. "NS32CG160-15/20/25 32-bit Integrated System Processor," National Semiconductor, Santa Clara, CA data sheet, 1990.
20. "NS32GX320-20/25/30 High-performance FAX Processor," National Semiconductor, Santa Clara, CA data sheet, 1990.
21. "NS32FX16-15/20/25 High-performance FAX Processor," National Semiconductor, Santa Clara, CA, data sheet, 1990.
22. G. E. Bernier, "Forth Based Controller," *WESCON '82 Conference Record*, Anaheim, CA, session 17B/4, pp. 1–6, 14–16, September 1982.
23. B. G. Lipták and K. Vencezel, eds. *Instrument Engineers' Handbook*, rev. ed. Radnor, Pennsylvania: Chilton Book Company, 1982.
24. C. T. Jones and L. A. Byran, *Programmable Controllers, Concepts and Applications*. Atlanta, GA: International Programmable Controls, 1983.

PROBLEMS

1. What is the disadvantage of a completely orthogonal instruction set?

2. List a RISC repertoire of instructions that you feel would be necessary in an MCU designed primarily to implement a programmable controller for process control.

3. If you were designing an instruction set for an MCU dedicated to controlling a video game, what specific instructions would you incorporate that would be considered unnecessary for a general-purpose MCU?

4. Discuss the effect of register size on the computational capability of an MCU and the number and type of native data types that can be made available.

5. Draw a ladder logic diagram to control the magnetron of a microwave oven. Assume that there is an interlock switch on the door, a start pushbutton, and a timer that is NC when set to a time other than zero.

6. List the advantages and disadvantages of interrupts that automatically stack the state of the machine upon entry and those that store nothing but the return address. How does this architectural difference affect the instruction set of an MCU?

7. Discuss the impact of memory-mapped I/O on instruction sets.

8. With reference to Table 2.1, why do you think the 8- to 16- bit MCUs do not have bit-test instructions, but the 32-bit MCU does?

9. With reference to Table 2.2, why do you think only the MCS-51 has a swap nibbles instruction?

10. Why are instruction sets designed to operate with different data types?

11. List several alternative structures for instructions with specific reference to the number and type of operands.

12. What would be the disadvantage of a machine designed to operate only with real-number data types (no integers or bytes)?

13. Even though there are 2^k possible instructions with an instruction length of k bits, why are fewer instructions usually implemented?

14. What are the disadvantages of using benchmarks to compare the performance of various machines, particularly if they are of different architectures?

15. What are several characteristics of control-oriented instructions that distinguish them from business- or science-oriented instruction sets?

16. What is the advantage of making a computer with a large address space, but at the same time limiting the number of address pins that leave the chip, thereby severely restricting the actual memory address space available to the computer?

17. What is the difference between shift and rotate instructions?

18. Discuss the relative merits of an instruction set that has specific instructions for bit manipulations relative to another instruction set that performs bit manipulation through logical operations of memory or register variables?

19. If one could implement only a single arithmetic operation in a computer, would it be better to implement two's complement subtraction or addition? Why?

20. What are the values and limitations of a single-operand instruction?

Controller Software Design

In Chap. 2, the architectural options used in the implementation of single-chip microcontrollers (MCUs) were introduced. The assumption is that any of the available MCUs have sufficient capabilities to perform any necessary computations and that the selection of which architecture or implementation to use is based on other considerations such as cost, availability, or lack of need to retrain engineers and technicians on a new MCU. In this chapter, microcontrollers are considered equivalent in the sense that at some level of abstraction, there exists a hypothetical MCU that is capable of performing the desired task. By taking this view, one can design the software independently of the machine on which it is going to be implemented. The key word here is *design* rather than *code*. Brooks[1] has offered a generally accepted partitioning of the time associated with the various tasks of computer software design as being "⅓ planning, ⅙ coding, ¼ component test and early system test, ¼ system test, all components in hand." The temptation is to start coding immediately and let the design evolve. This approach is invariably doomed for any project but the simplest. Experience has shown that Brooks' estimate of only one-third planning (designing) may be optimistically short. Part of the temptation to start coding from the start is the lack of availability of suitable design tools for the expression and logical progression of software development for controller design. This can be traced to the lack of a suitable higher-level model of the complete controller system and the familiarity with writing computation oriented software. This chapter presents two models for machine design—state tables and Petri tables—

and then integrates them into a single high-level design methodology that inevitably leads to the final one-sixth of the effort to code the software for a particular MCU.

Before continuing this discussion, some terminology must be clarified or the term *control* and its many variations and their use could lead to confusion. The term *controller* is used to here indicate the complete system, including the MCU and other electronics used to control a plant. The term *plant* is borrowed from automatic control and is meant here to imply the preexisting hardware that should function in a specified fashion. This does not mean that controllers only use the principles of automatic control; they can also implement timing functions, Boolean algebraic functions, or human-machine interface functions. The combined controller and plant is referred to here as a *system* or *microcontroller system*. The *single-chip microcontroller* is, or is part of, the controller, and it executes a computer program that *controls* the plant in the sense that it *interfaces* with actuators by generating *control signals* determined by the program it is executing. There is another level of reference in which the term *control* is used. This level concerns the program that is executing. The program execution is itself under an implicit type of control that is independent of the control function it is performing relative to the plant. The order of program execution can be determined in two ways. The first way is the *Von Neumann* approach, which means that all of the steps of program execution of the program are determined specifically by the program and in an order predetermined by the software programmer. This type of program is known as *control flow*. A second type of software program control that can be executed by a computer is called a *data flow*[2] program, since the data itself determines which instruction to execute next. In this chapter, liberties are taken with this definition of data flow to allow for consideration of external, asynchronous events as data that can control the flow of the computer program. The term *control* can be applied at the system as well as the computer program level, but it should be clear from the context which is meant.

Process control computers deal with four major function areas:[3] data acquisition and conditioning, direct digital control, supervisory control, and sequence control. Data acquisition and sequence control can be implemented as control flow machines by the very nature of their task and the usual requirements of collecting data at uniform intervals. Direct digital control and supervisory control, depending on the application, may have irregular demands placed on them that are neither predictable nor regular. These process control programs are representative of the dataflow paradigm.

Two models, the *finite-state machine* (FSM) model and the *Petri net* (PN) model, are presented here to address two different types of controller problems: those suitable for the application of the control flow (CF) approach and those suitable for the data flow (DF) approach. In the control flow approach, the steps of the control are deterministically and specifically laid out, with complete control of the timing and sequencing of inputs, outputs, and computations predetermined by the software designer. In the data flow approach, the data itself determines the order of computation and the order of execution of the controller program. In the context of controller design, a broad view of data flow is used to include the processing of interrupts. Since interrupts are in general nondeterministic, they must be processed as they arrive, subject to the operational constraints of the system. FSMs are a subset of general state machines and are introduced first since they are familiar to engineers and conceptually simpler to discuss. This is followed by the introduction of Petri nets, since they are able to handle interrupts and data flow concepts more effectively. A slightly restricted view of PNs is then presented using a subset of PNs as FSMs and the two concepts merged into a common method for designing controller software so one method can be used for the two different controller system models. The chapter concludes with a discussion of the use of higher-level languages and their integration with assembly language. Examples are presented to instantiate the methods presented.

3.1 FINITE-STATE MACHINE MODEL

Principles of automata theory have previously been applied to the design of computer software. The concepts are simple and straightforward and have been shown to be effective in the design and implementation of small, complex executive routines.[4] Even though the details of hardware implementation are hidden from the designer and the details of database concerns are not addressed, the control graph approach is attractive because of the ability to apply the principles of automata theory to determine the reachability of states, as well as apply the decomposition theorem to reduce the design and implementation into a series of orthogonal subprojects. The finite-state machine model is an example of these concepts from automata theory.

   The point of departure for controller software design is to move a metalevel above the MCU and consider the microcontroller system as a sequential machine since the discussion here is limited to a single-instruction, single-data (SISD) implementation. This FSM, specifically a Mealy machine, can be formally defined as an ordered quintuple of sets[6,7]

$$M = (S, I, O, \delta, \beta) \tag{3.1}$$

where S: finite, nonempty set of states
$\quad\ \ I$: finite, nonempty set of inputs
$\quad\ \ O$: finite, nonempty set of outputs
$\quad\ \ \delta$: $S \times I \to S$, the state transition function
$\quad\ \ \beta$: $S \times I \to O$, output function

This is not only the most general formulation but also a useful and intuitive formulation for the design of controller software since it is easy to visualize the controller as being in a particular state at each instant of time. The first step in designing an FSM is to specify the set of inputs and the values that they can take on as well as the set of outputs and their possible values. The complete set of states is usually difficult to specify at the outset of a machine design, but usually is developed as the design process progresses. The software design function starts to become difficult in the specification of the state transition function δ and the output function β.

3.1.1 The State Transition Function

It may not be obvious that any function, much less the state transition function, is actually a set, but it is not difficult to see when you consider that $\delta : S \times I \rightarrow S$ means a mapping from the cartesian product of the set of states S and the set of inputs I back onto the set of states S. Remember that the cartesian product of two sets is just another set consisting of ordered pairs of all elements of the first set with all elements of the second set. For example, if $S = \{s_1, s_2\}$ and $I = \{i_1, i_2, i_3\}$, then $S \times I = \{(s_1, i_1), (s_1, i_2), (s_1, i_3), (s_2, i_1), (s_2, i_2), (s_2, i_3)\}$. Continuing with this example, one can then set up a state transition function as in Table 3.1. Usually the tabular method is not used, but rather a state transition diagram. The example of the table as a state diagram is shown in Fig. 3.1. An example of a more complex state diagram typical of that used to document the internal operation of a microprocessor is shown in Fig. 3.2, which documents the internal operation of the 8085 microprocessor as a function of the clock period and the status of control signals. The clock is not necessary to the concept of a state diagram but only makes it easier to design the microprocessor as a synchronous FSM.

TABLE 3.1 A Combined Moore Machine and Mealy Machine Example of the State Transition Function and the Output Functions (Output Equivalent Machines)

Present State	Input	(State, Input)	Next State	δ	β Mealy	β Moore
s1	i1	(s1,i1)	s2	((s1,i1), s2)	((s1,i1), o1)	(s1, o1)
s2	i2	(s2,i2)	s2	((s2,i2), s2)	((s1,i1), o2)	(s2, o2)
s1	i3	(s1,i3)	s1	((s1,i3), s1)	((s1,i1), o1)	---
s2	i1	(s2,i1)	s2	((s2,i1), s2)	((s1,i1), o2)	---
s1	i2	(s1,i2)	s2	((s1,i2), s2)	((s1,i1), o1)	---
s2	i3	(s2,i3)	s1	((s2,i3), s1)	((s1,i1), o2)	---

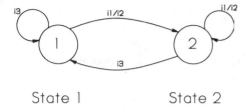

State 1   State 2

Figure 3.1 State diagram implementation
of state table. Circles represent states, and
directed arrows show transitions from one
state to another.

3.1.2 The Output Function

The same argument can be made for the output function β as was made for
the state transition function δ, that it also is a set. In this Mealy model, $S \times I$
$\rightarrow O$ means that the output is a function of both the current state and the
current input. There is an alternative model, the Moore machine, whose
output value is only a function of the current state. That is, $\beta:S \rightarrow O$ is the
output function. It is well known that any Moore machine can be converted
to an equivalent Mealy machine and vice versa,[6] and there are times when it
may be more convenient to make the initial formulation as one type of
machine or the other knowing that the conversion to the other type can be
easily made. There is a second difference between the two types of
machines—the type of output that is generally associated with each of
them.[7] The Moore machine usually has a steady-state, or level, output such
that as long as the machine is in a particular state, the output is a constant
value. Mealy machines, on the other hand, usually have pulsed outputs that
occur during the transition from one state to the next. There is no reason
why either a pulsed or level output cannot be associated with either type of
machine. For consistency throughout the rest of this book and because the
Mealy model can be designed using fewer states than the Moore model, the
Mealy model is used exclusively. An example of the output function for both
a Moore machine and a Mealy machine is shown in Table 3.1, along with the
state transition function.

3.1.3 FSM Table or FSM Diagram

Although it is easy to use a state diagram at the early stages of development
to map the flow of the machine into a visual representation, it is not useful for
any but the smallest of designs. The difficulty arises in the use of the state
diagram in several areas: documentation, readability, and conversion to
code. Documentation is difficult because of the directed graph nature of the
diagram. For example, if a state needs to be added, lines must to be erased,

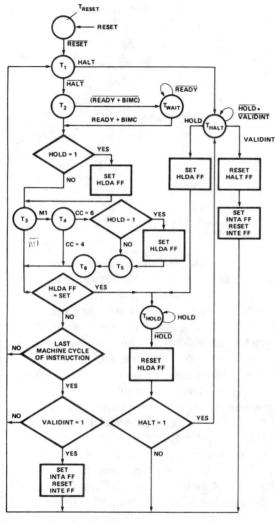

Figure 3.2 State diagram of internal operation of 8085 microprocessor. *(Courtesy of Intel Corporation)*

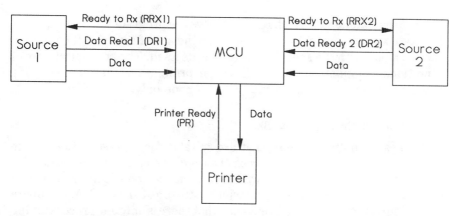

Figure 3.3 Block diagram of print spooler with two sources and one printer attached to an MCU.

new circles drawn, and new arcs drawn all without disturbing the rest of the diagram. Unless there is an automatic tool like some of the flowcharting software, the documentation soon becomes unreadable. As the program expands, the size of the page on which the state diagram is drawn comes into play. If the diagram needs to be continued onto another page, interconnecting lines must be cross-referenced, and the original intent of an easy to understand visual representation of the states ceases to be meaningful. In addition, there is no direct way to convert diagrams to executable code. For the lack of an automated way to update state diagrams, design documentation will always lag behind the actual code and, except at the earliest stages, will not reflect the actual program that is implemented.

Decision tables are an earlier tabular method for the design of software.[4] These have been applied to deterministic engineering problems to list confusing decision making problems. There are many forms of the decision table, but they all revolve around the notion that for every possible combination of inputs, the desired response of the system is known. Although there is a branch of automata theory dealing with stochastic machines, the consideration here is restricted to deterministic machines that can be properly modeled by decision tables. The particular form of decision table that is used here is a *state table* since it is intended to represent an FSM explicitly, although in a different form than a state diagram.

State tables[8] are an alternative to state diagrams that take as their point of departure the state transition function and output function as depicted in Table 3.1. A table of this type, with appropriate modifications introduced later, is easily maintained by simple word processing software. If a document is easy to maintain, it is more likely to keep abreast of the software development; and with the appropriate discipline, no changes will be made to the actual code implementing the states until the state table is modified. Since the state table is already a text file, it can be directly incorporated into the code as comments, which readily allows the programmer to implement the state as a separate function and have the specification of the function carried along with the specific section of code itself. The appropriate use of conditional compiler directives allows for printouts of the complete code along with comments, comments only, or code only.

3.1.4 Control Flow and the FSM

The types of control systems that are suitable for representation by the FSM model are those that progress in a regular manner from one identifiable state to another and those with inputs that can be selectively polled in each state to make a decision about which state to go to next. FSMs assume a nice, orderly progression of events and that there is only one process taking place—the one in the FSM. An example of this type of controller is a printer controller (spooler) that works with multiple parallel interfaces and handshaking signals. A simplified state diagram of this type of controller is shown in Fig. 3.4, given the assumption that there are only two sources of data. The printer controller itself is considered the sink for the data. Notice that the

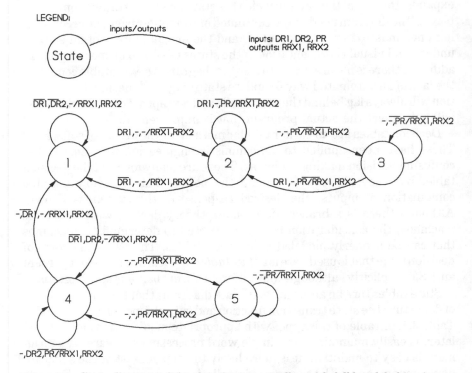

Figure 3.4 State diagram of simple printer controller with full handshaking interface.

TABLE 3.2 Free-Form State Table Equivalent of the Printer Controller State Diagram of Fig. 3.4

Present State	Action	Until Event	Output	Next State
1	Idle	~DR1,~DR2,-	RRX1,RRX2	1
		DR1,-,-	RRX1,~RRX2	2
		~DR1,DR2,-	~RRX1,RRX2	4
2	Read Source 1	DR1,-,~PR	RRX1,~RRX2	2
		-,-,PR	~RRX1,~RRX2	3
		~DR1,-,-	RRX1,RRX2	1
3	Data to Printer	-,-,~PR	~RRX1,~RRX2	3
		DR1,-,PR	~RRX1,~RRX2	2
4	Read Source 2	-,~DR1,-	RRX1,RRX2	1
		-,-,PR	~RRX1,~RRX2	5
		-,DR2,~PR	~RRX1,RRX2	4
5	Data to Printer	-,-,~PR	~RRX1,~RRX2	5
		-,-,PR	~RRX1,~RRX2	4

complete handshaking between the source and sink requires the source not to send a character until the sink (the controller) responds with a ready-to-receive signal. In this way, a single controller can service many different sources to the point of even maintaining buffers of data from all of the sources while only printing from one buffer until that one is exhausted. In this example, the controller has complete control of all of its incoming signals and actuator signals. It is designed to best utilize its resources and in doing so, cannot just respond to data. This type of controller could easily be implemented as a control flow program. A free-form equivalent of this state diagram is the state table of Table 3.2. The columns headings are interpreted as follows:

1. *Present state:* An arbitrary number associated with a particular state.

2. *Action:* The output function column β has been relabeled *action* since *output* is too restrictive in implying a signal that is used to activate an external device. *Action* as used here means anything that is required to be done when in this state, such as computing numerical values, transferring data from one section of memory to another, setting flags, or enabling external hardware, indicators, or actuators.

3. *Until event:* Conditions or inputs that must be satisfied before the FSM can transition to another state. The "~" means the complement of the indicated signal. For example, DR1 is "Data Ready One," and ~DR1 is "Not Data Ready One."

4. *Output:* Signal lines that are connected to the external environment and the values to which they are set. These can be either analog or binary (single-bit) values.

5. *Next state:* The state to which the FSM will transfer after the particular event occurs and the output values are set.

The manner of reading the table is that the FSM is at any given time in one of the numbered states in the leftmost column. The FSM will stay in this state until an event occurs. This event may be the change in a single input value or some combination of input values. After the event, the outputs are changed, and the FSM switches to the same or different state. If the machine is a Moore machine, the outputs are constant and only a function of the state. If the machine is a Mealy machine, as in this printer buffer example, the outputs can be either pulsed or steady until the next state transition, depending on the needs of the system.

3.1.5 A Limitation of the FSM Model

It would seem that the FSM is just what is needed to model any type of controller behavior, since at any given time the controller must be in some particular state, the inputs and outputs are clearly defined, and the state transition function and the output function can be completely defined, but that is not the case. There is a difficulty regarding the handling of interrupts and multiple asynchronous processes. There is no direct provision for these asynchronous service requests in the FSM formalism other than to list each interrupt request as a separate state. Furthermore, each state that can be interrupted must have associated with it its own set of interrupted states, or otherwise the removal of the interrupt request—the removal of the input that caused the FSM to go to the interrupt service routine—leads to an ambiguous return path. Of course, this is not a problem with actual code because of the automatic stacking of a return address upon calling an interrupt-service routine (ISR), but that also must be accounted for in the high-level design of the software if proper operation of the software is to be verified. As an example of this proliferation of states, and continuing with the foregoing printer example, assume that the handshaking between the sources of data and the microcontroller has been eliminated. Assume, furthermore, that the communications are serial and asynchronous and are handled through a universal asynchronous receiver transmitter (UART) that interrupts the MCU whenever it has received a character or completed the transmission of a character. Of course, the MCU could poll the devices, but for the sake of discussion assume that associated with both UARTs is a single ISR that both reads and writes to the UARTs as well as determines which UART generated the interrupt. The state diagram of this modified printer controller is shown in Fig. 3.5, and the associated state table shown in Table 3.3. The problem is evident in that each state that can be interrupted needs duplicate states so that as the number of "normal" states increases, the number of interrupted states increases proportionately. A better way to handle this is to introduce tokens, a concept that is defined shortly.

Another way of considering this proliferation of states is with the concept of *product machines*. Figure 3.6 shows two simplified state machines, one that represents the main program and a second that represents an interrupt service routine. These two machines can be formally combined into a single,

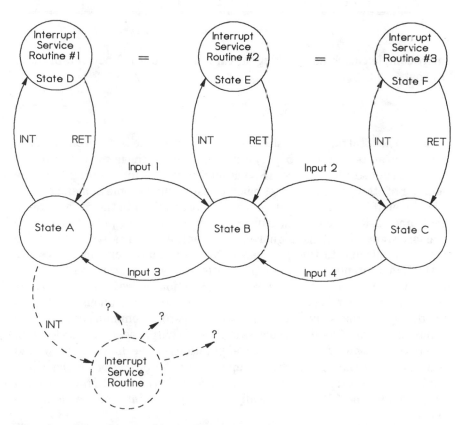

Figure 3.5 State diagram of printer controller operating with UARTS generating interrupts to a single interrupt service routine (ISR).

TABLE 3.3   State Table Representation of State Diagram of Fig. 3.5

Present State	Action	Until Event	Output	Next State
A	something	INT	--	D
		Input 1	--	B
B	something else	INT	--	E
		Input 2	--	C
		Input 3	--	A
C	something else	INT	--	F
		Input 4	--	B
D	service interrupt	RET	--	A
E	service interrupt	RET	--	B
F	service interrupt	RET	--	C

more complex machine by taking the cartesian product of their sets of states so that

$$S_{product} = S_{main} \times S_{ISR} \tag{3.2}$$

That is, the total number of states of the new machine is the product of the number of states in the main program multiplied by the number of states in the interrupt service routine (ISR). For any but the simplest programs, the number of states in the product machine can easily become unmanageable. This level of interrupt attentiveness is only required in the most difficult real-time tasks, such as measurement arrival time of pulsed electronic warfare signals in a dense signal environment. An alternative architecture that is applicable to this problem is to use a multiplicity of processors (MIMD) simultaneously. One of the processors can perform direct memory access (DMA) to move the data from the collection device into the processor memory, the second and third processors operate synchronously but independently on the same bus. The two processors communicate through common memory (mailbox), such as on alternative phases of the clock,[9] with one performing only quantitative computations on the data and the second interfacing to the operator and supervising the operation of the DMA controller. This effectively maintains the partition of the FSMs of the main program and the ISR by physically isolating them into independent systems.

Since there are many less time critical problems to which microcontrollers are applied, an alternative is to partition the software or the state diagram

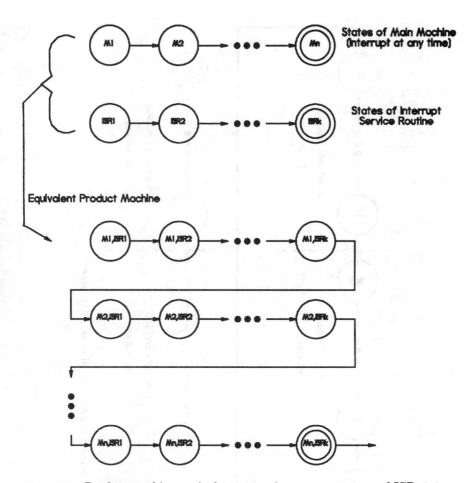

Figure 3.6   Product machine equivalence to main program states and ISR states.

into those regions in which interrupts are allowed and those regions in which interrupts are not allowed. Except for the nonmaskable interrupts, this can usually be done through software control of an interrupt enable bit. This approach is shown in Fig. 3.7, in which it is assumed that there are L distinct states in the ISR. There are E interruptable regions, each consisting of a different number of states. There are also D noninterruptable regions. The total number of states is not the simple product, but rather the sum of the noninterruptable states plus the product of the interruptable states and the number of states of the ISR.

$$\text{Number of states} = \sum_{i=1}^{D} K_i + \sum_{l=1}^{E} [N_l \times j] \tag{3.3}$$

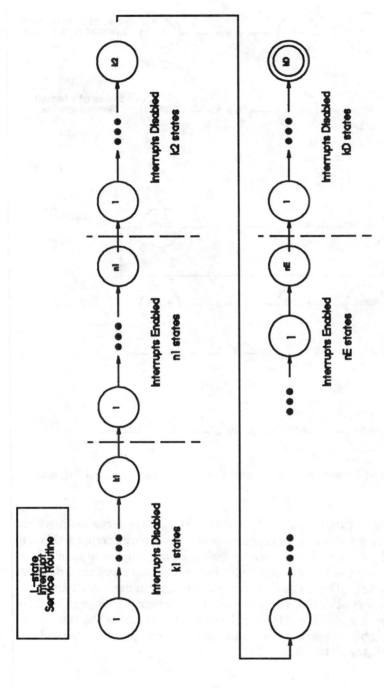

Figure 3.7 State machine with only some portions allowed to be interrupted.

If the number of states during which the main program is allowed to be interrupted is limited, then the total number of states of the resulting product machine can also be reduced. The penalty for this approach is the interrupt latency that is introduced. That is, the maximum time between the occurrence of an interrupt and its being serviced is determined by the longest sequence of states and their attendant execution times that can occur between the interrupting event and its being serviced. Both this maximum delay and the average delay may need to be calculated precisely in some circumstances. The maximum is easy, although tedious, to compute. The average delay time is somewhat more difficult since it is the expected value of the latency,

$$\mathscr{E}\{\text{interrupt latency}\} = \sum_{i=1}^{k} \text{Latency}_i \times P\{\text{Latency}_i\} \qquad (3.4)$$

that is needed.

An example that is expanded in the following IR tracker example is that of one MCU receiving a serial message from another device. Although each individual character generates an interrupt upon its receipt, it is only when the message is complete that it is of interest to the main program. A three state model for this type of ISR is shown in Fig. 3.8 after Lipovski.[10] In the

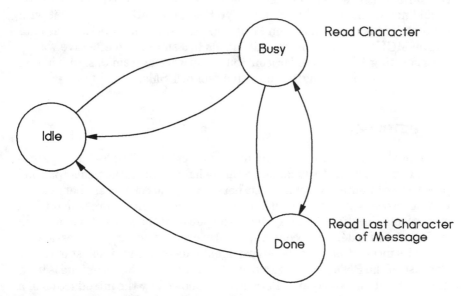

Figure 3.8 Three state model of interrupt service routine to read a string of characters.

idle state, the ISR has not received a character. When an interrupt occurs, the ISR transitions to the busy state. During the busy state, each character that generates an interrupt gets placed in a buffer until a termination character is received or the character count for the message is reached. When this occurs, the ISR transitions to the done state and then returns to either the idle state or to the busy state depending on whether another character has been received. The busy state only applies to the ISR, and while a character is being assembled through the serial port by the UART, the ISR returns control to the main program, which resumes its execution.

Rather than generate product machines from the ISR and main program, the approach taken here is to consider the main program as a foreground program that executes certain functions when semaphores have been enabled. The interrupts are considered as background processes that set semaphores as a result of the interrupt. This approach requires the introduction of Petri nets, discussed in Sec. 3.2.

Of all of the tasks associated with microprocessor and microcontroller system design, the development, writing, and operation of interrupt service routines seem to be the most dreaded. It is easy to see why this is so since the machine is not driven by the programmer, but by some temporally unpredictable interrupting device. Clearly, a method for documenting the behavior of ISRs, incorporating them into the high-level software design, and accounting for their being either enabled, disabled, or enabled only to a certain level needs to be inherent in the high-level design methodology. One should not abandon the highly intuitive FSM, but rather augment it with concepts of token passing from Petri nets. Considering interrupts as data and an MCU as a data flow machine leads to a simple and effective way of incorporating ISRs into microcontroller software system design while at the same time not losing the intuitive feel of FSMs.

3.2 PETRI NETS

Petri nets[11,12] are a graphical and mathematical modeling tool used extensively to model discrete dynamic system behavior, and in particular parallel processor computer systems. Petri nets handle the concurrent behavior of parallel processor systems by using tokens to represent the presence of data at particular locations in the system. The power of the method is more than sufficient to model microcontroller software design and is necessary to model its interaction with its environment except in the simplest of cases. Because of the PN's generalizability, only a subset of the technique is used here. A brief introduction to Petri nets is followed by the introduction of a variation called *colored Petri nets* (CPNs). FSMs are linked to Petri nets through an example of a coin counter for a candy machine, and the coin counter is extended to a CPN representation to introduce a tabular method similar to the state-table method previously introduced. Another example, a subset of an automotive engine control, is used to demonstrate the ease of incorporating interrupts and semaphores into the microcontroller software model using Petri tables (PTs). Petri tables are the desired high-level software design paradigm that is used for the rest of the software development examples in this book.

3.2.1 Formal Definition of a Petri Net

A Petri net[12] is a directed graph consisting two types of nodes, called *places* and *transitions*. Connecting the places and transitions are *directed arcs*. *Tokens*, of which there may be one or many but which are all of the same type, are passed along these arcs from transitions to places and from places to transitions. Places represent conditions. Transitions represent computations that need to be performed and/or actions that need to be taken. Transitions have associated with them a number of *input places* and a number of *output places*. In a graphical representation, places are usually

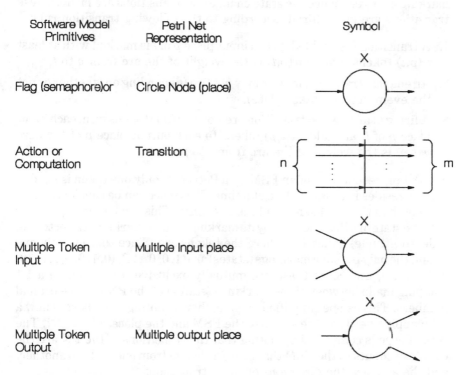

Software Model Primitives	Petri Net Representation	Symbol
Flag (semaphore)or	Circle Node (place)	
Action or Computation	Transition	
Multiple Token Input	Multiple Input place	
Multiple Token Output	Multiple output place	

Figure 3.9 Graphical representations of Petri net components.[13]

drawn as circles, transitions as bars or boxes, and arcs as interconnecting arrows with associated weights indicating the number of tokens, as shown in Fig. 3.9 (taken from Tabak and Levis[13]). If only one token is necessary, no label is associated with the arc. Tokens are represented as black dots. The formal definition of Petri nets is

$$PN = \{P,T,F,W,M_0\} \tag{3.5}$$

where $P = \{p_1, p_2, \ldots, p_m\}$, a finite set of places
$T = \{t_1, t_2, \ldots, tn\}$, a finite set of transitions
$F \subseteq (P \times T) \cup (T \times P)$, a set of arcs
$W : F \rightarrow \{1, 2, 3, \ldots\}$, a weight function
$M_0 : P \rightarrow \{0, 1, 2, 3, \ldots\}$, the initial marking
$P \cap T = \varnothing$ and $P \cup T \neq \varnothing$

A *marking* of a PN associates with each place a number of tokens. A PN structure with no marking is denoted by $N = (P, T, F, W)$, and one with an initial marking is indicated by (N, M_0). The *state* of a PN is specified by its marking, that is, at which places tokens exist and how many are at each of those places. A PN begins its operation with an initial marking, M_0. The marking changes, hence the state changes, as transitions are made. These transitions are made (fired) according to the following transition rule:[12]

1. A transition t is enabled if each input place p of t is marked with at least $w(p,t)$ tokens, where $w(p,t)$ is the weight of the arc from p to t.

2. An enabled transition may or may not fire (depending on whether or not the event actually takes place).

3. A firing of an enabled transition t removes $w(p,t)$ tokens from each input place p of t, and adds $w(t,p)$ tokens to each output place p of t, where $w(t,p)$ is the weight of the arc from t to p.

For the representation of an FSM by a Petri net, only one token is needed since it resides in a single place at a time. This place can be considered as a state both in the FSM sense and the PN sense. This can be seen by noting that the state of a PN is defined by its marking, and its marking is a vector (an ordered n-tuple) indicating where the tokens are. Since there is only one token, the only possible markings (states) are $(1, 0, 0, \ldots)$, $(0, 1, 0, 0, \ldots)$, $(0, 0, 1, 0, \ldots)$, etc. Since these are mutually exclusive, there can be a 1:1 mapping made between these markings (states) of the PN and the ordinal numbers. This can be simplified even further by noting that there is now a 1:1 mapping between the states of the FSM and the places of the PN. The equivalence is completed by noting that the transitions of the PN are the events or inputs to the FSM that cause a change from one state to another, and the arcs are the directions of these transitions.

3.2.2 An FSM Coin Counter in Petri Net Notation

A good example of an FSM is a coin counter for a candy machine that accepts four input symbols; nickels, dimes, 15¢ candy select button, and 20¢ candy select button. A Petri net representation of this machine is shown in Fig. 3.10, which is also easily readable as a state diagram. The single token, which indicates the present state, is shown as a black dot and moves from place to place depending on which transition is fired. The token in place P_1 enables the two transitions, T_1 and T_2. Neither of these transitions fires since each is lacking the proper external input (nickel or dime, respectively) to cause them to fire. If a nickel is deposited into the coin counter, then transition T_1 fires and the token is removed from place P_1 and moved to place

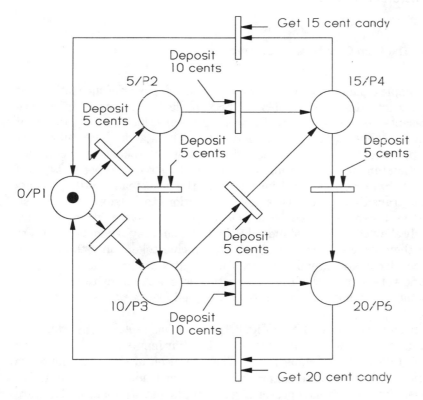

Figure 3.10   Petri net representation of FSM coin counter for candy machine.[12] Coin return transitions are omitted.

P_2. The other option is for a dime to be deposited, which causes transition T_2 to fire, moving the token from P_1 to P_3. The flow can easily be followed for the rest of the transitions until the proper candy button is pushed to enable the final transition for a 15¢ or 20¢ candy bar. This passes the single token back to place P_1, and the cycle starts again. It can be readily seen that the single token is a place marker for the current state as well as an enabling input for the transitions.

A characteristic of the FSM subclass of PNs is that each transition has exactly one incoming arc and one outgoing arc. The event that causes enabled transitions to fire is written alongside the transition. There is a severe limitation of FSMs: They allow for the representation of decisions but do not allow for the synchronization of parallel activities. Although an MCU is itself a SISD machine, when considered with the external processes that it is controlling, there are multiple processes operating in parallel. To consider this reality and account for it in the software design methodology, the concept of PN representation of FSMs can be expanded with the consideration of different types of tokens.

3.2.3   The Coin Counter as a Colored Petri Net

As a simple example that does not exploit the full power of the method but serves to introduce colored PNs, the FSM of Fig. 3.10 is now modified to be represented by a colored PN. Colored PN means that there are different types of tokens rather than a single type. The standard PN allows for multiple tokens, but they are identical in that any token(s) arriving at a transition can enable the transition. Colored PNs require specific types of tokens to be present at transitions to enable them. Instead of writing events next to transitions that cause enabled transitions to fire, events are considered as tokens and come from *source places*. *Source transitions* are already allowed for in the standard PN theory, but they cannot be used here since they are continually enabled, introducing tokens into the system. Source places are intended to be places that asynchronously introduce specific tokens into the PN. These tokens enable transitions and are consumed by transitions when they are fired or are passed on to other places.

Consider the colored PN of Fig. 3.11, which has replaced the transition activating inputs with source places for the four inputs and has included a unique token, the machine token. The operation of the coin counter is identical to the previous example, but the external inputs are considered as tokens. The four tokens representing the inputs that originate at places are as follows:

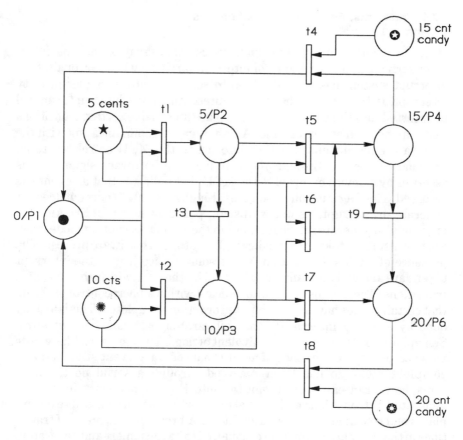

Figure 3.11 Colored PN representation of candy machine coin counter showing inputs as explicit source places.

| 5¢: | ★ | 15¢ bar: | ◉ |
| 10¢: | ✳ | 20¢ bar: | ◉ |

Initially the machine token, the solid dot, is at P_1 enabling T_1 and T_2. If a 5¢ token is placed in source place P_5, the transition is completely enabled, fires, and passes the machine token on to place P_2. Note that in firing, transition T_1 has consumed the 5¢ token from the source place P_5 which is now empty. So places P_1 and P_5 are input places for transition T_1. Place P_2 is an output place for transition T_1. The reason for making this conversion is not obvious in this simple example until it is pointed out that the assumption in the coin counter example is that only one coin is inserted at a time, so the two choices of transitions are mutually exclusive.

3.2.4 Interrupts, Flags, and Semaphores as Tokens

If one considers an interrupt driven system, there may be multiple interrupts arriving at the same time, or interrupts arriving while part or all of the interrupt system has been disabled to service a single, higher-priority interrupt. In these cases, the tokens representing these interrupts can still be inserted into the PN from a source place and enable certain transitions that may be fired at a later time. Another way of communicating with the hardware environment, whether internal to the MCU or related to an external process, is through *flags*. A flag is a hardware signal that is asserted by a device external to the MCU that indicates that an event has occurred. This flag is usually readable through a digital I/O port but does not generate an interrupt. Its status is usually polled by the MCU. Flags may or may not also generate interrupts, but in their noninterrupting mode, they can be considered tokens that indicate the status of a hardware process. The presence of a token is initiated from a source place just as the interrupt tokens are and will be available to be read by the program during specified transitions. There is also a need to indicate within a microcontrol program that certain values have been computed or actions completed. This is usually done by enabling internal (to the software) signals called *semaphores*. Semaphores are the software equivalent of flags. They are variables whose value asserts the occurrence of an internal software event such as that a complete message has been received through a serial port. These semaphores can easily be integrated into PNs by allowing them to be generated at transitions and exist also as tokens that are passed to output places and consumed by other transitions. A broad view is taken of transitions in that they are not merely considered to be consumers and generators of tokens in response to events and enabling tokens, but also the computational entities of the microcomputer software. It is during transitions that computations are performed, measurements made, external devices enabled, and interrupts enabled and disabled. Places are just that—indicators of breaks in the program where changes can be made in the control path as a function of enabling tokens. If there are no external event tokens, the normal control flow program executes by passing the machine token from place to place through transitions. The control is explicitly governed by the program design. The externally and internally generated tokens can alter the program flow, making the microcontroller behave at times as if it were a data flow machine.

Flags are usually hardware interface indicators that are read-only or perhaps are cleared automatically when they are read by the software. For example, an analog to digital converter might set an end-of-conversion (EOC) flag but not generate an interrupt. This may be either the way that the hardware is designed, or the interrupt may be temporarily disabled so the program can focus on a more relevant task, choosing to determine the status of the A/D converter at a time under its control. Once the A/D converter has made this conversion, there may be no temporal urgency to reading the data into the computer since the computer knows when the conversion was ordered and the value in the A/D result register will not change.

As an example of a semaphore, one might consider an internal variable that represents the mode of operation of an interrupt service routine (ISR) task scheduler. The tasks may be ordered such that each is completed in a deterministic order, the longest idle task gets serviced first, or the most important one gets serviced independent of the duration since servicing the other tasks. There are actually two types of semaphores here. The first indicates which method of scheduling is in effect; the other indicates the status of the task ordering, such as which is the most recently serviced interrupt.

3.2.5 Petri Tables as a Software Design Tool

As was pointed out in the description of FSMs and their associated state diagrams, the diagrams soon become unwieldy and difficult to maintain. An alternative to state diagrams or Petri net diagrams is to put the information

TABLE 3.4   Petri Table of Coin Counting Candy Dispenser Showing the Sparseness of Representation of a Simple FSM in This Format

NAME	PLACE	POSSIBLE TOKENS	ENABLING TOKENS	TRANS-ITIONS	OUTPUT TOKENS	ACTIONS	NEXT PLACE	ΔT
0¢	PA	•	• & ★	T1	•	-	PB	
		•	• & ✳	T2	•	-	PD	
5¢	PB	•	• & ★	T3	•	-	PD	
		•	• & ✳	T4	•	-	PC	
15¢	PC	•	• & ◉	T8	•	Dispense 15¢ Candy	PA	
		•	• & ★	T6	•	-	PE	
10¢	PD	•	• & ★	T5	•	-	PC	
		•	• & ✳	T7	•	-	PE	
20¢	PE	•	• & ◉	T9	•	Dispense 20¢ Candy	PA	

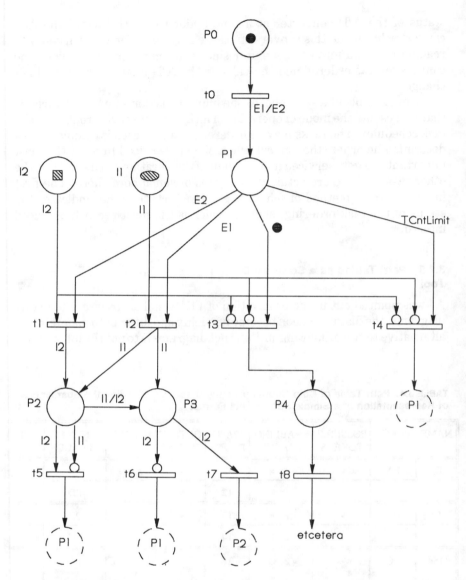

Figure 3.12 Partial Petri net of automotive engine controller with source places I1 and I2 representing interrupts.

in a tabular form, which we call *Petri Tables* (PTs). As a first example, the coin-counting, candy-dispensing colored Petri net of Fig. 3.11 is used, although may table entries are blank. The table of the coin counter is shown in Table 3.4. This tabular form can be easily automated and can be used for development of the microcontroller software since each place (state) and its associated transitions can be coded as separate functions or procedures depending on the language of choice.

As a second example, the partial Petri net representation of an automotive engine controller is shown in Fig. 3.12. In this figure, the concept of source places is introduced to represent the origination of tokens due to two interrupts: I_1, which represents a real time clock; and a lower-level interrupt, I_2, which represents a flywheel sensor producing an interrupt with each rotation. Between these two sensors and software accumulators, the speed of the engine can be measured either directly in terms of revolutions per unit time, or indirectly by counting the time between successive pulses from the flywheel. *TCntLimit* is a semaphore that is maintained in the software and becomes active when the number of pulses reaches the number desired for performing the speed calculation. This semaphore originates at transition T_6 and is a result of the accumulations generated due to interrupt I_1. Also introduced here is the *inhibit arc* represented by the *negating circle* on the input to a transition. There are no complement tokens passed; thus this is the mechanism used to ensure that a transition does not fire under the presence of certain tokens. It is easy to see how the graph of a Petri net can get quickly out of hand, especially if all of the possible interrupt scenarios are taken into account at each transition. This makes it more obvious why the tabular method is a more useful way of implementing a high-level description of the microcontroller software. This example of an engine controller is shown in Table 3.5.

There is one significant difference between the Petri table of the coin machine and that of the automotive engine controller: the lack of the unique machine token explicitly listed in each row under enabling tokens. It is assumed that this is a required token for all transitions since an MCU is an SISD machine, and no confusion can result. Each of the columns is intended to be used in the following fashion:

1. *Name:* A brief verbal description of the state of the machine (its marking)

2. *Place:* A unique, numerical identifier for the place

3. *Possible tokens:* Those tokens such as interrupt tokens, interrupt enable tokens, and semaphores (the unique machine token is assumed here for all places), that can be stored in this place

4. *Enabling tokens:* Those tokens that are required to enable each transition listed in the column to the right

5. *Transitions:* The unique label associated with a particular token that is enabled by the tokens in the column to the left

6. *Output tokens:* Those tokens that the particular transition in the column to the left passes on to the next place(s)

7. *Actions:* Mathematical computations to be performed, semaphores to be

TABLE 3.5 Petri Table of Subset of Automotive Engine Controller Showing Usage of Interrupt and Semaphore Tokens

Name	Place	Possible Tokens	Enabling Tokens	Trans-itions	Actions	Output Tokens	NEXT PLACE	ΔT
Init	P1	E1,E2, Tcntlimit	E2 & I2	T1	Disable I2	I2	P2	
			E1 & I1	T2	Disable I1	I1	P3	
			~Tcntlimit & ~I1 & ~I2	T3	Measure Temp, Store	-	P4	
			Tcntlimit & ~I1 & ~I2	T4	Compute Speed	-	P1	
ISR2	P2	I1,I2	I1 & I2	-	-	I1, I2	PA	
			I2 & ~I1	T5	Incr. Shaft Cntr	-	P1	
ISR1	P3	I1,I2	I2	T7	Incr. Timer	-	P2	
			~I2	T6	Incr. Timer	-	P1	
Temp Meas.	P4	E1,E2	~I1 & ~I2	T8	Measure Airflow	-	(etc.)	
			E1 & I1	T9	Disable I1	I1	P3	

set, or other actions to be performed during the transaction in the column to the left

8. *Next place:* The next place (state) that is the result of the enabled transition

9. ΔT: The estimated or measured duration of transition execution.

In real-time programming, maximum operating speed can be realized by taking advantage of the fact that an operation takes a known time. This approach is opposed to the use of semaphores to signal the beginning and end of a process. For example, a read of a disk sector into a buffer, once started, cannot be stopped. If the buffer into which the data is being read is not completely empty, the data will overrun the previous data. To reconcile these conflicting goals of speed of execution and rigid semaphore maintenance to ensure no timing errors, the ΔT column is maintained. The table is designed to accommodate Wirth's[14] three steps toward making real-time programming manageable. Briefly, the three steps are as follows:

1. Initially make all process interactions explicit using semaphore synchronization signals.
2. If a hardware signal is not available, derive an analytical time constraint.
3. Verify that these time constraints are met by the microcontroller.

Initially, the ΔT column contains estimates of critical transitions, and later, as the program becomes refined or explicit semaphores are replaced with timed control, these estimates provide the raw data for that reformulation.

Another advantage of the Petri table is that the control problem to be designed is effectively partitioned and orthogonal, and the subunits of code can be directly associated with a particular set of high-level functions. Dangerous interactions among various segments of code are minimized since all of the information transferred is in the form of tokens. The intent is to do away with the graphical representation as an aid to microprogram software development and convert it to a scalable, easily maintainable, orthogonalized representation of the software. At the same time, both control flow and data flow (interrupt flow) methods are preserved and can be programmed simultaneously since the action of every token and desired combinations of tokens is accounted for in the Petri table.

When the Petri table is in a specified place, more than one transition might be enabled. In this case, priority of execution is arbitrarily assigned to the first listed transition. In any real system, this should not be a problem, but the implementation in code and the desire to transfer explicit information to the coder through the medium of the Petri table requires this priority to be defined. This is the price of using an SISD machine to control multiple concurrent processes.

3.3 INTEGRATED SOFTWARE DESIGN MODEL

The previous section introduced a method for the high-level design of microcontroller software based on Petri nets. A subset of the PN concept was chosen as being similar to FSM methods but allowing for the explicit introduction of interrupts, flags, and semaphores. This concept was also chosen because of the growing body of theory about PNs, which will eventually allow for algorithmic checking of Petri net representation of control systems to determine whether there are any unreachable places or terminating places. Although the concern here is with single MCU design, the introduction of PNs and PTs allows for the integration of multiple MCUs through the introduction of tokens that are passed from one MCU to another. At the PT level of description, there is no distinction between one machine or process and another, and all communication takes place through token passing.

3.3.1 Summary of Design Steps for Petri Tables

Although the process of Petri table development can be stated in a few steps, the development of this, as any high-level design, is an iterative process. One does not sit down and design the complete control software as a single,

TABLE 3.6 Petri Table Worksheet

NAME	PLACE	POSSIBLE TOKENS	ENABLING TOKENS	TRANS-ITION	ACTIONS	OUTPUT/ CHANGED TOKENS	NEXT PLACE	ΔT

TABLE 3.7 Token List Worksheet for Petri Table Design

(S)emaphore/ (I)nterrupt/ (F)lag	Label	Meaning

continuous process. One of the benefits of PTs is the ability to add new transitions easily without disrupting the rest of the design. Since each transition is enabled by a specified ensemble of tokens, it is only necessary to ensure that no two transitions have the same enabling tokens. The following discussion refers to two accomanying worksheets, since the majority of the software design process consists of "filling in the blanks," whether this is done manually in duplicated worksheets or automatically in a word processor or dedicated PT design program. The first worksheet is the blank Petri table worksheet of Table 3.6. The second is the blank token list worksheet of Table 3.7. The token list is a convenient way of maintaining correspondence between the token mnemonics that will naturally develop in the program design, to minimize the repetitious writing of long descriptive names for the tokens. The mnemonics take the place of unique single character symbols since they have some of the characteristics of the interrupt, flag, or semaphore with which they are associated. This token list also allows for a quick check for duplicate token names, since as the size of the program goes, so will the length of the mnemonic list. The steps in generating a Petri table representation of a control problem follow:

1. Determine significant places associated with the control process and the completion of some phase of the control process.

2. Determine source places in terms of interrupts, semaphores, and flags.

3. Determine tokens that allow the transition from each place to another place.

4. Transfer the information from the previous steps to the PT to effect the control strategy.

5. Trace all the paths through the PT to ensure desired operation, and modify as necessary.

6. If required, enter timing values in the rightmost column.

The timing values are entered to determine the latency of interrupts or minimum and maximum times between critical events. These may be either estimates or allowed values, depending on the stage of software development. In the early stages of software design, the timing column provides goals for the actual coding; during the final stages it can indicate regions of slack or excess where improvements in the temporal operation of the program can be made.

3.3.2   Example Problem: UAV Controller/Autopilot (UAV CAP)

To show how the PT process can be applied effectively, the method is used to design the control software for an autopilot for an unmanned aerial vehicle (UAV). Since it is customary to have some idea of the hardware to be used in the design of such a system, a Motorola MC68HC11 microcontroller is assumed, but detailed knowledge of the operation of this device is not necessary at this stage. A block diagram of the complete system is shown in Fig. 3.13, where it can be seen that the MCU will receive a binary-valued signal from a receiver on the airplane. This binary-valued signal contains three channels of information encoded as pulse widths (PW). The duration of a complete frame containing the three channels is 16 ms. Each of the three channels is encoded as a pulse width continuously variable from 1.0 to 2.0 ms

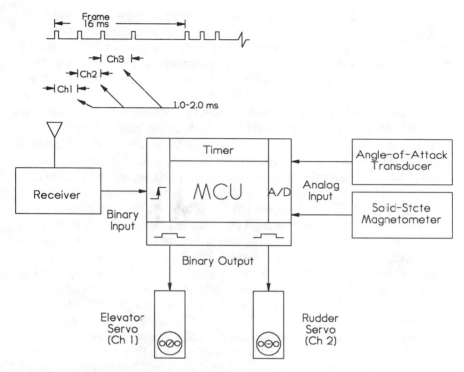

Figure 3.13 Block diagram of unmanned aerial vehicle autopilot.

duration. A pulse is defined as the time between rising edges of the signal. The duration of the high portion of the signal is irrelevant and can be ignored. The channels are transmitted in order and follow exactly after one another. If all three channels are at their maximum value of 2.0 ms, then the information contained in a frame is contained in the first 6.0 ms at the start of a new frame, and the remaining 10 ms contain no information. This 10 ms "dead time" can be used to synchronize the MCU to the incoming signal since the time preceding the channel-one rising edge is the only time in which the signal is low for more than 2.0 ms. Channel pulse width is measured from rising edge to rising edge, and a new frame is preceded by a low level of greater than 2.0 ms duration.

The three channels of information transmitted to the UAV contain the following:

Channel 1: Elevator angle or desired angle of attack (AOA), depending on mode of autopilot

Channel 2: Rudder angle

Channel 3: Autopilot mode (elevator angle control or AOA attitude hold).

Channels 1 and 2 are continuous valued channels, where the pulse width, which varies from 1.0 to 2.0 ms, is the desired elevator angle (or AOA) and rudder angle, respectively. Channel 3, although continuous from 1.0 to 2.0 ms, is to be interpreted by the MCU as a binary-valued signal with a PW $\leqslant$ 1.5 ms interpreted as no autopilot mode ($\sim$APM) and a PW $>$ 1.5 ms interpreted as autopilot mode (APM).

Internal to the UAV are two sensors that provide aircraft attitude (not in the inertial sense of the word) and magnetic heading information. The first sensor measures the angle of attack of the wing. The method of operation of this sensor is not important here, but it consists of an analog output from 0 to 5 volts that is a monotonically increasing function of the angle of attack with 0 representing the minimum AOA and 5 volts representing the maximum measurable AOA. The AOA is used by the autopilot to maintain the UAV at either its maximum endurance AOA or maximum range AOA, depending on the type of mission of the UAV. The second sensor is a solid-state magneto-meter (SSM) that measures the magnetic heading of the aircraft. Again, the exact method of performing this function is not of importance here,[15] but the sensor is assumed to produce an analog voltage proportional to heading such that the range 0 to 5 volts represents 000 to 359 degrees. The SSM is used to stabilize the heading of the UAV and can be used for dead-reckoning navigation over short distances. Alternatively, it could be programmed to cause the UAV to orbit continually in a fixed pattern. To make the discussion aeronautically complete, it is assumed that the UAV has enough inherent aerodynamic stability such that expensive inertial frames of reference are not required for it to maintain upright flight.

To complete the implementation of the controller/autopilot (CAP), two actuators are needed, and these consist of DC electric motor servos. There are two servos, one for the elevator (CH_1) and the other for the rudder (CH_2). The control signals to these are the same as described earlier in that they are pulse widths of from 1.0-2.0 ms duration at logic levels. While it may look as if the encoded signal described above may be used to drive the servos, the encoded signal must be separated into the channels; only the channel-1 PW is sent to the channel-1 servo, and the channel-2 PW is sent to the channel-2 servo.

With this simple servo consisting of one binary input, two analog inputs and two binary outputs, a UAV can be controlled in two modes. The first mode is a nonautopilot mode in which the control signals received by the airborne receiver are decoded and sent individually to the two servos. The second mode of operation, the autopilot mode, combines the AOA sensor information with commanded AOA data from the receiver, to control the elevator servo by adjusting the pulse width sent to that servo. The autopilot mode also uses magnetic heading information from the SSM sensor to stabilize the UAV in yaw and maintain a constant heading or some other desired aircraft heading change such as a racetrack pattern.

To complete the requirements for the development of the system, it is further assumed that the MCU contains an internal timer. The availability of a timer is typical on MCUs, the MC68HC11 is particularly useful for implementing this system. The timer can operate in a variety of modes, but the modes assumed here are as follows:

1. A continuously running clock that generates real-time interrupts (RTI) at regular intervals

2. A register that captures the time of the clock when a rising edge on an external input occurs

3. Registers that can be set to values that will generate interrupts when those counters reach the set value

3.3.3 Petri Table for UAV Controller/Autopilot (UAV CAP)

A Petri table has been generated for this controller/autopilot and is shown in Tables 3.8 to 3.10. To familiarize the reader with the reading of a PT, various places and transitions are described. The initial place is *Restart* and is assigned an arbitrary place number of P_0. No tokens are required because of the earlier assumption of there being only one machine and, hence, an assumed machine token located in each of the enabling token blocks. The extension of this technique to multiple interacting MCU machines would necessitate the inclusion of an explicit machine token if the PT represented the operation of all machines. The first transition, T_1, is somewhat hardware specific in that the MC68HC11 requires specific registers to be instantiated with their values during the first 64 clock cycles of operation after restart. This may or may not be required with other MCUs. Since the end of configuration of the hardware is an identifiable place, it is given a name P_1, after which the system hardware and software self-tests are specified. At this point, the PT could be expanded to list each of the individual tests as a transition, specify a place after each one to indicate which test has been completed, and have individual responses to the passing or failing of a specific test. This level of detail is not included here because it is the highest level of description of the software and controller operation. Clearly, each of the individual transitions can be decomposed to its own PT with no loss of documentation clarity or control as long as its entry and exit points are the same as those of the higher-level PT. Be that as it may, transition T_2 is placed in this master PT since it is a meaningful, well-defined transition that is not in need of detailed design at this time.

There are two possible semaphore tokens that can originate at transition T_2, and they are mutually exclusive. If the self-test is passed, then *PST* (passed self-test) is enabled. If the test fails, then *FST* (failed self-test) is enabled. These tokens also appear in the token list of Table 3.11. The

TABLE 3.8 First Part of Petri Table Representation of Unmanned Autonomous Vehicle Control System

Name	Place	Possible Tokens	Enabling Tokens	Trans-ition	Actions	Output/ Changed Tokens	Next Place	ΔT
Restart	P0	PST	~PST	T1	Configure Hardware	--	P1	
Hdware Confg'd	P1	PST	~PST	T2	Self-test			
					If passed	PST	P2	
					If not passed	~PST	P0	
Passed Self-test	P2	PST	PST	T3	Init Var Init Tokens Config rising edge timer Clear interrupts Enable RE Timer int	~OPCH1IE ~OPCH2IE ~RTI ~AOA ~SYNCH RETIE	P3	
System Initialized	P3	RETIE RETI SYNCH	RETIE & RETI	T4	Store time	~RETI RETIE	P4	
		OPCH1IE OPCH1I	OPCH1IE & OPCH1I	T9	Clear Ch1 o/p bit	~OPCH1IE ~OPCH1I	P3	
		OPCH2IE OPCH2I	OPCH2IE & OPCH2I	T10	Clear Ch2 o/p bit	~OPCH2IE ~OPCH2I	P3	
		AOA RTI	RTI & APM	T11	Enable A/D ch1	~RTI AOA	P3	
		APM AD1	AOA & APM	T12	Read AD1			
					If AD1	AOA APM	P7	
					If ~AD1	AOA APM	P3	
			~APM	T13	Set output compares to stored values Enable o/p compares	OPCH1IE OPCH2IE	P3	

proliferation of tokens and whether a single token and its complement are used to indicate the state of a semaphore is a matter of personal choice. When memory is scarce, the implementation of the PT in software may force the coder to combine these two tokens into one; however, that is not germane to this level of discussion and is a decision best left to the coder. At this level, one is more concerned with the proper logical flow of the PT than with the details of implementation in software. Note that in the leftmost column of the token list, Table 3.11, the type of token is indicated since its type determines whether it is setable (a semaphore), readable only (a flag), or able to interrupt the process (interrupt or trap) if the appropriate interrupt enable token is activated.

TABLE 3.9 Continuation of UAV CAP Petri Table

Name	Place	Possible Tokens	Enabling Tokens	Trans-ition	Actions	Output/ Changed Tokens	Next Place	ΔT
Rising Edge Rx'd	P4	RETIE RETI SYNCH	RETI & RETIE & ~SYNCH	T5	Store time Compute pulse width			
					If 0 < PW < 1.0 ms	~RETI	P4	
					If 1.0 ≤ PW ≤ 2.0 ms	~RETI	P4	
					If 2.0 ≤ PW ms	~RETI SYNCH	P4	
			RETI & RETIE & SYNCH	T15	Store time Compute pulse width			
					If 0 < PW < 1.0 ms	~RETI	P4	
					If 1.0 ≤ PW ≤ 2.0 ms	~RETI	P5	
					If 2.0 ≤ PW ms	~RETI SYNCH	P4	
Channel Data Rx'd	P5	CH1 CH2 SYNCH	SYNCH & ~CH1 & ~CH2	T6	Store Channel 1 PW	CH1	P4	
			SYNCH & CH1 & ~CH2	T7	Store Channel 2 PW	CH2	P4	
			SYNCH & CH1 & CH2	T17	Store Channel 3 PW	~SYNCH ~CH1 ~CH2	P8	

TABLE 3.10 Continuation of UAV CAP Petri Table

Name	Place	Possible Tokens	Enabling Tokens	Trans-ition	Actions	Output/ Changed Tokens	Next Place	ΔT
CH3 Decoded	P6	—	--	T8	Read internal counter Set Ch1/2 output bits	OPCH1IE OPCH2IE	P3	
AOA Read	P7	APM AOA	APM & AOA	T14	Store AOA Compute Control Values Store Channel PWs Set output compares to stored value Enable o/p compares	~AOA OPCH1IE OPCH2IE	P3	
CH1/2/3 Rx'd	P3	--	--	T16	If CH3 ≤ 1.5 ms	~APM	P6	
					If CH3 > 1.5 ms	APM	P6	

TABLE 3.11   Token List for Unmanned Aerial Vehicle Control and Autopilot Example

(S)emaphore/ (I)nterrupt/ (F)lag	Label	Meaning
S	PST	Passed self-test
S	FST	Failed self-test
S	OPCH1IE	Output Channel 1 Interrupt Enable (compare counter interrupt enabled)
S	OPCH2IE	Output Channel 2 Interrupt Enable (compare counter interrupt enabled)
I	RTI	Real-Time Interrupt (generated every 1/10 second)
S	SYNCH	Synchronized (PN synchronized to framing signal)
S	RETIE	Rising Edge Timer Interrupt Enable
I	RETI	Rising Edge Timer Interrupt
S	CH1RX	Channel 1 data received
S	CH2RX	Channel 2 data received
S	AOA	Angle-of-Attack measured
F	AD1	A/D CH1 done, (Flag bit read in control register)
S	APM	Autopilot mode
S	CH1	Channel 1 pulse width received
S	CH2	Channel 2 pulse width received

The last major transition takes place at T_3 in which the software and hardware are initialized for the operation of the control system. As outputs of this transition, six tokens are originated or at least initialized to specific negative values. For example, the "~" before token *OPCH1IE* of the "Output/Changed Tokens" column means that the output-channel-1 interrupt-enable token is made FALSE. That is, if the output-channel-1 interrupt occurs, it will be ignored by the software because that particular interrupt token has not yet been enabled.

Once the system has been initialized, place P_3 is the focus of operation of this implementation. It is by no means the only method of implementing the desired autopilot control functions, but has been chosen because it incorporates elements of semaphores, flags, and interrupts. The transitions associated with the output of place P_3 were not all written at one time, but evolved as the rest of the table was developed. This is typical, and it should be expected that a full understanding of the control program only develops as it is instantiated in the Petri table. However, because of the discipline imposed by the requirement for specific tokens to be available to enable transitions, this evolutionary process is easily traceable, modifiable, and readable.

Interrupt-driven transition. Transition T_4 is an example of an *interrupt-driven transition*. One of the capabilities of the MC68HC11 is to define, through the programming of a status register, the edge of an input that causes a counter's value to be read, stored, and an interrupt generated. For T_4 to be enabled, two tokens must be active: the rising-edge interrupt enable (RETIE) token and the rising-edge interrupt (RETI). All interrupts were cleared (e.g., RETI) in the previous transition, T_3, and an output token of that transition was RETIE. Provided that none of the other enabling tokens fire a transition, the MCU will stay in place P_3 until the rising-edge timer interrupt occurs. When the interrupt occurs, the value of the master counter that is internal to the MC68HC11 is stored, the rising-edge interrupt cleared, and the RETIE token passed on, since the firing of a transition assumes that all enabling tokens are consumed. To remind the programmer of the status of these tokens, the complemented RETI is listed in the changed tokens list as well as the still active RETIE. This example interrupt-driven transition may be either vectored or polled with little effect on the Petri table logic.

Semaphore-driven transition. After the first rising edge is detected from the receiver, as indicated by place P_4, it is necessary to determine whether this is useful information. Initially, the MCU has no information about where its received signal is within a data frame. This status is indicated by the complemented SYNCH semaphore that will be used to demonstrate a *semaphore-driven transition*. In P_4, when the second rising edge occurs, if the SYNCH token is not there, T_5 fires, and the time between rising edges is computed. It can easily be seen from the actions column that pulse widths of less than 1.0 ms are an error condition and remain in P_4. Pulse widths between 1.0 and 2.0 ms are also error conditions since there is no way to know whether the information is from channels 1, 2, or 3. If the PW is greater than 2.0 ms, then it can be safely assumed that the time between rising edges was the time after channel 3 and that the next rising edge will be the beginning of the PW for channel 1. Because of this, the SYNCH token is activated by T_5. The rising edge interrupt is cleared, and the rising edge interrupt enable is maintained.

After the MCU is synchronized to the incoming pulse train, the next rising edge enables T_{15}, which stores the time of the rising edge and computes the time between the rising edges—the pulse width for channel 1. Again the pulse width is checked, rejected if it is less than 1.0 or greater than 2.0, or accepted if it is between 1.0 and 2.0 ms, and transferred to place P_5. At place P_5, more semaphores (CH_1, CH_2) are used to determine which transition fires to store the channel data in its proper place. Since no interrupt or interrupt enable tokens were consumed in the process, the only tokens that need to be set are those for the appropriate channel when its data is stored and the SYNCH token complemented after the third channel.

Now that all three channel data are available for processing, the PT proceeds to place P_8, which determines which mode the autopilot is in, as explained previously, and sets or clears the APM semaphore token. Once the mode is determined, both the elevator and rudder channel outputs are enabled (set high) while the control loop computations are made for the attitude and heading control loops (if the controller is in autopilot mode). This immediate setting of the output bits can be done since the duration of the pulses to the servos is at least 1.0 ms, and this allows sufficient time for the necessary computations. Control is then returned to place P_3 which continues to test for enabling combinations of tokens.

If it is assumed that autopilot mode is not enabled, then T_{17} fires, which reads the stored channel 1 and channel 2 data, adds them to the internal counter values that were stored during transition T_8, and writes them to the output compare timer registers, simultaneously enabling their interrupts. This method of timing the duration of the high level sent to the servos is unique to the M68HC11, but equivalent functions are available in other MCUs or could be emulated in software. In the M68HC11, the master counter is never stopped, but certain timer registers can be loaded with values that, when the master counter reaches that value, will generate an interrupt. Two of the six available output compares are used in this controller.

Flag-driven transition.   Based on the dynamics of the UAV, the control loop may or may not need to be operated at the same rate as the incoming data (1/16 ms) Hz. For the sake of this example, assume that the control loop values only need to be computed at a 10-Hz rate and that the real-time interrupt (RTI) generates an interrupt every 100 ms. Since one of the inputs to the control loop is the AOA value, and the A/D in the MC68HC11 does not generate an interrupt at the end of the A/D conversion, the final type of token is introduced, and it consists of a *flag* token as well as a *flag-driven transition*. Even though an interrupt is not generated, a hardware bit (CCF—conversion complete flag) is set in the ADCTL register of the MC68HC11,[16] which indicates that the conversion is complete. The flag token can be handled in the following manner. If the controller is in autopilot mode, then APM will be TRUE, and the PT will be continually cycling through place P_3 by design. When a real-time interrupt occurs, token RTI becomes TRUE, enabling transition T_{11}. T_{11} sends the appropriate commands to the on-chip A/D to start the conversion as well as setting the AOA semaphore token, finally returning to place P_3. Now that RTI has been cleared, T_{11} is no longer enabled, but T_{12} is causing the MCU to read the flag bit (AD_1 token) from the A/D control register. Depending on the results of this read, the PT continues looping through place P_3 or continues on to place P_7. In place P_7 the digitized value of the analog AOA sensor is read and stored. Since the PT cannot arrive at this state unless it is also in autopilot mode, the AOA is used to compute the appropriate control signals (PW) for the elevator channel, sets the output compares in the timer, and enables the O/P compares, finally returning to place P_3.

3.3.4 Example Problem: Infrared Sensor Target Tracker

As a second example using the Petri table approach, the simplified land-based infrared (IR) target tracking system of Fig. 3.14 is introduced. The components of this system are as follows:

1. *IR sensor:* This sensor receives energy in the near or far infrared region and generates an image similar to that of a standard television, although

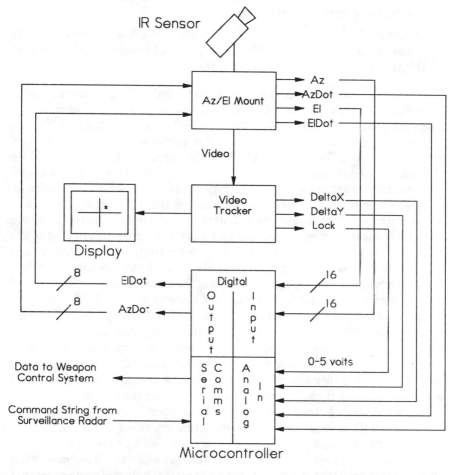

Figure 3.14 Diagram of infrared sensor target tracker showing the sensor, the azimuth/elevation mount that points it, the automatic video tracker that generates error signals, and the microcontroller.

usually with more or less resolution in terms of both the number of scan lines and the number of quantization levels. Its exact operation is not of interest to us since the signal is processed by a video tracker.

2. *Video tracker:* This subsystem analyzes the video (implies large bandwidth) signal from the IR sensor and generates error signals that can be used to drive the Az/El mount.

3. *Az/El Mount:* This subsystem positions the optical axis of the IR sensor to point it in any desired direction in azimuth or elevation. The input signals to the Az/El mount are velocity signals rather than position signals. These control signals are generated by the microcontroller.

4. *Microcontroller:* The MCU that measures position and velocity signals, receives commands from a surveillance radar subsystem, and computes the control signals to drive the mount.

5. *Display:* The IR tracker is usually supplied with a display that shows the video signal along with superimposed gates, which are meant to bracket the target visually as it is being tracked.

The various signals depicted in Fig. 3.14 are typical and may be either analog or digital, although the resolution of the azimuth and elevation signals may be as many as 20 + bits to obtain the required degree of Az/El positioning accuracy. It would also be typical to have both a coarse and fine Az/El control as well as position feedback, but this would complicate the example. Note that Azimuth (Az) and Elevation (El) are in an absolute coordinate system, whereas the error between the boresight of the IR sensor and the target is given as error in the X (horizontal) direction and in the Y (vertical) direction. These are relative errors and are all that can usually be obtained by a passive sensor. The video tracker supplies an additional signal that indicates the quality of its ability to discriminate the target from noise or background clutter: the lock signal. This is used to indicate to the microcontroller when the video tracker is supplying useful data to the microcontroller. The control signals used to drive the Az/El mount are velocity signals. While that may seem awkward at first, it is natural as a motor turns when power is supplied and even joysticks attached to the common video games and the sticks of aircraft control rates of movement rather than position. Two signals pass through the serial communications section. One is from a hypothetical surveillance (search) radar which initially locates the target and passes this Az/El information to the IR tracker so it can acquire the target more easily. The other signal is for communications to a weapon system providing more accurate Az/El information to it than can be provided by the radar.

There are two assumed modes of operation, both under control of the search radar. The first is a position mode (PM) in which the MCU makes the IR tracker point at the Az/El passed to it through the serial communication channel from the search radar. This mode is used to allow for the acquisition

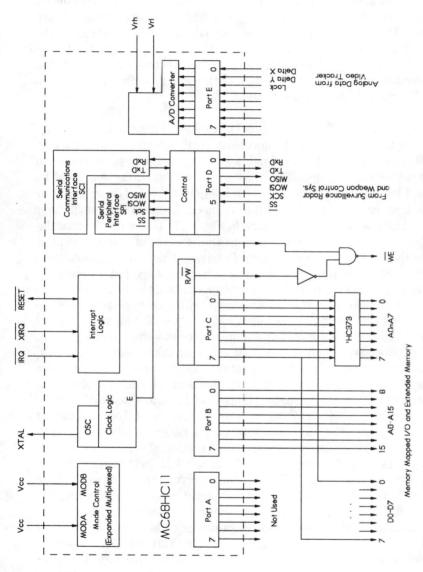

Figure 3.15 Block diagram of implementation of IR Tracker using MC68HC11 MCU.

and fusion of sensor data from multiple sensors, in this case a radar and an IR sensor. The second mode of operation is the track mode (TM). In this mode, the Az/El signals from the search radar serve for the initial pointing of the IR sensor, but when the video tracker acquires the target, its error signals are used by the MCU to generate the control signals to maintain IR track of the target. Although not expanded further in the following Petri table representation, the video tracker signals could be passed through the MCU to the weapon control system for integration with the search radar signals (e.g., to ensure the identification of the target).

IR tracker Petri table. To demonstrate the top-down approach afforded by the Petri table approach to the design of MCU software, the Petri net of Fig. 3.16 is first presented. This Petri net demonstrates the basic operation of the IR tracker without detailing many of the subfunctions that must be performed but that do not interact at this highest level of operation. For example, transition T_5 (which has a dashed box inside of it) has been further expanded into the same Petri net form in Fig. 3.17. This sort of encapsulation is possible when there are functions that are not affected, except at entry, by the tokens of the higher level Petri net. This step of generating a Petri net is not a required part of the design methodology, but is presented here to show the link between the control system and the Petri table. You may find it helpful to start the system design with a Petri net; however, as was explained previously, the net by itself is only useful for small control systems. The Petri net of Fig. 3.16 also introduces a variation on a concept earlier introduced—namely, places that originate tokens asynchronously, although not continuously. Place P_6 represents the arrival of a message from the surveillance radar. When a message arrives (SRMsg token activated) and the interrupts are enabled (IntEn token), transition T_9 is enabled and the message is received byte by byte until it is completely assembled, at which time T_9 passes the SRMsgRx token to P_7. Likewise, places P_4 and P_5 are semaphore-originating places whose generation of semaphores is controlled by whether than have received a PM or ~PM, in the case of P_4, or, TM or ~TM, in the case of P_5. If the mode received is position mode, then transition T_4 passes a PM token to place P_4 which causes it to generate PM tokens until the ~PM token is received from transition T_6.

  Some assumptions have been made in the design of the expanded Petri net of Fig. 3.17. As with the earlier example, a Motorola MC68HC11 is assumed and is used in the generation of the code for this example in Chap. 4. A more detailed block diagram of the I/O pin assignments for the MC68HC11 is shown in Fig. 3.15 and is necessary for the development of the software for this example Chap. 4. Normally, this processor would not be used for the IR tracker because of its limited computational power and basic 8-bit architecture. It also does not have enough digital I/O and would have to be operated in an expanded mode, wherein some of the digital signals would be

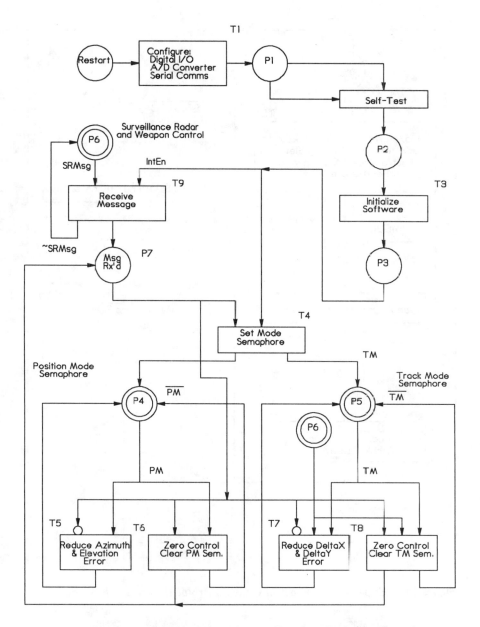

Figure 3.16 Petri net representation of control system for infrared tracker.

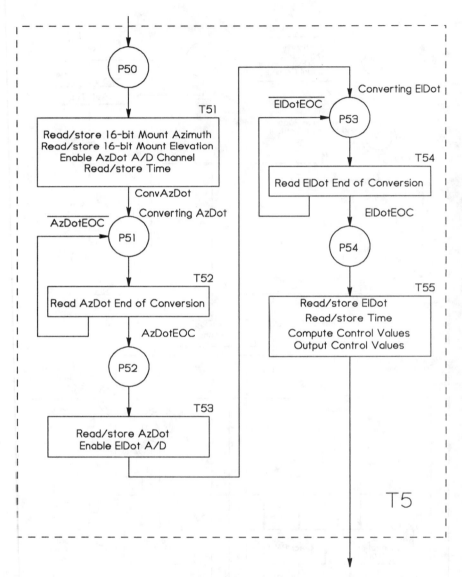

Figure 3.17 Expanded Petri net of transition T5 of IR Tracker System.

read by the MCU as memory-mapped I/O. Rather than worry about the details of another MCU when our purpose is to explain Petri tables, the MC68HC11 is used in the expanded mode. Of course, it is not necessary at the Petri table level to be too concerned about the MCU under consideration, but one can incorporate MCU-specific information at this level when it is known and convenient. For example, the MC68HC11 has some parameters that must be initialized within the first 64 clock cycles if they are to be altered, and this is accounted for in the first transition. The details of the expanded T_5 might also change for different processors. For example, the MC68HC11 does not generate an interrupt when the A/D converter completes a conversion, and one must either wait a minimum amount of time for the conversion to be completed or poll the conversion-complete flag (CCF) to determine the end of a conversion. In the case of an interrupt-driven A/D conversion complete, the polling of P_{51}/T_{52} would be replaced by an enabling of T_{53} by interrupt enable (IntEn) and the convert azimuth rate (ConvAz-Dot) tokens.

Liberty is taken here with transition T_{56} which says "Compute control values." No effort is made here to detail the various control laws that can be implemented to direct the movement of the Az/El mount. Whether they are linear or nonlinear or the order of the control law is not of interest. There are any number of excellent books on digital control that treat this subject extensively. The control values can either be computed algorithmically, or through a lookup table when the computations are extensive and memory is cheap. Another issue in this implementation is that of the temporal regularity of the control signals. Some type of synchronization of them may be necessary to ensure regular intervals between the times of repetitive sampling and compution of the control values.

The Petri table equivalent of the IR tracker Petri net is shown in Tables 3.12 and 3.13, with the expansion of transition of T_5 shown in Table 3.14. The associated token list is in Table 3.15, which is not complete since it will be used as the starting point for several homework problems at the end of this chapter.

3.4. SUMMARY

In this chapter, the concept of finite-state machines has been reviewed as an introduction to the expansion of the concept to Petri nets. The review included several examples of finite state machines as used in the specification of control problems as well as the control of microprocessors themselves (8085). The state diagram was shown in its equivalent form of state table, a method for designing control systems. Petri nets were introduced because of their more consistent and logical handling of interrupts, semaphores, and flags. Petri nets were shown to be a super-set of finite-state machines, and an equivalence relationship developed. Several examples of Petri nets were presented to show the representational conversions from Petri nets to Petri tables. A more extensive example of a control system for an unmanned aerial vehicle was introduced, and a Petri table representation was developed and explained. This example will be utilized further in Chap. 4 to demonstrate the conversion from Petri tables to high-level and low-level software as well as the integration of the two into a single control program.

TABLE 3.12   First Part of Petri Table Representation of IR Tracker Control System

Name	Place	Possible Tokens	Enabling Tokens	Trans-ition	Actions	Output/ Changed Tokens	Next Place	ΔT
Restart	P0	PST	~PST	T1	Configure Hardware: Digital I/O A/D Converters Serial Comms	--	P1	
Hdware Confg'd	P1	PST	~PST	T2	Self-test			
					If passed	PST	P2	
					If not passed	~PST	P0	
Passed Self-test	P2	PST	PST	T3	Init Var Init Tokens Clear interrupts	RETIE	P3	
System Initialized	P3	Ready SRMsgRx	Ready & SRMsgRx	T4	Set Mode Semaphore			
					If Position Mode	PM ~TM	P4	
					If Track Mode	TM ~PM	P5	
Position Mode Semaphore	P4	PM	PM & ~SRMsgRx	T5	Read Az/El Read AzDot/ElDot Read Time Compute Cont Signals Output Cont Signals	PM	P4	
			PM & SRMsgRx	T6	Output Zero Control Signals	~PM	P7	
Surveill. Radar Msg Rx'd	P7	SRMsgRx	SRMsgRx	T4	See above	--	--	

TABLE 3.13   Continuation of IR Tracker Petri Table

Name	Place	Possible Tokens	Enabling Tokens	Trans-ition	Actions	Output/ Changed Tokens	Next Place	ΔT
Surveil. Radar Message Semaphore	P6	SRMsg IntEn	SRMsg & IntEn	T9	Receive Message Decode String and store			
					If Message Complete	SRMsgRx	P7	
					If Message Not Complete	~SRMsgRx	P6	
Track Mode Semaphore	P5	TM SRMsgRx IRTrkLock	TM & ~SRMsgRx & IRTrkLock	T7	Read IrTrkLock Read DeltaX Read DeltaY Read Time Compute Control Output Control	TM	P5	
			(TM & ~SRMsgRx & ~IRTrkLock) + (TM & SRMsgRx)	T8	Read IrTrkLock Zero Control Output Control	~TM	P7	
IR Tracker Lock Semaphore	P8	--	--	T7/T8	See above			

TABLE 3.14   Expansion of Transition T5 of IR Tracker Petri Table

Name	Place	Possible Tokens	Enabling Tokens	Trans-ition	Actions	Output/Changed Tokens	Next Place	ΔT
Position Mode and No Message	P50	PM SRMsgRx	PM ~SRMsgRx	T51	Read/Store 16-bit Mount Az Read/Store 16-bit Mount El Enable AzDot A/D Channel Read/Store Time	ConvAzDot	P51	
Converting AzDot	P51	ConvAzDot	ConvertAzDot	T52	Read AzDot End of Conversion Flag			
					If AzDot EOC	AzDotEOC	P52	
					If Not AzDot EOC	~AzDotEoc	P51	
AzDot Conversion Complete	P52	AzDotEOC	AzDotEOC	T53	Read/Store AzDot Enable ElDot A/D Channel	ConvElDot	P53	
Converting ElDot	P53	ConvElDot	ConvElDot	T54	Read ElDot End of Conversion Flag			
					If ElDot EOC	ElDotEOC	P54	
					If Not ElDot EOC	~ElDotEOC	P53	
ElDot Conversion Complete	P54	ElDotEOC	ElDotEOC	T55	Read/Store ElDot Read/Store Time Compute Cont. Values Output Control Values			

TABLE 3.15   Token List for IR Tracker Example

(S)emaphore/ (I)nterrupt/ (F)lag	Label	Meaning
S	PST	Passed Self-Test
S	Ready	Microcontroller initialized and passed self-test
I	SRMsg	Message available from Surveillance Radar
S	IntEn	Interrupts are Enabled
S	PM	Position Mode (mount points to Az/El position commanded by SR message
S	TM	Track Mode (mount nulls error in X and Y position of IR video image. DeltaX and DeltaY are produced by separate video tracker.)
F	IRTrkLock	Signal from video tracker which indicates it has locked onto the target and is generating correct DeltaX and DeltaY signals.
S	SRMsgRx	A message has been received from the Surveillance Radar and is ready for implementing.
S	ConvAzDot	A/D is converting Azimuth rate of change
S	ConvElDot	A/D is converting Elevation rate of change

REFERENCES

1. F. P. Brooks, Jr., *The Mythical Man-Month*, as extracted in *Datamation*, December 1974, republished in *Milestones in Software Evolution*, P. W. Oman and T. G. Lewis, eds., pp. 35–42, Los Alamitos, CA:IEEE Computer Society Press, 1990.
2. P. C. Treleaven, D. R. Brownbridge, and R. P. Hopkins, "Data-driven and Demand-driven Computer Architecture," *Computing Surveys*, 14(1):93–143, March 1982.
3. J. Gertler and J. Sedlak, "Software for Process Control—a Survey," *Automatica*, 2(6):613–625, 1975, as reprinted in R. L. Glass, *Real-Time Software*. Englewood Cliffs, NJ: Prentice-Hall, pp. 12–44, 1983.
4. L. J. Peters, *Software Design: Methods and Techniques*. New York: Yourdon Press, 1981.
5. M. J. Flynn, "Some Computer Organizations and their Effectiveness," *IEEE Transactions on Computers*, c-21(9):948–960, September 1972.
6. M. W. Shields, *An Introduction to Automata Theory*, Oxford: Blackwell Scientific Publishers, 1987.
7. J. Hartmanis and R. E. Stearns, *Algebraic Structure Theory of Sequential Machines*. Englewood Cliffs, NJ: Prentice-Hall, 1966.
8. ———, "Programming Techniques," *New Logic Notebook*, 1(7), March 1975, published by Microcomputer Technique, Inc., Reston, VA.
9. K. J. Hintz, "Multiplexed-Bus Dual Microprocessor System," master's thesis, University of Virginia, Charlottesville, 1979.
10. G. L. Lipovski, *Single- and Multi-Chip Microcomputer Interfacing*. Englewood Cliffs, NJ: Prentice-Hall, 1988.
11. J. L. Peterson, *Petri Net Theory and the Modeling of Systems*. Englewood Cliffs, NJ: Prentice-Hill, 1981.
12. T. Murata, "Petri-Nets: Properties, Analysis and Applications," *Proceedings of the IEEE*, 77(4):541–80, 1989.
13. D. Tabak and A. H. Levis, "Petri Net Representation of Decision Models," *IEEE Trans. System Man., Cybernetics*, SMC-15(6):812–18, November-December, 1985.
14. N. Wirth, "Towards a Discipline of Real-time Programming," *Communications of ACM*, 20(8), August 1977, as reprinted in R. L. Glass, *Real-time Software*, Englewood Cliffs, NJ: Prentice-Hall, pp. 128–42, 1983.
15. M. J. Gilstead, "Magnetic Heading Sensor for the HERO 2000 Robot," senior project report, George Mason University, May 5, 1989.
16. *M68HC11 Reference Manual*, #M68HC11RM/AD. Englewood Cliffs, NJ: Prentice-Hall, 1989.

PROBLEMS

1. Draw the state diagram and transition table for a Mealy machine that controls a clothes dryer. The inputs to the machine are run/stop, too hot/too cold, and dry clothes/tumble with no heat. The outputs of the machine are heat on/heat off, motor on/motor off, and clothes dry buzzer.

2. Do Prob. 1 as a Moore machine.

3. Explain the difference between an event-driven and a control-driven machine.

4. Describe the colored Petri net of Fig. 3.11 using the formal Petri-net set-theoretic form.

5. Redesign the colored Petri net of Fig. 3.11 to accept pennies (1¢).

6. Design a colored Petri table representation of a machine to control a computer keyboard. Assume that any key pushed generates an interrupt. Each key (including key combinations such as Control-C or Alt-F4) is represented by a

unique 8-bit code that is passed unchanged to another device (the computer). A special key combination, Control-Break, must be sensed as it passes through the controller. When it is sensed, the controller generates an output in parallel with the character that it passes on to the computer. In addition to the keys that generate interrupts, the computer has an output that must be read by the keyboard controller that tells it not to accept any more key inputs. The internal buffer for the keyboard controller can hold up to 25 characters (type-ahead buffer), and when that number is exceeded, the key must be ignored and a tone sounded to indicate that it is full.

7. Design a Petri table to control a radar speed detector for speed-limit enforcement. The inputs to the controller are whether a car is present, the speed of the vehicle from the radar, mode, and a speed at which to sound a buzzer, which is read from one input port set by the person monitoring the radar. Internally, the controller is to keep a log of the average speed of the passing cars and compute the standard deviation of these speeds. The mode input selects whether to set off the alarm when a passing vehicle exceeds the speeds set by the person monitoring the system or whether to set off the alarm only for those cars that exceed three standard deviations from the average speed.

8. Design a Petri table to control a local telephone exchange within an office. Allow for three phones that can signal an off-hook condition and two incoming lines with ring indicators. The outputs are control signals to connect any of the phones to either of the incoming lines, with automatic switchover to the line that is not busy.

9. Modify the previous solution to allow for party line calls in which any or all of the phones can be simultaneously connected to the active incoming phone line.

10. Write a Petri table similar to that of Table 3.14 to implement transition T_7.

11. Modify the Petri table of Table 3.12 to include the transmission of a message to the weapons control system. The message should contain the Delta X and Delta Y of the target as well as the actual value of the Lock signal from the video tracker.

12. Assume that the IR tracker is mounted on a moving platform such as a ship or an aircraft and that there is a stabilized platform available to use as an attitude refererence. Modify the IR tracker diagram of Fig. 3.14 to incorporate such a device. Assume that the stabilized platform has outputs of Heading, Pitch, and Roll. These are each 18-bit digital signals. Which transitions would have to be modified to take this new data into account?

13. Assume that the display of Fig. 3.14 is not driven directly by the video tracker, but rather must be driven through remote signals generated by the MCU. This new display requires as its input two analog signals representing the angle Θ that the target is offset from the top of the screen as well as the radius ρ. These signals are scaled as follows: 0 to 360 degrees = 0 to 5 volts, and center of screen to edge of screen = 0 to 5 volts. Modify Fig. 3.14 accordingly as well as the Petri net of Fig. 3.16 as well as the Petri table of Table 3.12.

14. Write a Petri table for receiving a message that has as its first character an ASCII character for the number of characters to follow (1–9). Detect an error if

a zero length is received and set a message-error semaphore. At the end of the reception of the completed message, set a received-message semaphore.

15. Assume that a product machine is used to implement an MCU program and that it is interruptable during 15 percent of the states that are uniformly distributed throughout the main program. If the probability of receiving an interrupt is 0.1 during each main program state, there are 1000 states, and the complete program executes in 1 s, what is the expected latency?

Microcontroller Software Implementation

4.1 SOFTWARE DEVELOPMENT PROCESS

In Chap. 3 a new method of high-level software design was introduced with no consideration of how the design would be implemented in software. This chapter shows how individual transitions in the Petri table can be converted into executable code in both assembly language and a high-level language, and it discusses elements of real-time kernels. An example of both C and assembly language code for the MC68HC11 MCU, which was written without the benefit of a Petri table, is also shown. These examples demonstrate how either or both languages can accomplish the same task in an MCU. The target processor for which the assembly language is written and the C language compiled is the Motorola MC68HC11. The PC-based C cross-compiler, cross-assembler, linker, and file conversion programs that are used for the examples in this chapter are available from INTROL Corporation.* Other development products are available from INTROL and other corporations specializing in software and hardware design tools for microcomputers. The differences among the various vendors are not of interest here, and the tools used effectively deal with all aspects of software design

---

*INTROL Corporation, 647 W. Virginia Street, Milwaukee, WI 53204; used with permission.

for microcontrollers. INTROL also has assemblers for a variety of MCUs and microprocessors that can be run on a variety of hosts from PCs and workstations to mainframes. The availability of a large variety of target processors is a key ingredient in vendor selection because of the time involved in learning the syntax and operation of a new programming environment.

The key topics addressed in this chapter, which differ from normal software development, are as follows:

1. Interfacing assembly language and high-level language (HLL) in the same program

2. Linking of different segments of code to different types of memory

3. Substitution of custom object files for library files

4. High-level language (HLL) implementation in MCUs without operating system support

5. Conversion of Petri tables to executable code.

At first inspection, the path for software development for an MCU will seem convoluted and unnecessarily tedious, but this is only because your first exposure to programming was in a well-structured, platform-independent environment. When you first learned C, PASCAL, FORTRAN, or another high-level language, the machine for which it was to compile was the furthest from your consideration, and that is as it should be for programming that is computation oriented with inputs from a terminal or file and output to the real world through the same "standard" devices. The moment you start working with small machines that are intended to interface intimately with the real world, it is unlikely that any two MCU systems will be the same. Certainly, the MCUs around which they are built may be the same, and many of the registers and I/O ports may exist at the same locations in memory, but how they are used, and the variety of programming options for these devices and their I/O ports and memory, make almost every system unique. Because of this variety of usage, much more detail must be specified in the software design, not only in the executable code, but in the command file to the linker that specifies where particular sections of code are to be placed.

4.1.1 Software Development

The process advocated in this chapter takes advantage of the highest-level implementation of the control software, with reduction to assembly code occurring only at a later stage, and only when necessary. This process is shown in Fig. 4.1. The benefits of this approach are as follows:

1. Minimum time to a development model for use by engineers designing other parts of the system.

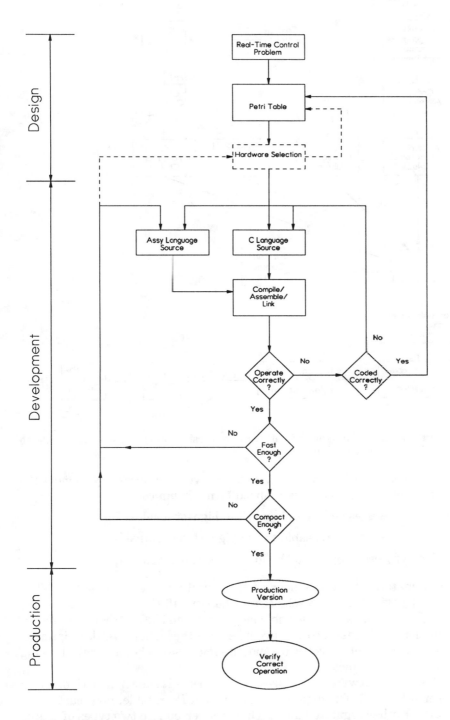

Figure 4.1 Process of top-down, sufficient, software development for an MCU.

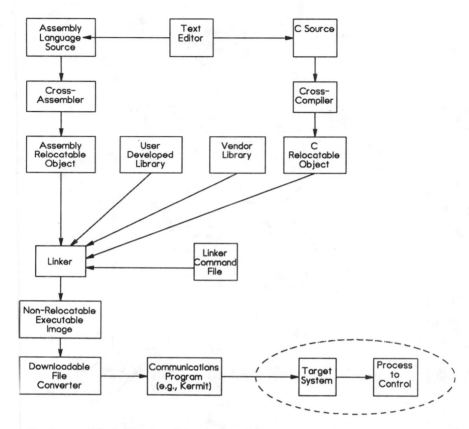

Figure 4.2 Block diagram of software conversion process from source file to downloadable, executable software.

2. No more effort is expended than that which is necessary to meet the control system goals, be it speed or minimum code.

3. Software is easiest to debug at the highest level.

4. The software is traceable to the highest conceptual level.

5. The process supports a "build a little, test a little" approach.

The temptation for an engineer is to start writing code to "get it done." The difficulty with this approach occurs when (not if) the code does not work. Troubleshooting at the assembly language should be restricted to a determination of whether it correctly implements the higher level description of the logical flow of the software and, in some cases, whether critical timing parameters are met.

A sample software assemble/compile process is shown in block diagram form in Fig. 4.2. The high-level design, be it Petri table, flowchart, state table, or other organized approach, is converted into two types of source files. The first is the *assembly source*, which in our examples has the suffix .*s11*, with the *s* indicating assembly language and the *11* indicating for the 6811 target processor. The second source is the high-level source, in this case C, with the suffix .*C* to indicate its type. How the code is apportioned among these two languages is discussed later in this chapter. The assembly source is then *assembled* to a relocatable *object file* with the suffix .*o11* appended to it. The C source is compiled to an assembly language equivalent resulting in an .*s11* file, which is then assembled automatically as part of the compiler's function to its .*o11* linkable representation. Two object files now exist, each of which is linkable (not executable) but not yet linked to the other or to the library functions.

At this stage, nothing is essentially different from a normal programming environment. Normally, a linker would be called that would combine the object files into a single *relocatable* execution image file along with relocation information. *Relocatable* is the key word since a normal program is linked so it can be executed under the control of an operating system. Because of the dynamic allocation of memory to executable programs by the operating system, it is not known at link time where the executable code will be placed in memory for execution. The specific addresses for jumps and data accesses are also not known. These are resolved at load time by the operating system. In the usual case of an MCU implementation, there is generally not an operating system but rather a small kernel for control without the extensive utilities that are taken for granted in a normal programming environment. Because of this, the relocatable object files are linked such that all external references are resolved and all addresses specified, assuming that the program will be loaded for execution at a particular location. In the case of the INTROL and other compilers, the necessary addresses are specified in a command file to the linker that has the .*ld* suffix. An example of a linker command file is discussed in detail later in this chapter.

The linker has now produced an absolute executable image of the program, but it is still residing on the host system, which in this case is a PC. The program must now be transferred from the PC to the target system through a file transfer utility. A typical utility is KERMIT,[1] which is primarily of value because of its conversion of all binary codes during a file transfer from one machine to another into printable ASCII characters so a network or monitor program does not inadvertently receive a control character that changes the operation of the network. By virtue of its translation to printable codes, KERMIT is not the fastest file transfer protocol currently available. While it is not the fastest program for file transfer, it is freeware and can transfer data among a large variety of machines. While a file transfer program is available, the target machine is still lacking a real operating system, so it is generally unable to receive and store a normal file. Usually, the executable binary image is transferred from the host to the

Type	Record Length	Address	Code/Data	Checksum

Figure 4.3 Motorola S-Record format showing the five data fields.

TABLE 4.1 Field Definitions for Motorola S-Record Format

Field	Printable Characters	Contents
Type	2	Type of S-Record (S1, S2, etc.)
Record Length	2	Character pair count in this record, excluding the type and record length
Address	4,6, or 8	2, 3, or 4-byte address where data is to be loaded
Code/Data	0-2n	Zero to n bytes of executable code, memory loadable data, or descriptive information
Checksum	2	LSB of the 1's complement of the sum of the values represented by the pairs of characters making up the record length address, and the code/data fields.

target machine in some standard format with embedded error checking and a rigorous protocol, so a very simple (and small) section of code in the target can receive the data stream and place it directly in its executable location in memory while at the same time providing a modest amount of error checking. A typical format for this data record is the Motorola S-Record format shown in Fig. 4.3. The five fields of this record[2] are explained in Table 4.1. The important points of this format are that (1) it consists of all printable characters so it is easy to display and, if necessary, edit; (2) it contains explicit address information of where to put the executable code; and, (3) it contains a checksum to ensure odd number of bit error detecting transmission and reception. Notice that errors are only detected but not corrected. It is assumed that the transmission medium is essentially error free and that the rare event of incorrect transmission only needs to be detectable with a subsequent manual redownloading of the file.

4.1.2 Include and Header Files

In principle, the generation of the source file, the assembling/compiling, linking, conversion, and downloading are straightforward. The difficulty lies in the many details that must be specified to customize the software to the particular hardware system for which the software is being written. Much of this detail is handled in the vendor-supplied *include* files, which list the standard mnemonic definitions of registers. A portion of the INTROL-supplied H11REG.H for C language development, START.S11 for assembly language development, and the linked assignment of numerical values to

```
/* vers=1.01 */
/*
 *      definitions of 6811 on-chip registers.  These are defined
 *      in 'start.s'.
 */

#ifndef      H11REG_h

#define      H11REG_h

extern volatile unsigned char H11PORTA;    /* i/o port A */
extern volatile unsigned char H11PIOC;     /* parallel i/o control register */
extern volatile unsigned char H11PORTC;    /* i/o port C */
extern volatile unsigned char H11PORTB;    /* i/o port B */
extern volatile unsigned char H11PORTCL;   /* alternate latched port C */
extern volatile unsigned char H11DDRC;     /* data direction for port C */
extern volatile unsigned char H11PORTD;    /* i/o port D */
extern volatile unsigned char H11DDRD;     /* i/o data direction for port D */
extern volatile unsigned char H11PORTE;    /* input port E */
extern volatile unsigned char H11CFORC;    /* compare force register */
extern volatile unsigned char H11OC1M;     /* OC1 action mask register */
extern volatile unsigned char H11OC1D;     /* OC1 action data register */
extern volatile unsigned H11TCNT;          /* timer counter register */
extern volatile unsigned H11TIC1;          /* input capture register 1 */
extern volatile unsigned H11TIC2;          /* input capture register 2 */
extern volatile unsigned H11TIC3;          /* input capture register 3 */
extern volatile unsigned H11TOC1;          /* output compare register 1 */
extern volatile unsigned H11TOC2;          /* output compare register 2 */
extern volatile unsigned H11TOC3;          /* output compare register 3 */
extern volatile unsigned H11TOC4;          /* output compare register 4 */
extern volatile unsigned H11TOC5;          /* output compare register 5 */
extern volatile unsigned char H11TCTL1;    /* timer control register 1 */
extern volatile unsigned char H11TCTL2;    /* timer control register 2 */
extern volatile unsigned char H11TMSK1;    /* main timer interrupt mask 1 */
extern volatile unsigned char H11TFLG1;    /* main timer interrupt flag 1 */
extern volatile unsigned char H11TMSK2;    /* misc timer interrupt mask 2 */
extern volatile unsigned char H11TFLG2;    /* misc timer interrupt flag 2 */
extern volatile unsigned char H11PACTL;    /* pulse accumulator control register */
extern volatile unsigned char H11PACNT;    /* pulse accumulator count register */
```

Text Box 4.1   Partial header record for C files that assigns specific addresses to mnemonics for hardware registers of the MC68HC11 MCU.

the mnemonics are shown in Text Boxes 4.1, 4.2, and 4.3, respectively. This is a good example of a fundamental principle of software writing, which is to minimize the number of occurrences of actual values in the code itself. In almost every instance, a value should be replaced by its mnemonic label whose actual value is defined in an include or header file. The reason for this

```
|
*
*       start - startup code        6811 version (standalone version)
*

*       standalone definitions

              opthdr                 vers=$Revision: 1.4 $

              lib             config.lib
              lib             vector.lib

              if              buffalo

*
*       BUFFALO dependant definitions
*

AUTOLF                equ            $00A9           auto line feed flag
_vec:         equ            $00C4          address of revector table

              else
              import               H11RAMREG
              endif

*       define on chip registers

              section                .H11REGS

H11PORTA:     ds.b            1       i/o port A
              ds.b            1       reserved
H11PIOC:      ds.b            1       parallel i/o control register
H11PORTC:     ds.b            1       i/o port C
H11PORTB:     ds.b            1       i/o port B
H11PORTCL:    ds.b            1       alternate latched port C
              ds.b            1       reserved
H11DDRC:      ds.b            1       data direction for port C
H11PORTD:     ds.b            1       i/o port D
H11DDRD:      ds.b            1       i/o data direction for port D
H11PORTE:     ds.b            1       input port E
H11CFORC:     ds.b            1       compare force register
H11OC1M:      ds.b            1       OC1 action mask register
H11OC1D:      ds.b            1       OC1 action data register
H11TCNT:      ds.w            1       timer counter register
H11TIC1:      ds.w            1       input capture register 1
H11TIC2:      ds.w            1       input capture register 2
H11TIC3:      ds.w            1       input capture register 3
H11TOC1:      ds.w            1       output compare regieter 1
H11TOC2:      ds.w            1       output compare register 2
H11TOC3:      ds.w            1       output compare register 3
H11TOC4:      ds.w            1       output compare register 4
H11TOC5:      ds.w            1       output compare register 5
H11TCTL1:     ds.b            1       timer control register 1
H11TCTL2:     ds.b            1       timer control register 2
H11TMSK1:     ds.b            1       main timer interrupt mask 1
H11TFLG1:     ds.b            1       main timer interrupt flag 1
```

Text Box 4.2 Include file for assembly language source files that defines the physical addresses of 6811 registers.

```
.H11REGS      4096 (0x1000), section: 9
H11PORTA      4096 (0x1000), section: 9
H11PIOC       4098 (0x1002), section: 9
H11PORTC      4099 (0x1003), section: 9
H11PORTB      4100 (0x1004), section: 9
H11PORTCL     4101 (0x1005), section: 9
H11DDRC       4103 (0x1007), section: 9
H11PORTD      4104 (0x1008), section: 9
H11DDRD       4105 (0x1009), section: 9
H11PORTE      4106 (0x100a), section: 9
H11CFORC      4107 (0x100b), section: 9
H11OC1M       4108 (0x100c), section: 9
H11OC1D       4109 (0x100d), section: 9
H11TCNT       4110 (0x100e), section: 9
H11TIC1       4112 (0x1010), section: 9
H11TIC2       4114 (0x1012), section: 9
H11TIC3       4116 (0x1014), section: 9
H11TOC1       4118 (0x1016), section: 9
H11TOC2       4120 (0x1018), section: 9
H11TOC3       4122 (0x101a), section: 9
H11TOC4       4124 (0x101c), section: 9
H11TOC5       4126 (0x101e), section: 9
H11TCTL1      4128 (0x1020), section: 9
H11TCTL2      4129 (0x1021), section: 9
H11TMSK1      4130 (0x1022), section: 9
H11TFLG1      4131 (0x1023), section: 9
```

Text Box 4.3   Actual values assigned to mnemonics
relating to 6811 register locations as a result of
linking the object modules.

is simple. Assume that it is necessary to change a particular location of a
hardware port. If this address has been referred to many times throughout
the program by its actual value, then each file that contains it must be
"searched and replaced" to address the port correctly after the revision.
This leads to two possibilities, both of which are bad. In the first case, it is
easy to miss a single occurrence by not editing all files. The second
possibility is replacing too much: that is, perhaps the address is the same
hexadecimal value as another value that is used as a literal in a computation.
Search and replace will also replace this value with the new hardware port
address unless each occurrence is examined before replace. It involves a
little more effort up front, but it pays off in the long run for all numerical
values to be defined by labels and then the labels used throughout the
program development. If a value is changed, then only one occurrence of the
value needs to be changed in the include file and the compile process redone.
Experience has shown the value of this practice. The effect of the linker on
program segment location is discussed later.

4.1.3 HLL/Assembly Language Program Development

One of the major reasons for encouraging the use of an HLL such as C for real-time event controllers is the relatively rapid prototyping capability with subsequent refinement to tailor the resulting code to final product constraints.[3] It is typical for the development hardware to have more capability, at least in memory and possibly in program execution speed, than the final product. If this is the situation, then it is more important to determine the correct logic of program operation and speed of execution as early as possible rather than coding the program in the most size- or speed-optimal manner. A high-level language allows that capability. Once the logic of the program and the operation of the system with the process to be controlled has been verified, it is time to reduce the size of the code to fit the ROM allotted or speed up the execution of certain critical sections of code. The first step is to identify time critical elements of the code that require faster operation. Once these segments have been identified, then a close scrutiny of the code developed by the optimizing compiler may reveal some areas that the programmer can optimize. For example, if the code being evaluated contains arithmetic routines and it is known that negative numbers will never be encountered, then sections of the code that check for negative and/or establish the sign of the result of the computation can be eliminated. This both speeds up the code and at the same time reduces the

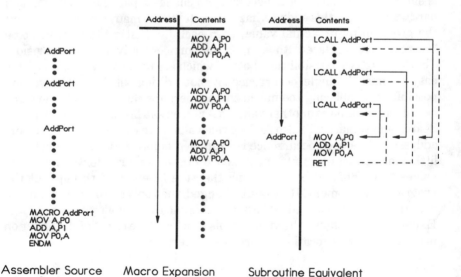

Figure 4.4 Comparison of macro expansion and subroutine usage in assembly language coding.

total quantity of code. It may also be that no amount of "kluging" the code will make it run as fast as required, but a completely new approach in assembly language will. That is the time to write assembly language code.

The approach that began in Chap. 3 of defining the problem completely and explicitly through the use of Petri tables is logically continued through its implementation in a high-level language. If this implementation is sufficiently fast and sufficiently ROM efficient, then the process can stop here. If, however, correct or timely control of the process demands, or the economics of the system indicate, the HLL can be further reduced by hand correcting the assembled HLL. In the extreme, modules can be replaced with original assembly language code. This rational approach to software design for controllers will ensure the minimum time to prototype, minimum amount of nonrecurring engineering (NRE), and a sufficient system. Later in this chapter a section of the UAV is implemented in C, another real-time controller in both C and in assembly language, and the two linked together into a single program to demonstrate the concept of stratified software development. A second example focuses on interrupt-service-routine (ISR) programming in C and assembly language.

4.2   REAL-TIME PROGRAMMING REQUIREMENTS

Nothing discussed thus far has limited the applications of the Petri table method or the software development cycle to real-time systems, although that has been the hidden agenda. Real-time systems and batch-processing systems have different requirements[4] but do have one thing in common: They interact with some "process" even if that process is a human.[5] Since this discussion focuses on MCUs, the applications that are considered are small, although the principles can be extended to more comprehensive real-time systems. An example of a real-time control problem suitable for implementation in an MCU is shown in Fig. 4.5, which is a block diagram of an automotive engine control system. Since applications vary significantly, nothing is said about sizing the MCU to the task at hand other than to mention that the MCU must be able to perform all of the required input, output, and computations such that the execution time is much shorter than the time constant of the process under control. For example, if the interaction is with a human operator, a delay of 100 ms will hardly be noticed, and if it is a relatively predictable delay, an operator may be willing to wait several seconds for a response to an input to the system. Physical processes, such as chemical reactions or electromechanical machines, require inputs at a sufficiently high rate that the process itself acts as a low-pass filter for smoothing out the responses to the control signals. There is little benefit to be gained by controlling the plant at a rate more frequent than several times the inverse of the plant time constant unless precise control is required. In

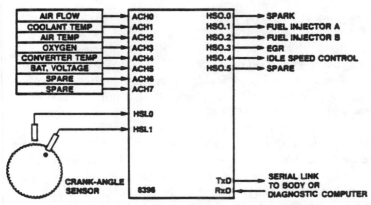

Figure 4.5 Intel 8096 I/O port assignments for a typical automotive engine control application.

the case of the automotive example, timing of outputs is critical since the MCU causes the spark plugs to fire in the engine. A four-cycle engine operating at 6000 r/min requires one ignition every two crankshaft revolutions or 50 ignitions per second. Each revolution takes 0.010 seconds, which makes a 1-degree timing error equate to an error of 0.010 seconds/360 degrees or 27.78 μs. If a 1-degree tolerance on ignition timing is required to prevent detonation and yet ensure economic operation of the engine, the error in outputting the ignition signal must be less than 27 μs. Assuming a 1-μs instruction execution time, this means that at most the latency in the determination of the occurrence of the correct firing time and the outputting of the ignition signal can be 27 instructions. Further assuming that the firing time it computed, loaded in a timer compare register, and an interrupt generated at the correct time, the interrupt-service machine must respond to the interrupt, stack the state of the machine, determine which timer has interrupted it, and load the appropriate registers all within 27 clock cycles. All of this must be done while nothing else is allowed to interrupt the MCU. In this case, one cannot count on the low-pass filtering properties of the process to smooth out the control signals, and the MCU must respond at an exact time or the process will not be controlled properly. In another example of an engine controller, a Motorola MC6801 was used in a Bendix fuel injection system.[6] In this application, the fuel was required to be injected into a 4-cylinder engine every 5 ms when it is running at 6000 r/min. While the exact timing of fuel injection is not as critical as ignition timing, this shows the maximum amount of time available for computations between cylinder firings assuming that a complete computation is performed. An alternative would be to perform computations in accordance with an engine response time constant in the background and to use these computations to update values that are used by an interrupt-service routine to set the actual times of firing. This could be done on a per-cylinder basis or by computing a single value of ignition timing and fuel injection for all cylinders.

The various levels of process controllers can be divided into a commonly accepted terminology introduced by Intel.[4] These levels are as follows;

1. *Event controllers*: Embedded systems controlling a dedicated process. These typically have a kernel operating system in ROM, no secondary storage, and fast response times. Applications are microwave oven control, VCR control, or automotive engine controllers. Response times are less than 100 s.

2. *Data controllers*: Embedded systems controlling one or more processes. These controllers are distinguished by the fact that they are connected by a network and require more operating system services. Examples of these types of controllers are network controllers or military command and control systems. Response times are on the order of 100 to 500 s.

3. *System controllers*: Not embedded, but part of a host system controlling other embedded controllers. An example of this type of controller is a network protocol switching node. Response times are generally greater than 500 s.

From this classification, it can be seen that the general term *event control-lers* can be applied to the example UAV controller and IR Tracker presented in Chap. 3 and expanded later in this chapter. Another term that can be used to distinguish among the three classes of controller applications (for it is not the MCU but the task to which it is applied), is *hard real time* as opposed to *soft real time*. Particular hard real-time applications often force implementation-dependent decisions about whether to implement repetitive computations over a limited range, to a lookup table if time is critical and ROM available, or about whether to perform the computation when time is not of a premium but ROM is.[6]

Even though there is a variety of hard real-time tasks that can be implemented in MCUs, a fundamental piece of software must be included in every one of these: the kernel, which controls which process, place, or state the machine is in. This can be seen also in the Petri table example of Chap. 3. While it was typical in the recent past for MCU system developers to write their own kernels, a number of vendors have made available real-time kernels around which systems can be built.[4,7] One reason for using these preprogrammed kernels, at least as a starting point for system development, is the increasing sophistication of the newest MCUs with cache and on-chip memory management, particularly if the MCU is operating in a multiprocessor environment where it must compete for and utilize common memory.

Another concern in the choice of MCU and kernel is whether the MCU is a RISC or CISC architecture. Associated with the RISC architecture is an unusually large number of registers and a load-store-only memory access instruction set. With this increased number of registers comes the requirement to store all of them whenever there is a context switch. This increased context to save translates to increased latency when servicing interrupts or switching processes under a multiprocess kernel. For example, a C executive takes 17 μs to perform a context switch on a 25 MHz 68020, while this same context switch takes 29 μs on a 20-MHz AM29000.[7] While these are not MCUs, the example points out the effect of RISC/CISC on context switch latency. One solution to this problem is implemented in a PL/M-51 compiler for the MCS-51 microprocessors based on their unique multiple register bank architecture. In this compiler, a procedure can be associated with a particular register bank, and hence only the program counter, rather than all registers, needs to be saved on the stack during a subroutine call.[8] One can also take advantage of some of the newer MCUs that allow sections of their register space to be "locked" so necessary values for the ISR do not need to be retrieved from memory each time the ISR is invoked. This minimizes the latency time between an interrupt and the completion of its ISR.

4.2.1 High-Level Languages

While the availability of kernels may ease the development of applications, the development cycle can be speeded up even more by the use of a high-level language. High-level languages can take many forms, not all of which are applicable to MCUs, from interpreters (such as BASIC) to fifth generation languages such as PROLOG and LISP or the transputer language OCCAM. The scope of this discussion is limited to several language types with no loss of generality.

There are three requirements, any of which is sufficient, to qualify an HLL for use in an MCU system:[9]

1. The ability to define new high-level I/O structures

2. Direct access to computer memory and hardware for reading and writing

3. The ability to call user-written assembly or machine language routines.

Several more requirements must be added to this list which, while not sufficient, are nonetheless necessary conditions for the ready use of an HLL in an MCU system. [11]

4. *Compatibility*: If a new dialect of a language must be developed, then some, if not all, of the benefits of an HLL are lost.

5. *Direct connection to interrupts*: If it is necessary to write all ISRs in

assembly language and link them in, then a large portion of the software will have to be done at this level. HLL access to interrupts is a reasonable extension to an HLL without making it a dialect.

6. *Optimization*: The lack of an optimizing compiler to produce efficient code both in terms of memory usage or speeds (preferably selectable depending on application) will negate much of the benefit of working in an HLL.

7. *Development environment*: The interfacing of HLL object code and assembly language and its subsequent linking and downloading should be as painless as possible. The machine-specific characteristics should all exist in a command file, which is all that needs to be modified to compile the code for a different system configuration.

8. *Reentrancy*: Because of the asynchronous real-time nature of controller usage, executing code must be easily interruptable. The passing of arguments on the stack and returning values in registers allows for complete reentrancy.

For example, there are implementations of BASIC stored in ROM for some microprocessors that can appear to execute high-level instructions through calls to BASIC procedures. This approach has three drawbacks. The first is the expense of the ROM to store the interpretation of each BASIC instruction. The second is the slow execution speed because each instruction has at least the overhead of a subroutine call to a general implementation of the BASIC instruction. That is, there is no opportunity to optimize the execution of the overall program. For example, BASIC typically operates at a rate 400 to 2000 times slower than an equivalent machine-coded program.[9] The third objection is the large amount of storage that must be set aside for the executable program since it is not coded in machine language but rather in the HLL, which must be interpreted. The only thing that can be said in its defense is its ease of implementation for small control or computation programs with modest temporal requirements.

Another popular alternative to assembly language is a self-defining language such as FORTH. FORTH has been used to implement multi-computer real-time controllers.[10] FORTH is attractive to many programmers because of its flexibility and the fact that a new implementation of FORTH can be coded in typically less than 1000 bytes of machine code. It also meets all three of the requirements listed previously. The power of FORTH is that there are only a few high-level instructions, and all other instructions are coded in these. As soon as a procedure is coded, it immediately becomes another function of the language usable in defining more complex instructions. With this approach, there is a tendency to code a little, test a little. This incremental approach to software development is preferred by some. Since FORTH procedures are defined in terms of more primitive FORTH procedures and the basic instructions are limited, there is less of a requirement to fine tune the compiler for specific machines. Of course, one does not have the advantage of any optimization, because all instructions are interpreted, if not directly, recursively through the simplest FORTH instructions. FORTH is not as slow as BASIC, but is typically 10 to 100 times slower than tight machine code.[9]

A next level of HLL is C. This language is sometimes referred to as a portable assembly language, and sometimes as an implementation of a higher-level syntax expressing the PDP-11 assembly language,[11] for it offers many capabilities that other HLLs do not. The language is also disliked by some because it allows the programmer too much freedom and because it is easy to get oneself into trouble when using it. For example, it does not do the same type of explicit type checking which PASCAL does, and one can inadvertently set a character equal to a floating-point number with no warning being generated at compile time. Pointers can be used before they are initialized, leading to bizarre effects that are difficult to locate, if easy to correct. But this power is necessary when dealing with embedded systems since they are much more I/O intensive and less computation intensive. For example, if one is trying to convert data types, as in the case in a network protocol converter, FORTRAN is not a good choice since the only way to do this is to declare a common variable of one type and then access it through a pointer as a different type. It works, but it certainly is not in the spirit of the language. C, on the other hand, will let one declare a variable of one type and access it implicitly or explicitly as another type. Also, as seen in the IR tracker example, structures can be defined that refer to memory mapped I/O and allow for easy access and manipulation of the associated data or hardware registers.

The high-level language of choice appears to be C, although it is difficult to argue definitively why this is so. Many developers use it, and perhaps it is because of its natural link to UNIX systems. C may form the intellectual as well as the software bridge between high-level machines and event controllers since it is not only possible, but appropriate, to use C for programming in both environments. To substantiate this choice further, each of the eight previously enumerated desired properties of an MCU HLL is discussed here:

1. *I/O structures*: C has the ability to define new high-level I/O structures by virtue of its *struct* specifier. These new types can also be used as elements of other, higher-level structures.

2. Direct access to memory: C can directly access memory and hardware for both reading and writing by declaring a pointer and assigning it a value that is the address of a memory location or a hardware I/O port.

3. *Assembly language link*: A nominal extension to C implemented for an MCU is the ability to call user written assembly or machine language

routines. Typical C compilers not intended for MCUs usually have this as a normal part of the language,[12,13] while it is not specified in ANSI C.[14]

4. *Compatibility*: Compatibility is not a problem since extensions to C that aid in the development of MCU systems usually are directed toward direct attachment of interrupts and handling of on-chip hardware functions. These extensions have no meaning outside of the target MCU unless the source is transported to another MCU with similar capabilities. These enhancements would then be supported by the appropriate target-dependent implementations of the C calls.

5. *Direct connection to interrupts*: See item 4 above.

6. *Optimization*: It is assumed that the C compiler does perform optimization or can, through the use of command line switches, perform various levels of optimization (not a function of a compiled language, but the compiler itself).

7. *Development environment*: The only difference from normal high-level software development environments is the need for a linker command file that explicitly specifies addresses and object files to be linked into an executable image (not a property of a language, but the programming environment).

8. *Reentrancy*: C functions can be programmed to be reentrant since variables declared within a function are allocated on the stack.

It can be seen from the foregoing that the C programming language does indeed meet the requirements that have been established through experience and therefore is a suitable HLL for programming event controllers.

4.2.2 Macro Expansion and Functions

Before beginning the discussion of assembly language and HLL programming, there are a few general points that are applicable to both environments. As always, the hardware constraints of a particular event controller problem usually dictate whether compact code or execution speed is the dominant consideration. *Compact code* here means the length of the total executable program rather than the tight coding of an individual function or subroutine. Compact code and fast execution are not only not synonymous, but usually are conflicting programming goals. To illustrate the point, assume that it is necessary repeatedly throughout a program to read data in from one I/O port, P_1, and add it to another port, P_2, and then output the sum to port P_1. In the Intel MCS-51 this addition can be done with the sequence of assembly language instructions

```
AddPorts:      MOV A, P1      ;Read data from Port 1 into accumulator
               ADD A, P2      ;Add data from Port 2 to accumulator
               MOV P1,A       ;Output sum to Port 1.
```

This sequence of code can be handled in several ways that result in two possibilities. First, at each requirement for the addition in the program, the code can be duplicated. This can be done more easily by defining a *macro*. A macro is a sequence of code that replaces every occurrence of the macro's label at the time the code is assembled. For example, AddPorts could be written as a macro by adding system-dependent keywords such as the following:

```
AddPorts        MACRO
                MOV A, P,1      ;Read data from Port 1 into accumulator
                ADD A, P2       ;Add data from Port 2 to accumulator
                MOV P1,A        ;Output sum to Port 1.
                ENDM
```

Every time it is desired to output the sum of the ports, the macro label *AddPorts* is inserted in the code. This practice makes code more readable and at the same time minimizes the possibility of coding errors by letting there be only one copy of the code that is replicated at each occurrence of the macro label at assembly time. Aside from these programming advantages, macros have the distinct advantage of providing the fastest executable code since there is no need to context switch or jump to a subroutine to output the sum of the two ports. The expense, however, is in the total size of the executable code since the macro is expanded into its equivalent instructions at each of its occurrences.

The second way is to program this code sequence as a subroutine and end it with a return from subroutine as in

```
AddPorts:       MOV A, P1       ;Read data from Port 1 into accumulator
                ADD A, P2       ;Add data from Port 2 to accumulator
                MOV P1,A        ;Output sum to Port 1.
                RET             ;Return from subroutine.
```

There is not much difference in the appearance of the three code examples, but in the case of a subroutine, there is only one occurrence of the executable code whereas in the other two examples, the direct coding and the macro expansion, there are multiple copies. The advantage of the subroutine call is that it occupies the minimum program space. This is in addition to the advantage of the macro of having only one copy of the code, which helps reduce coding errors. The disadvantage of subroutine calls is the temporal overhead required each time the subroutine is called. If the subroutine is not located within 2 kbytes of the PC (Intel 8051) requiring the use of the LCALL instruction to call the subroutine, each call will require two machine cycles to execute the LCALL and two cycles for the RET, return from subroutine, instruction execution. With a 12-MHz clock, the minimum instruction cycle time is $1.0\,\mu s$ for a total of $4\,\mu s$ overhead for each subroutine call. Remembering the earlier example requiring a maximum of $27\text{-}\mu s$ delay, it can be seen that these apparently small overheads can add up to make a hard real-time system work properly or not. At least to a first approximation, the fastest programs will be those that use macro expansions to generate inline code at the expense of executable program size. The smallest programs will be those that maximize the usage of subroutine calls. The macro expansion process and the subroutine equivalent are shown in Fig. 4.4.

The terms *function* and *procedure* are sometimes used interchangeably, although there is a distinction between the two. Functions always can return a value, procedures do not. The importance of these two terms is in the implementation of code in either HLL or assembly language. While PASCAL has both functions and procedures, C has only one alternative, and that is a function. Lest you think that C has deprived you of some capability, notice that a procedure is a degenerate form of a function. That is, a function can return a value of *void*, which is the same as returning no value.

In the case of assembly language, the choice of using procedures or functions is determined by whether the code is linked only with other assembly language routines or whether it is meant to interface with C. Since the assembly language routine is usually defined in the C program as an external function with its return type also defined, there is little question what, if any, type of value is to be returned to the calling program. Depending on the C cross-compiler, the manner of returning the value may differ. One would expect that the values are returned either on the stack in a compiler-specific order or in one or more of the target machines registers. In the case of the INTROL 68HC11 cross-compiler and cross-assembler, 16-bit integer values are returned in the D-register and 32-bit integer values are returned in the D- and Y-registers, with the Y-register containing the most significant 16 bits. Return value types smaller than 16 bits are converted to an *int* and returned in the D-register. Floating-point values are returned the same as 32-bit integers.[15]

The same programming principle can be applied to C programming with the #*define* directive equating to a macro and a function equating to a subroutine call. The define directive has general format[14]

```
#define identifier token-sequence
```

with

```
#define OutFormat "%5d %5ld %2d %2d %5.3f %e %e %e\n"
```

as a specific example. Every time this string (including the quotation marks) is to be used, it need not be typed but rather the label *OutFormat* can be used. A different form of the define directive can be used to implement functions. For example[14]

```
#define ABSDIFF(a,b) ((a)>(b) ? (a)-(b) : (b)-(a))
```

defines a macro that returns the absolute value of the difference in its arguments. This is not a function call since it is only a replacement of an identifier by a particular sequence of mathematical operations at the time the program is compiled.

The C equivalent of an assembly language subroutine call is just a function itself. A function is defined by its return value type, the function name, and the name and type of parameters that are passed to it. The general form of a function is[14]

```
return-type function-name (parameter declarations, if any)
        {
                declarations
                statements
        }
```

with a specific example being

```
int GetIntegerFromFile(FILE *FileP)

{
char    TempString[32];
int     IntegerValue = 0;

    fgets(TempString, 31, FileP);                    /* skip text     */
    fscanf(FileP,"%d",&IntegerValue);                /* read integer  */
    fgets(TempString, 2, FileP);                     /* skip crlf     */
    return(IntegerValue);
}
```

which is a function to read an integer value from an input file that is pointed to by the pointer *FileP*. The actual value of the integer is returned by the function as indicated by the function return type of *int* and the last statement in the function return(IntegerValue);. This function can be called from anyplace in the program without incurring increased program size, as would be the case if the *#define* macro had been used.

4.2.3   Assembly Language Programming

As indicated earlier, assembly language programming will be limited in system development to those cases where an HLL is unable to accomplish the task either in terms of speed or code size. The rudiments of assembly language programming are not covered here as there are adequate texts dedicated to that purpose.[16-18] There are some fundamental practices that should be followed to generate effective code. First, code must be documented. Documentation consists of multiple levels, including traceability to the particular place it came from in the Petri table, to what the segment of code is designed to do (including assumptions), and finally to what each line or small group of lines does. This does not mean that the comment on each line is a restatement of the instruction. The line comments should answer the question "Why?" this line, not "What?" this line. The traceability back to the highest level of documentation is necessary for orthogonal debugging. The first symptom of a program not operating correctly is a high-level failure that causes one to check the operation of the program starting at the highest level of documentation, in this case the Petri table. Only when it is determined that the operational logic is correct should one go to the code to determine whether it was implemented correctly. Notice that the debugging of the code is concerned only with the correct implementation of the higher level documentation, not the correctness of the high-level documentation. There is no tendency so strong as the one to start changing the machine code when an event controller program does not work correctly. This is not the correct approach.

The absolute minimum number of numerical values should reside in the source code itself. While the source programs are longer, they are more readable and less likely to contain coding errors. It is easy to spot the difference between InitialValue and FinalValue, but much more difficult to detect the difference between the numbers 12 and 17 if they were the actual values of the two parameters. When you are reading the program looking for errors, the connotative meanings of Initial and Final values are more likely to attract your attention if the wrong variable name were used than if 12 were substituted for 17. This numerical error detection requires a mental conversion from the number to what the number represents before you can determine whether or not it is the correct value.

4.2.4   C Language Programming

No attempt is made here to teach C language programming. There are a number of texts devoted to teaching C programming[19] as well as documents that go into great detail about the language and its structure.[14] The focus in this section is on those elements of C that bear on the subject of its integration with assembly language programming as well as implementation in an environment that has, at best, a bare-bones operating system. It must be remembered that C's origins are in the UNIX operating system and, as such, it assumes the presence of system calls for many operations such as input/output. The hierarchy of I/O descends from the formatted output and input functions printf( ) and scanf( ) down to single character I/O with putchar( ), putc( ), getchar( ), and getc( ). Since printf( ) and scanf( ) are usually defined in terms of lower level functions such as gets( ) and puts( ), only these need to be coded in target assembly language if they are not already available from the software vendor for the target machine. These functions may need to be written in assembly language and/or modified to write to ports at nonstandard addresses.

Just as with C implemented on a variety of machines to operate under an

```
/*      The following are true for all implementations of Introl-C     */

#define CHAR_BIT      8               /* bits in char                   */
#define CHAR_MAX      127             /* maximum char value             */
#define CHAR_MIN      0               /* minimum char value             */
#define SCHAR_MAX     127             /* maximum signed char value      */
#define SCHAR_MIN     (-127)          /* minimum signed char value      */
#define UCHAR_MAX     255             /* maximum unsigned char value    */
#define SHRT_MAX      32767           /* maximum short int value        */
#define SHRT_MIN      (-32767)        /* minimum short int value        */
#define USHRT_MAX     65535U          /* maximum unsigned short value   */
#define LONG_MAX      2147483647L     /* maximum long int value         */
#define LONG_MIN      (-2147483647L)  /* minimum long int value         */
#define ULONG_MAX     4294967295UL    /* maximum unsigned long value    */

/* maximum values for doubles and floats                                */
#define DOUBLE_MAX   1.79769313486231570e308
#define FLOAT_MAX    ((float)3.402823466385288598e38)

/* largest powers of 2 exactly representable as double or float         */
#define FLOAT_MAXPOW2     ((float)(11 << 23))
#define DOUBLE_MAXPOW2    ((double)(11 << 30) * (11 << 22))


/*      The following are dependent on the use of IEEE format floating  */
/*      point values at run time and are common to all Introl-C         */
/*      implementations                                                  */

#define FLT_RADIX     2             /* radix of float exponent          */
#define FLT_ROUNDS    1             /* add rounds (1) or chops (0)       */
#define FLT_MAX_EXP   38            /* maximum power of 10 in float      */
#define FLT_MIN_EXP   (-38)         /* minimum power of 10 in float      */
#define FLT_DIG       7             /* digits of precision in float      */
#define LDBL_RADIX    2             /* for long double                   */
#define LDBL_ROUNDS   1
#define LDBL_MAX_EXP  308
#define LDBL_MIN_EXP  (-308)
#define LDBL_DIG      15
```

Text Box 4.4   Limits.h header file for INTROL-C implemented for several Motorola MCUs.

operating system, the dynamic range of different types varies with the target processor. The definitions are usually found in a header file called LIMITS.H and may differ significantly from those specified as the minimums in Kernighan and Ritchie (K&R).[14] For example, an inspection of the LIMITS.H header file of Text Box 4.4 supplied with the INTROL-C cross-compilers shows that some of the definitions are exactly the same as the K & R minimums except that the smallest values can be one more negative than the K&R specification. Some other values, as shown in Text Box 4.5, vary among the various target processors. For example, the maximum value of an *int* on the 68HC11 target is 32767 (**SHRT_MAX**), whereas on the 68020 an *int* can take on values up to 2147483647 (**LONG_MAX**). Note, incidentally, the use of the C convention of all capital letters to denote a constant.

To account for the fact that MCUs generally have some of their code and constants stored in EPROM, INTROL has incorporated the type modifier *const*. This modifier causes the data to be placed under a different location counter so it can be compiled for the memory addresses associated with the EPROM or ROM. A second type of modifier is used to account for the fact that some memory locations are actually hardware ports. *Volatile* is used to turn off register optimization for that variable so writes and reads of that variable will, in fact, take place at the specified address. This modifier does not turn off register optimization for all variables, but only the ones specified, thereby allowing for the maximum optimization possible consistent with the desire to write to specific hardware addresses. Both of these modifiers are part of the ANSI standard but not in the original K&R version of C.[14] A further discussion of the specifics of RAM initialization and MCU-specific considerations can be found in Smith.[20]

If you prefer not to write a separate assembly language source code, assemble it, and link it to the C object code, the function asm("<string>") is available for inline assembly language code. For example, the previous example of summing two ports could be coded in INTROL-C as

```
/*      The following are dependent upon the particular Introl-C      */
/*      implementations                                               */

#if     defined(__CC09__)|defined(__CC11__)|defined(__CC03__)|\
        defined(__CC01__)|defined(__CC86__)
#define     INT_MAX     SHRT_MAX      /* maximum int value            */
#define INT_MIN     SHRT_MIN     /* minimum int value            */
#define UINT_MAX    USHRT_MAX    /* maximum unsigned int value   */
#elif   defined(__CC68__)|defined(__CC20__)|defined(__CC32__)|defined(__CC62__)
#define INT_MAX     LONG_MAX     /* maximum int value            */
#define INT_MIN     LONG_MIN     /* minimum int value            */
#define UINT_MAX    ULONG_MAX    /* maximum unsigned int value   */
#endif


/*      The following are dependent upon the particular Introl-C      */
/*      implementations                                               */

#if     defined(__CC09__)|defined(__CC11__)|defined(__CC03__)|defined(__CC01__)
#define DBL_RADIX           FLT_RADIX
#define DBL_ROUNDS          FLT_ROUNDS
#define DBL_MAX_EXP FLT_MAX_EXP
#define DBL_MIN_EXP         FLT_MIN_EXP
#define DBL_DIG             FLT_DIG
#elif  defined(__CC68__)|defined(__CC20__)|defined(__CC32__)|defined(__CC62__)
#define DBL_RADIX           LDBL_RADIX
#define DBL_ROUNDS          LDBL_ROUNDS
#define DBL_MAX_EXP LDBL_MAX_EXP
#define DBL_MIN_EXP         LDBL_MIN_EXP
#define DBL_DIG             LDBL_DIG
#endif

#endif
```

Text Box 4.5 Target-dependent type definitions for INTROL-C cross-compiler.

```
asm("AddPorts:\MOV A, P1\ADD A, P2\MOV P1,A");
```

This would result in a compilation in which this assembly language would be inserted in the *.s11* file generated by the compiler before it is automatically assembled into the relocatable object file.

There is usually a preprocessor associated with the C language that, at a minimum,[14] allows for macro definition and expansion to ease the task of writing code. For example, there is a standard set of directives that control the operation of the compiler but have nothing to do with the syntax of the C language itself. These directives are all preceded by the octothorp, #, and are case sensitive just as are keywords in C. These directives are as follows:

#define *identifier token-sequence*

Replace subsequent instances of *identifier* with the *token-sequence* surrounded by white space.

#define *identifier(identifier-list) token-sequence*

In this second form, if there is no white space between *identifier* and (, then the identifier is taken as a macro with parameters given by the *identifier-list*.

#include <*filename*>

Replace this line with the contents of *filename*, which will be found in a standard directory.

#include "*filename*"

As above, but *filename* is found as quoted.

#ifdef *identifier*

Allows for conditional compilation where the lines following are compiled if *identifier* has been previously defined by a *#define* directive. Other conditional compile directives are #ifndef, #else, and #elif.

#endif

Directive ending sequence of lines to be included in conditional compile. Used to terminate #ifdef, #ifndef, #else, or #elif.

#line *constant "filename"*

Sets the line counter in the preprocessor to a value where the directive occurs and identifies the line number with a specified file. Usually used to eliminate confusing numbering caused by include files, and tie line numbers of the listing file to line numbers of the source file.

#pragma *token-sequence*

Causes the preprocessor to perform an implementation-dependent function. If *token-sequence* is unrecognized, it is ignored.

The sequence of operations required to proceed from a C language source file to an executable file for the target processor consists of the following, using the INTROL cross-compiler as an example:

1. Compilation of the C source program, such as HELLO.C, using the CC11.EXE program

2. Assembly of the output of the CC11 program, such as HELLO.S11, by the macro assembler AS11.EXE program

3. Linking of the resulting object module, such as HELLO.O11, with other object modules that have been assembled and/or object modules that are contained in standard or special libraries

4. Conversion of the linked program, such as HELLO.OUT, to a hexadecimal format by the hex conversion program IHEX.EXE for downloading to the target processor

5. Downloading the program to the target through the use of a terminal program such as KERMIT.

The operations of these programs and their interaction are explained more fully in the following sections, in which example programs are compiled, assembled, linked, and converted. Associated with each of these programs is a command line that specifies not only the program to execute, but which source files to operate on and which optional measures to employ while performing the operation.

The C compiler can have associated with it a variety of options specified by a hyphen, -, followed by a letter, number, or other optional argument. For example, the INTROL C compiler, CC11.EXE, takes a command line of the form

```
cc11 <filename> [{<option>}]
```

The metacharacters <>, {}, and [] are not to be included in the command line itself, but only serve to delimit the required or optional parameters. Items not included in these metacharacters are to be typed literally; items in <> are required and are selected by the user; items in [] are optional; items in {} can occur one or more times. For example, to compile one of the examples shown later, the command line would be

```
cc11  smrtcomp -l
```

which will take the file named SMRTCOMP.C (the .c suffix is assumed by the compiler if none is supplied), compile it, and produce an object file with the name SMRTCOMP.O11. Because of the -l option, a listing file will also be produced with the filename SMRTCOMP.LST. The optional listing file is not the only option that can be invoked by the command line. Other options for the INTROL compiler are as follows:[15]

$-a<ch> = <string>$: ch is a single character from the set, $t\,d\,b\,s\,1\,2\,c\,p$, and $<string>$ represents a section name. Sections are discussed in more detail later in this chapter since the topic is important to implementations of code in target processors. This option forces the compiler to put generated output of a specific type into a specific named section.

-b = *<directory>*: Specify the directory where subsequent passes of the compiler can be found.

-c: Use the default option file labeled OPTCC11 if it exists.

-cn: Inhibit the search for the default option file.

-c = *<filename>:* Get the compiler options from the file named *filename*.

-g: Various *g* options affect the internal operation of the compiler such as the order of filling bit fields, whether a char is signed or unsigned, or the internal format for floating-point values (IEEE or otherwise).

-h[<#>]: # is an optional digit from 0 through 9 that specifies the level of compiler optimization to be used.

-i = *<directory>*: Search *directory* for include files that cannot be found in other standard directories.

-k: Print the name and version number of each compilation pass. This is discussed in more detail later in this chapter.

-l[= *<filename>]:* Causes the compiler to generate an assembly language listing of the code and optionally put it in the file named *filename*. If *filename* is not present, the listing file is the same root name as the source file with the suffix *.lst* added.

-m<name>[= *<string>]*: equivalent to *#define* directive on the command line.

-n: Do not load the next sequential compilation pass.

-p: Generate *comm* assembler directives instead of define storage.

-o = *<filename>*: Output the object file to an explicitly named file.

-r: Retain the *.s11* file produced by the compiler rather than erasing it as would normally be done.

-s: Allow nested comments.

-t = *<directory>*: An explicit directory for the compiler's temporary files rather than the current directory.

-w[#]: # is an optional digit from 0 through 9 that specifies the level of warnings to be issued by the parser.

-x: Do not produce an object file.

-y[= *<n>]*: Truncate all identifiers to length *n*.

It can be seen from this list that complete control of the compiler can be exercised through the use of these command line options. Because of the difficulty of remembering them all as well as the errors can be propagated through repeated typing of the command lines, it is typical either to specify the options in the compiler command file, OPTCC11, or specify the compiler command line as well as the linker and hex converter command lines in a batch file, such as

```
c:\introl\bin\ccll smrtcomp -k -l
c:\introl\bin\ildll smrtcomp.o -a -m -u -y -gc:\introl\kjh\kjh -f smrtcomp.lnk -o
   smrtcomp.out
c:\introl\bin\ihex -k -v -g m -o smrtcomp.out
```

The first line invokes the CC11 C compiler, the second line invokes the ILD11 linker, and the third line invokes the IHEX conversion program to put the code in a downloadable format. The linker options and the operation of the linker are detailed in a later section.

Hidden in the CC11 compiler are actually three separate compilation programs that are normally transparent to the user. These three programs, or passes, are named ICC, IMP, and CG11. The first pass, ICC, stands for INTROL C compiler and is the front-end language parser. It produces a file that is subsequently piped to the language improver, IMP, which is an optimization program that improves the output of the parser to reduce the size and increase the execution speed of the final executable program. It is only in the last of the passes, CG11, where the target-specific considerations are taken into account. The code generator, CG11, takes the output of the improver pass and symbol table of ICC and combines them into an assembly language output file for subsequent assembly by the AS11 macro assembler. The code generator pass also continues the optimization of the program by implementing processor-specific optimizations. The final pass of the compiler calls the AS11 assembler to generate the relocatable, linkable object code.

4.2.5 Comparison of C and Assembly Language

The comparison of C and assembly language can be done on several levels. The first level of comparison is productivity, as measured in time to produce error-free executable code. Since C is working a metalevel higher than assembly language and programmers do not have to account for the contents of each register and status register bit, but can concern themselves more with the logic of the program flow and proper computation, C has the advantage in the timely production of useful code. A second level is the accessibility of the hardware resources of the computer to the programmer. Of course, everything is available at the assembly language level, but so are most things available to C. For example, if one wants to read or write from a memory location or a hardware port that is memory mapped, a pointer can be declared appropriate to the type of access desired. For example, if the hardware port were the 8-bit control register for the A/D converter on the M68HC11 MCU, a directive may be used to define the address of the port,

```
#define H11ADCTL 0x1030
```

and a pointer to an unsigned character be declared,

```
unsigned char *ADAddress;
```

with a subsequent assignment within the initialization function of the program of the identifier to the pointer, as in

```
ADAddress = (unsigned char *)H11ADCTL;
```

It may seem at first that the typecasting in this line is unnecessary, but ADAddress is a pointer and, as such, cannot be assigned an integer value even though the programmer knows it is all right. The compiler has caught this as an error and issued a warning. To assure the compiler that it is intended to initialize the pointer to this integer value, the identifier is typecast as a pointer to an unsigned character. Now the 8-bit A/D control register can be accessed by dereferencing its pointer, as in

```
*ADAddress = ADCTLWD;        /* initialize A/D converters  */
```

which writes the value defined by ADCTLWD to the hardware address pointed to by ADAddress. To determine when the A/D conversion is complete, one must test the most significant bit of the A/D control register for a 1. This bit position is the conversion complete flag (CCF) of the A/D control register. This is done in C using a while operation testing on the bitwise ANDing of CCF (which has a value of 0x80) and the dereferenced address of the A/D control register, as in

```
while (!(*ADAddress & CCF));     /*Wait for conversion complete      */
```

Once the conversion is complete, any of the A/D values can be read from their respective hardware locations by dereferencing their pointers, as in

```
MagneticReference = (int)*ReferenceAddress;      /* read      */
```

An interesting construction is shown in the preceding line. It was desired to use the value of the 8-bit A/D converter at ReferenceAddress for numerical computations, so the 8-bit value was typecast to an integer before its assignment to the integer variable MagneticReference. Of course, the value at ReferenceAddress is still available as an unsigned character as if it were at any other memory location. The complete C source from which these lines were taken is contained in App. I, Smart Compass. The Smart

Compass example shown in the appendix is modeled after an original design by Doug Garner, who designed the system using an Intel 8748 MCU.[21] The Smart Compass is an electronic compass based on a flux gate magnetometer that consists of two orthogonal windings around a saturable core. This is sometimes referred to as a solid-state magnetometer since there are no moving parts. The two windings sense the earth's magnetic flux in a vector form. The basic operation of the flux-gate magnetometer is that a reference signal saturates the toroidal core, thus shutting out any outside magnetic flux. When the core unsaturates, the earth's magnetic flux is gated into the core, generating a time-varying magnetic field that induces an EMF into the transverse and longitudinal coils.[22] The Intel 8748 version uses an external A/D converter, whereas the M68HC11 version uses the on-chip A/D. The circuitry associated with the M68HC11 Smart Compass is shown in Fig. 4.6. This example is used in later sections to show the complete integration of C and assembly language and the various alternatives for programming an MCU from only assembly language to only C as well as a combination of the two.

Regarding the compactness of executable code, the smallest program that can be written to implement a particular real-time controller is one written in assembly language, since some compromises must be made in the translation from a high-level language to machine code. For example, you may need to take the square root of a number. Since you know that the operand can never be negative in your application, you do not need to code the section that checks for a negative operation in order to not take the square root of a negative number. The HLL implementation must, since it

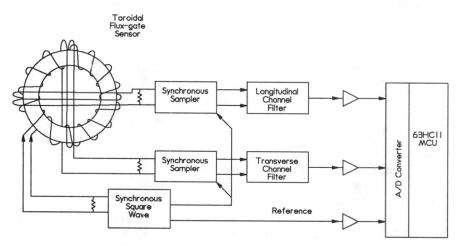

Figure 4.6 M68HC11 Smart Compass diagram showing flux-rate circuitry and interface to MCU.

does not have this *a priori* knowledge of the sign of the operand, include the negative checking, resulting in longer executable code both in size and execution time.

4.3　CONVERSION FROM PETRI TABLE TO SOFTWARE

One of the advantages in using Petri tables to design MCU software is its ability to orthogonalize the two phases of software construction. The Petri table is used to construct and verify the logical operation of the stages of the program, and once this is done, it is a simple matter to implement each transition of the Petri table in code. It does not matter whether the code is a high-level language or assembly language. When debugging, this orthogonality allows one to begin by asking whether the problem is in the Petri table or in the implementation of it. This may seem an obvious question, but there are many people who debug their logic in the code itself with no high-level design documentation. This section takes the Petri table for the UAV of Chap. 3 and implements it in C, although it could equally well have been implemented in assembly language. A partial C source for the UAV example is shown in App. II, UAV Code.

4.3.1　UAV Petri Table Implementation in C

The unmanned autonomous vehicle (UAV) example from Chap. 3 is used here to show how straightforward it is to convert from a Petri table to executable code. As previously, only sections from the appendix are used here to show the various features. Note first that each transition is translated directly into a C function. For example, the simple transition T_6 translates to

```
void TransitionSix(void)
{
ChannelOnePulseWidth = TemporaryPulseWidth;
ChannelOne = TRUE;
}
```

in which the value of the channel-1 pulse width is stored in a more permanent location and the semaphore that indicates that the channel-1 data has been measured, ChannelOne (CH_1 of the Petri table), has been set to TRUE. In general, each transition (function) when called performs some operation(s) and sets or clears the appropriate semaphores before it returns. These semaphores, or tokens, that are used by the executive to determine which is the next transition that is enabled, and hence which is the next transition function to call. Notice that no transition function directly calls another transition function and the tokens are set or cleared by a transition that, exactly as in the Petri table, enables a transition to fire.

There are also some functions that are not transitions, and this is for the convenience of coding. Often used sequences of code that are incidental to the transitions are put into their own function or logical blocks that perform a single function isolated from the calling function. This aids in debugging and effectively decouples segments of the program. For example, since the M68HC11 time is measured by capturing a value of a 16-bit counter that is never cleared or set, a pulse width may span from one count sequence to another, as in its rising edge arriving at 0FF00H and the falling edge occurring after the overflow of the counter at 0010H. It seems appropriate to modularize this computation in a function as follows

```
unsigned int ReadPulseWidth(int CounterBeginValue)
{
unsigned int LocalPulseWidth;
   CounterEndValue = ReadAndSwapEndianBytes(TimerOneAddress);
   if (CounterEndValue >= CounterBeginValue)
         LocalPulseWidth = CounterEndValue - CounterBeginValue;
   else
     LocalPulseWidth =
               MAX_COUNTER_VALUE - (CounterBeginValue - CounterEndValue);
   return(LocalPulseWidth);
}
```

This function first finds the current value of the master counter, CounterEndValue, by calling a function that reads the counter and then compares it with the value of the counter when the timing began, CounterBeginValue, to determine whether there has been an overflow. If there has been one, then the order of subtraction is reversed and the value is subtracted from 65535, the maximum value of the counter.

The only other section that needs to be considered is the main function, which embodies the executive. A section of main follows, which shows the initialization that consists of only setting the passed self-test variable to FALSE. After that, the values of the tokens control the execution of the program. The for(;;) is a forever loop and continually cycles through the testing of the various tokens to determine which transition is enabled.

```
int main(void)

{
PassedSelfTest = FALSE;
while(!PassedSelfTest)
   {TransitionOne();                     /* configure hardware          */
    PassedSelfTest = TransitionTwo();    /* Self-test                   */
   }
TransitionThree();                       /* Initialize all variables    */
for (;;)
   {
   if(Synchronized & !ChannelOne & !ChannelTwo)
        TransitionSix();
   else if(Synchronized & ChannelOne & !ChannelTwo)
        TransitionSeven();
   else if(Synchronized & ChannelOne & ChannelTwo)
        TransitionSeventeen();
```

```
else if(!Synchronized & RealTimeInterrupt & RealTimeInterruptEnable)
    TransitionFive();
else if(Synchronized & RealTimeInterrupt & RealTimeInterruptEnable)
    TransitionFifteen();
...
```

4.3.2 Magnetometer Assembly Language Example

An example of a well-structured assembly language program for an MCU that is both I/O and computation oriented is shown in App. I, Smart Compass. This example predates Petri tables, but demonstrates some practices of assembly language programming that make implementation and debugging straightforward. Again, sections are extracted from the complete listing as examples. First, notice that there are none, or very few, numbers embedded in the program. All numbers are defined at the beginning through the use of the EQU assembler directive as

```
ADCTL      EQU    $1030     *A/D Control register
ADRTRAN    EQU    $1031     *A/D Result -Transverse
ADRLONG    EQU    $1032     *A/D Result -Longitudinal
ADRREF     EQU    $1033     *A/D Result -Reference
ADCTLWD    EQU    $10       *A/D Control word
CCF        EQU    $80       *Conversion complete mask
```

The addresses of the registers are first specified, such as the address of the A/D control register, ADCTL, as hexadecimal 1030. Other values such as the bit that indicates that the A/D conversion is complete, and bit 7 of the ADCTL register, are represented by the identifier, CCF, which is equal to hexadecimal 080. The value of this approach is that if the software is compiled for different hardware configurations, such as A/D result registers 0, 1, and 2 rather than 2, 3, and 4, the numbers need only to be changed once, and that is the one in the EQU. If the numbers are embedded in the code and a change like this needs to be made, one occurrence will probably be missed and the program will appear to work correctly, but in fact will give erroneous answers.

Because of the combination of ROM and RAM in an MCU system, the assembler may have to be told more than once where to place code. This is done in more sophisticated assemblers through the use of sections, but in this program the assembler directive ORG was used, as in

```
ORG RAMSTART
```

which begins the sequence of code that allocates variable storage space. Later in the program, the compiler directive ORG is used again for the section of code that is to be placed in nonvolatile storage. For example,

```
ORG    PROMSTART      *Set address for compilation
LDS    #STACKSTART    *Initialize stack
```

```
        JSR    INIT                *Initialize vars
main:   JSR    IPVECTORS           *Read in field component
        JSR    CALCOCTANT          *Calculate which octant
        ...
```

Of course, this would not work but for the fact that RAMSTART and EPROMSTART were previously defined through the use of the EQU directive as

```
RAMSTART      EQU    $0000 *Start of RAM
EPROMSTART    EQU    $F800 *Start of EEPROM
```

To show equivalent hardware interfacing and computations to those mentioned earlier in the example of C programming, the following sequence of code starts the A/D conversion, waits for the CCF flag to be set indicating the end of A/D conversion, then reads the three magnetometer values, and computes the two absolute values. Notice the similarity to the preceding C example, but the increased difficulty in reading the code.

```
IPVECTORS    LDAA    #ADCTLWD         *Put control word in control register
             STAA    ADCTL            *start A/D
             LDX     #ADCTL           *Wait until conversion is complete
NOTRDY       BRCLR   0,X,CCF,NOTRDY
             LDAA    ADRTRAN          *Put Transverse in ACCA
             SUBA    ADRREF           *Subtract Reference
             BHS     TPOSITIVE        *If positive, jump
             NEGA                     *Else, take magnitude
TPOSITIVE    STAA    ABSTRAN          *Store | Transverse |
             LDAB    ADRLONG          *Put Longitudinal in ACCB
             SUBB    ADRREF           *Subtract Reference
             BHS     LPOSITIVE        *If positive, jump
             NEGB                     Else, Take magnitude
LPOSITIVE    STAB    ABSLONG          *Store | Longitudinal |
             RTS
```

4.4 INTERFACING C AND ASSEMBLY LANGUAGE

The Smart Compass C example of App. I, Smart Compass, was written to demonstrate both the use of C for interfacing to a hardware port and the replacement of that function with an external assembly language function to perform the same task. As the program stands in the appendix, it contains the line

```
/*#define ADinAssembly*/
```

which (as seen from the /* */ around it) shows that ADinAssembly is not defined. This definition and another,

```
#ifndef ADinAssembly
#define ADinC
#endif
```

are used for a conditional compile for this demonstration. Because ADinAssembly has not been defined, the #ifndef will be TRUE and result in ADinC being defined. These two identifiers will be used in subsequent places in the source code to determine whether code is to be included. For example, the entire C function, InputVectorsC( ), is included in the compile because of the preceding #ifndef ADinC as

```
#ifdef ADinC
    void InputVectorsC(void)
```

Yet later, the assembly language section that equates the exported variables (AbsTran, AbsLong, and MagRef) from the assembly language subroutine into the C function is excluded from compilation, as in

```
#ifdef ADinAssembly
    InputVectors();
    AbsoluteTransverse          = AbsTran;
    AbsoluteLongitudinal        = AbsLong;
    MagneticReference           = MagRef;
# endif
```

Variables are made exportable from an assembly language program if their identifiers are followed by a colon (:). This indicates to the assembler that their identifiers should be in a symbol table that is exportable to the linker rather than just used internally by the assembler. An example of this from the final assembly language program of App. I, Smart Compass, the program SC.S11, is

```
AbsTran:    RMB    1    *exportable Magnitude of Transverse vec.
AbsLong:    RMB    1    *exportable Magnitude of Longitud. vec.
MagRef:     RMB    1    *exportable magnetic reference value
```

In this example, the values of the variables were passed from the assembly language subroutine back to the calling C program through named variables. There are times when this may be neither convenient nor desirable, such as when writing reentrant or recursive subroutines. Reentrant and recursive functions are an assumed part of C, but are not a given in assembly language. For example, the use of named variables at fixed locations in memory means that a subroutine cannot call itself since the second invocation of the subroutine will overwrite the variables with the new value and the original value will be lost. Again, using the INTROL cross-assembler and cross-compiler as an example, a calling convention has been established for the transferring of data from the C program to the assembly language subroutine and back. In the INTROL 68HC11 compiler, the calling function passes parameters to a called function by pushing them onto the stack.[15] In this way, the use of fixed memory locations is avoided. After the function call, the calling function increments the stack pointer to where it was before the parameters were pushed (the stack grows from higher to lower memory in the MC68HC11). The actual locations of the parameters on the stack are shown in Fig. 4.7. The first parameter is passed in the D-register (A and B) if it is 16 bits or smaller. If it is a 32-bit parameter, such as a long integer number, the most significant part of it is passed in the Y-register with the least significant part in the D-register. Additional parameters are passed on the stack in the big-endian format since the stack grows downward. Since the calling function restores the stack pointer to its value before the parameters were pushed, it is not necessary for the assembly language program to take care of this. It is the responsibility of the called function to clean up the stack and return with the stack pointer unchanged from its value when the called function was entered. Since functions only return a single value, this value is returned in the D-register if it is 16 bits or in the D- and Y-registers if it is 32 bits long.

There is one other way in which values can be passed and that is by being declared in the C program and read and manipulated in the assembly language program. For this to occur, the assembly language must use the IMPORT directive to specify that the identifiers location will be resolved at link time.

After an assembly language program is written, it must be assembled into

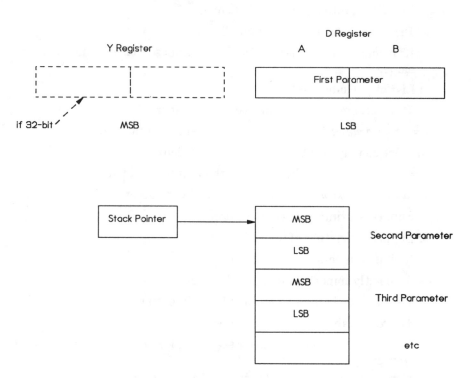

Figure 4.7 Location of parameters on stack as passed by the calling function.

a relocatable object file before it can be linked with the C program. Just as in the C cross-compiler, the command line for the AS11 assembler can take command line options in addition to the name of the assembly language source file to assemble. A typical command line might be

```
c:\introl\bin\as11 sc
```

Again, if the .s11 suffix is not present, it is assumed. The general form of the assembler command line is

```
as11 <file>{<options>}
```

where the options are

-*a:* All symbols except those with a question mark are placed in the object file.

-*b:* Prints all conditional directives in output file.

-*c:* Output listing to console rather than to a file.

-*d:* Inhibits printing of macro definitions.

-*e:* Inhibits printing of macro calls.

-*f:* List all conditionally excluded code.

-*g:* Prints data for data definition lines.

-*h:* If an undefined opcode is defined, it does not search for a macro by that name.

-*i:* List all include files.

-*k:* Print assembler name and version number.

-*l* = <*filename*>: Explicitly name the listing output file.

-*m:* List code generated by macro expansion.

-*o* = <*filename*>: Explicitly name the object output file.

-*p*<*digit*> = <*value*>: Pass a numbered parameter to the assembler.

-*r:* Suppress printing of cross-reference.

-*s:* Suppress printing of symbol table.

-*t:* Inhibit branch size optimization.

-*u:* Implicitly import all undefined symbols.

-*v:* Use line feeds instead of form feeds in output.

-*x:* No relocatable object output file produced.

-*y* = <*pathname*>: Pathname for searching for macro files after current directory.

-*z:* Delete the original source file (an extremely dangerous option).

Because of the number of options, it is typical to set up a batch file to do the assembling. Notice that unlike the C compiler, a command line option file is not automatically searched, making a batch file the only alternative to repetitive typing of the various options.

4.4.1 Example Linker Command File

Now that all of the C source files have been compiled and the assembly language object files have been assembled, it is necessary to link together the relocatable object files produced by them. This linking process not only fixes the location at which a program must run, but also resolves identifiers which are external to each of the modules. Some of these unresolved identifiers are in other object modules and some are contained in the C standard libraries supplied with the cross-compiler. In addition to resolving these undefined identifiers, the linker must be told the address at which the code is to be executed. Because of the nature of MCUs, there will be various types of memories at different locations in the address space, each with its own type of volatility. ROM and RAM are the two obvious examples, but EEPROM also must be considered due to its quasi-volatility. ROM, of course, cannot be written to by the executing program and hence cannot be a place for variables. On the other hand, RAM powers up in an unknown state and hence cannot start executing a program. One way of handling this selective compile for different locations is through the use of the ORG directive in assembly language programs. Unfortunately, there is no equivalent C directive since C was intended to be a platform-independent language with an operating system sitting between it and the hardware. Assignment of sections of code to different memory locations is accomplished through the use of *sections*. In assembly language code, the sections must be explicitly stated, and this replaces the ORG directive. In C programs, the compiler determines which type of code should go where and inserts the appropriate *section* directives into its output before passing it to the assembler for object file generation. Up to this point, the actual values of the sections are unknown, because absolute locations are of no value at the object stage of the compilation process. The linker is the only program that must know for which addresses to assemble the various types of code. The sections are, therefore, specified in a linker command file. A sample linker command file that has been slightly modified from INTROL's sample is shown in App. III, Linker. Segments from that appendix are used in the following discussion.

There are several types of sections as used by the INTROL compiler:

.text: Executable code

.data: Initialized external and static data

.bss: Uninitialized external and static data

.strings: Strings

.mod1: Noninterruptable executable functions

.mod2: Interrupt handlers not directly callable by C

.const: Symbols defined with the *const* type modifier

.base: Base page addressed symbols

Although these are automatically assigned to the different types of code generated by the C cross-compiler, they can all be used when writing assembly language code to force the linker to compile the code for a desired memory location. An example of section usage is

```
section .base bss origin 0;                    /* uninitialized base page storage    */
section .text origin 0xC000;                   /* exec. code put in 8K RAM of EVB    */
section .bss origin endof (.text) bss comms;/* foll. by uninit storage              */
```

which locates the *.base* section at 0000H, the executable code at C000H, and the variables immediately following the executable code. Errors in memory usage can also be checked if these checks are written into the linker command file. For example,

```
check endof(.text)    >  0xDFFF fatal "Code area too large";
check endof(.heap)    >  0xDFFF fatal "Program too large for RAM";
check sizeof(.base)   >  256 fatal "Base page too large";
check endof(.bss)     >  0xDFFF fatal "Uninitialized variable space too large";
```

checks to make sure that code was not linked that exceeds the memory available on the target system and prints the appropriate warning message to alert the programmer. The linker command file can also check to ensure that there is sufficient heap and stack space left after linking to make sure that the program is useful, as in

```
check __stackend - __stackstart < 64 fatal "Stack too small";
```

which warns you if there are not at least 64 bytes available for the stack.

Another purpose of the linker command file is to specify which object files other than those specified on the command line are to be linked as well as those to substitute for library object modules. For example, you may wish to use a version of OFMT that does not have output formats for long or floating-point numbers. The line

```
'c:\introl\kjh\shrtofmt.o'                     /* KJH: shorter version of printf    */
```

inserted before the normal library call

```
-lcio                                          /* C i/o library                     */
```

ensures that the abbreviated version will be used, thereby occupying less space of compiled code.

4.4.2 Smart Compass Example of C and Assembly Language Linking

In the case of linking the main Smart Compass C program and the assembly language program that reads the analog values at the inputs of the A/D converter, there are several ways in which the linking can be done. First, the two object files can be specified on the command line to the linker, as in

```
ild11 smrtcomp.o sc.o -gkjh -f smrtcomp.lnk -a -m -u -y
```

which invokes the linker to combine the two object files SMRTCOMP.O11 and SC.O11. The -*g* option says use the linker command file KJH.LD which can be found in the current directory. The -*f* option sends the ILD listing file to SMRTCOMP.LNK which is a valuable file for debugging since (with the -*m* option) it gives the actual address of all functions and (with the -*y* option) symbols, as well as (with the -*u* option) the names of all library functions that were used. The -*a* option creates an absolute file which will also indicate errors if there are any undefined references.

A second way to link the files, which is usually used when there are many object files that have not yet been put into a library, is to specify them in the linker command file itself. For example, the linker could have been invoked with

```
ild11 smrtcomp -gkjh
```

and the linker command file, KJH.LD, contain the line

```
'sc.o'.
```

Either of these methods is suitable and the one that is used is determined by the desires of the programmer. Usually as the number of object files increases, they tend to be put in the linker command file, with only the latest one that is being developed included on the command line. Although it is not discussed further here, most development systems have library programs that combine the object files into libraries so individual object files do not need to be listed, but only the name of the library that contains them.

4.4.3 Interrupt Handling in the IRTracker

The IR tracker example introduced in Chap. 3 is included to give additional insight into the conversion from Petri tables to executable C code. In particular, the manner of handling interrupts and the linking of interrupt-service routines (ISR) within C is demonstrated. All comments in this section are related to the example of C code as shown in App. IV, IR TRACKER. In this example, the interrupt handler is considered to be a background process that receives the interrupt, processes it, and sets or clears the appropriate semaphores. These semaphores are then used to enable or disable transitions. The firing of these transitions is controlled by the looping executive of the main function of the C program. Since the semaphores that enable each transition are tested in order, after a transition fires, the loop continues until it reaches the end and then restarts through the list at the beginning. This is only one example of how the execution of transitions can be prioritized. With this method, no single transition can monopolize the processor, since as soon as it fires, all of the others must be tested before that one can fire again. More complex executives could implement any of a number of other priority or queuing schemes merely by rewriting the main function with no change to the rest of the code.

Nesting of interrupts is not a problem since an interrupt of a higher priority can still interrupt another ISR. The use of C and its true function design, in which local variables can be allocated on the stack, allow for the writing of reentrant ISRs as well. The example of the IR tracker is not reentrant because it is assumed to read a character and put it in a buffer faster than the time it takes for the next character to be transmitted through the serial link to the SPI. As was shown earlier, the ISR could also have been written directly in assembly language and linked to the main program.

Through the use of the ATTACH ( ) C function, inherent in the INTROL cross-compiler, an ISR can be written in C or assembly language and associated directly with a particular hardware interrupt. For example, there are two ISRs in the IR tracker example. The SPIInterruptServiceRoutine processes the interrupts generated by the MC68HC11 serial peripheral interface (SPI), which is an on-chip synchronous serial communications device for communicating with other processors or peripherals. For our purposes, it is assumed to be a slave. When a character is received by the SPI, an interrupt is generated since the interrupt was enabled by writing 0C0H (SPI_ENABLE_INT) to the SPI control register (*SPIControlAddress) through its pointer (SPIControlAddress). The MSB enables the SPI, and Bit-6 enables the interrupt. The other bits can be set to zero in this example. The ISR first determines whether it is in the middle of receiving a message (the busy state) or whether it has received a complete message. It does this by reading the end-of-message (EOM) semaphore, since this would be TRUE if a complete message had been received and another one not started yet. If so, the message length and message index are zeroed and EOM set to FALSE. The message length is the second character received in the message and indicates where the checksum byte is for error checking. The message index is used to keep track of how many characters have been received and put into the message buffer, IncomingPositionMessage[ ]. The EOM semaphore is a distinct semaphore used by the ISR to maintain its own state, whereas the SRMsgRX semaphore is a main-program semaphore associated with its transitions. When a special byte is received, the START_OF_MESSAGE byte, the function considers the next byte as indicating the length of the following message and stores it in the variable MessageLength. This allows for variable-length messages. MessageIndex is then used both as an index for characters into the array, IncomingPositionMessage[ ], as well as an indicator of whether the message is complete. When the message is complete, then the now completely assembled message is transferred to the PositionMessage[ ] array and the SRMsgRx semaphore set TRUE. The timer interrupt-service routine is left as a homework problem, but it is attached in the same manner as the SPI ISR and is considered to be operating in the background, with its contribution being made available in the long integer variable RealTime.

4.4.4 Miscellaneous C/Assembly Language Interactions

The use of structures is also introduced in the IR tracker example. For example, the state of the tracking mount (i.e., its angular position and rate in both azimuth and elevation) as well as the times at which the first and last measurements were made are all stored in CurrentMount, which is of type MountStateType.

```
struct MountStateType
{
int Azimuth;
int Elevation;
int AzDot;
int ElDot;
long int StartTime;
long int EndTime;
};
struct MountStateType CurrentMount;
```

Time is part of the state because the angular position measurements are read from memory-mapped digital shaft encoders, but the angular rates are read as analog voltages from tachometers. Since the A/D conversions take a small but measurable time, and interrupts could occur during the measurement process, the storage of these times allows one to determine whether the time between the beginning and end of the measurements makes them unusable or requires corrections to the measured values. This is only one of several structure-type variables,

Another use of structured variables is to access hardware registers. It is assumed in the IR tracker example that the A/D converters that generate the control signals for the azimuth and elevation axes are located at memory-mapped locations. Specifically, they are assumed to be at sequential locations beginning at 06080H.

```
#define MOUNT_CONTROL_ADDRESS    0X6080    /* Mount D/A for control    */
```

A pointer type is also declared that points to a structure.

```
struct MountControlType
   {
   unsigned short int Azimuth;
   unsigned short int Elevation;
   };
struct MountControlType *PhysicalMountControlAddress;
```

The net result of this is that to output a command to the elevation axis, one must initialize the pointer at the beginning of the program and write

```
(*MountControlAddress).Elevation = 5;
```

to output the control value of 5. This exact scheme could be used to address all of the control registers of the MC68HC11 so long as the correct types are used in the declaration of the structure. The correct type is important since some of the registers are only 8 bits wide and others, such as the counter, are 16 bit. Starting in the middle of the register structure, one could declare the OC1M and OC1D as unsigned characters since they are both 8 bits wide, but declare TCNT as an unsigned integer to allow a single access to read the counter into a variable rather than two accesses to individual bytes with the attendant complication of having to combine them into a single unsigned integer later.

```
struct H11RegisterType     /* declares a structure equal to the regs    */
{
...
unsigned char OC1M;
unsigned char OC1D;
unsigned int TCNT;
unsigned int TIC1;
...
}
   struct H11RegisterType *H11Regs;   /* declares a pointer to the regs */
   #define H11RegLocation 0x1000     /* physical location of registers /
   H11Regs = (struct H11RegisterType *)H11RegLocation; /* inits pointer*/
```

Remember that when the directive defines H11RegLocation as 01000H, C will assume it is of type *int*, and because of this, its use to initialize the pointer to a structure must be specifically typecast as a pointer to the desired structure. Now that this is done, each individual register can be accessed by dereferencing its pointer, as in

```
TimeNow = (*H11Regs).TCNT;
```

Along these same lines, and what results from the INTROL compiler, is that the 68HC11 registers are declared in a portion of assembly language as

```
                 section            .H11REGS

H11PORTA:        ds.b          1   i/o port A
                 ds.b          1   reserved
H11PIOC:         ds.b          1   parallel i/o control register
   .
   .
   .
H11OC1M:         ds.b          1   OC1 action mask register
H11OC1D:         ds.b          1   OC1 action data register
H11TCNT:         ds.w          1   timer counter register
H11TIC1:         ds.w          1   input capture register 1
```

Initially one might think that these labels are pointers, but in fact the labels are treated as any other variable identifier. In other words, to read the input capture register 1, one only needs to write

```
TimeNow = H11TIC1;
```

rather than

```
TimeNow = *H11TIC1;
```

The section H11REGS was defined in the linker command file to be 01000H which is equivalent to setting the ORG in an assembly language program to 01000H. The result of this is that the storage is defined to exist at the register locations.

4.5 SUMMARY

This chapter presented the fundamentals of assembly language and C language programming as they apply to MCU programming. The conversion from Petri tables to executable code was shown to be simple and straightforward and orthogonal to the logic flow that is verified in the Petri table. The C language was presented as an excellent high-level language for the simple implementation of real-time event controllers, at least as a first approximation of the final code. It was argued that software development should proceed expeditiously through the use of high-level software on a well-configured development system to determine the functionality of the program, and only after it has been shown to be operational at this high level, apply other techniques to reduce either the size of the code or increase the speed of execution. Finally, a second example of the Smart Compass was presented exclusively in both assembly language and in C as well as in parts in each of these languages to show how they can be integrated. The INTROL cross-compiler and cross-assembler were used extensively as an example system to demonstrate the principles of writing, compiling, linking, and downloading executable code into a target MCU system.

REFERENCES

1. Kermit reference.
2. *M68HC11EVB Evaluation Board User's Manual*. Motorola Semiconductor Technical Data Publication 68HC11EVB/D1, Phoenix, AZ, 1st ed., September 1986.
3. I. Olsen, "C: The Link for Embedded Controllers and Productivity in Control Environments," *Electro 88*, 54(1), Boston, MA, May 10–12, 1988.
4. T. Williams, "Real-time Operating Systems Struggle with Multiple Tasks," *Computer Design*, pp. 92–108, October 1, 1990.
5. D. A. Mellichamp, ed., *Real-Time Computing With Applications to Data Acquisition and Control*. New York: Van Nostrand Reinhold, 1983.
6. S. Lamb and M. Pauwels, "Real-Time Programming with Single Chips," *WESCON '82 Conference Record*, Anaheim, CA, session 5A/2, pp. 1–4, 14–16, September 1982.
7. R. Wilson, "Real-time Executives Take on Newest Processors," *Computer Design*, pp. 88–105, February 1, 1989.
8. R. Jaswa, "PL/M-51: A High-Level Language for the 8051 Microcontroller Family," *WESCON '82 Conference Record*, Anaheim, CA, 17B/1, pp. 1–6, 14–16, September 1982.
9. R. M. Dumse and D. E. Smith, "High Level Language Solutions for Dedicated Applications," *WESCON '82 Conference Record*, Anaheim, CA, session 17B/1, pp. 1–14; 14–16, September 1982.
10. G. H. Smith, "A New FORTH-based Development System," *WESCON '82 Conference Record*, Anaheim, CA, session 17/C1, pp. 1–14, 14–16, September 1982.
11. D. LaVerne, "C in Embedded Systems and the Microcontroller World," *Electro 88*, session 50/4, Boston, MA, May 10–12, 1988.
12. *Turbo C++ User's Guide*. Borland International, Inc., 1800 Green Hills Road, Scotts Valley, CA, 1990.
13. *Microsoft C Advanced Programming Techniques*. Microsoft Corporation, Redmond, WA, 1990.
14. B. W. Kernighan and D. M. Ritchie, *The C Programming Language*. 2nd ed. Englewood Cliffs, NJ: Prentice-Hall, 1988.
15. *Introl-C MS-DOS Host Guide, Version 3.05*. INTROL Corporation, Milwaukee, WI, 1989.
16. W. C. Lin, *Computer Organization and Assembly Language Programming for the PDP-11 and VAX-11*. New York: Harper & Row, 1985.
17. G. J. Lipovski, *Single- and Multiple-Chip Microcomputer Interfacing*. Englewood Cliffs, NJ: Prentice-Hall, 1988.
18. J. B. Peatman, *Design with Microcontrollers*. New York: McGraw-Hill, 1988.
19. R. Johnsonbaugh and M. Kalin, *Applications Programming in ANSI C*. New York: Macmillan, 1990.
20. L. Smith, "Efficient RAM Utilization by C Compilers for Microcontrollers," *Electro 88*, session 50/3, Boston, MA, May 10–12, 1988.
21. D. Garner, "Building the Smart Compass," *Forum*, Oshkosh, WI, 1986, as updated by status report of June 5, 1987.
22. M. J. Gilstead, "Magnetic Heading Sensor for the HERO 2000 Robot," senior project, George Mason University, May 5, 1989.

PROBLEMS

1. Convert transition T_{17} of the UAV Petri table example to C and assembly language. Make all of the appropriate declarations for the assembly language. Include section declarations.

2. Discuss the advantages of using a high-level language for the initial coding of a real-time event controller.

3. Write a Petri table for a three-floor elevator and implement it in C.

4. Write an assembly language macro to sum the analog values available at the four A/D inputs of the MC68HC11 MCU. Include the code to start the conversion process as well as check for conversion complete. Repeat the example, but compute the average of four readings taken on the same A/D channel. Program the *ADCTL* register correctly to make multiple samples on the same input.

5. Define a C macro that swaps the two bytes in a variable of type *unsigned short int* (16 bits).

6. Write a short C program and assembly language that communicate by passing two 16-bit parameters on the stack. Assume that the assembly language program returns a 32-bit result.

7. With reference to the IR tracker example and App. IV, IR TRACKER, write a Petri table for the serial peripheral interface interrupt-service routine.

8. Modify the Petri table generated in Prob. 7 to ensure that the software synchronizes correctly with the incoming message.

9. Write a Petri table and C function to compute the checksum of the incoming message of the IR tracker and return **FALSE** if the checksum is not correct.

10. Is there a better way to do the A/D conversions in the IR tracker example rather than waiting for *CCF* after each A/D conversion? (This question is specific to the capabilities of the MC68HC11 MCU.)

4-Bit and 8-Bit Microcontrollers

Four- and eight-bit MCUs, because of their simplicity relative to the 16-bit and 32-bit progeny, hold the distinction of being the first commercially produced MCUs. In spite of their relative simplicity, 4- and 8-bit MCUs still command a significant portion of the MCU market, if not in dollars as shown in Fig. 5.1,[1] then at least in number of units used. These simplest of MCUs are still perfectly capable of controlling many devices, from washing machines to videocassette recorders (VCRs) as they have been doing so for many years. Although they are limited in their processing capability because of their word size and low speed, their low-cost and built-in peripheral controllers (display drivers, serial interfaces, and A/D converters) make them more than adequate for a number of volume production applications in which a cost-effective solution requires a minimum chip count.

This chapter surveys 4- and 8-bit MCUs, highlighting the architectural characteristics of the various families as well as the differences among the families. Because of the tremendous variety of manufacturing processes, packaging options, environmental specifications, and detailed technical information available about the individual devices, the devices are only viewed from this architectural and feature perspective, and the interested reader is encouraged to contact the individual manufacturers for the most current electrical and environmental specifications. In addition to the survey of MCU family characteristics, one 8-bit MCU, the Motorola

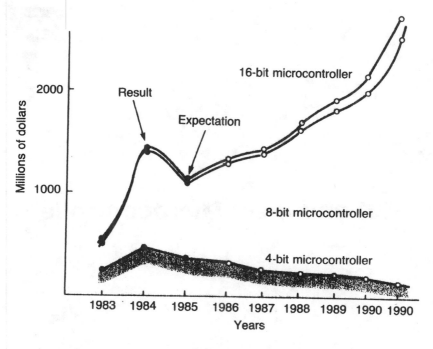

Figure 5.1 MCU shipment forecast for 4-, 8-, and 16-bit MCUs.[1]

MC68HC11, is discussed in more detail since it is representative of the capabilities of 8-bit MCUs. Included in this chapter are MCUs from Intel, Motorola, NEC, and Texas Instruments.

The distinction among MCUs as to 4, 8, 16, or 32 bits is determined by the width of the internal data path and not the internal register size, memory width, or external data path width. This distinction must be made since many MCUs internally move data over one width of data path and yet regularly perform arithmetic on register operands twice that width. In the TMS1000, a 4-bit MCU, the data path is 4 bits wide, but the program ROM is 8 bits wide. In the Motorola MC68xx series, there are two 8-bit registers, *A-register* and *B-register*, that can be arithmetically manipulated individually or simultaneously as the *D-register* to perform 16-bit operations. The only constant for classification of MCUs appears to be the internal bus width, and this is the classification used in the following discussion.

5.1 4-BIT MICROCONTROLLERS

Because of their intended use in large-volume markets, 4-bit MCUs tend to be characterized by design tradeoffs that minimize the recurring production costs at the expense of nonrecurring engineering (NRE) costs. This is because NRE is amortized over the total number of MCUs implemented in a particular application. The cost of the MCU itself, however, must be accounted for in the recurring costs of production, which dictate the minimum number of chips, interconnects, and peripheral hardware. The result of this type of tradeoff sometimes results in less than convenient programming of the MCU. For example, in the TMS1000 family, the program counter is not sequential since it is more silicon-effective to implement a pseudo-random number generator than a sequential counter.[2]

The discussion of 4-bit MCUs begins with the Texas Instruments TMS 1000 family which was first released to original equipment manufacturers (OEM) in 1974[2] and continues to be used today. The discussion expands to include the NEC μPD75xx series and finishes with the National Semiconductor COPS 400 series.

5.2 TEXAS INSTRUMENTS TMS1000 FAMILY MEMBERS

The TMS1000 family is characterized by a 4-bit data path from the RAM to the ALU as well as to the *X-register* and *Y-register*. The *R-output* latch is also addressed by 4 bits, allowing for 16 different 11-bit or 13-bit outputs. There are 4 bits of binary input referred to as *K-inputs*. The program counter itself is 6 bits wide, with a 4-bit page address that accounts for the ability to access 1024 8-bit-wide instructions.

5.2.1 TMS1000 Family Members

As with all MCUs, many of the family members consist of variations in the amount of ROM and/or RAM. The TMS100 family[3] varies from 512×8 bits of ROM to 4096×8 bits and from 32×4 bits of RAM to 128×4 bits. The more significant variations are among I/O characteristics in that they can vary from 9 output lines to 16 output *R-lines* and from 8 to 10 *O-lines*. The *R-lines* are latched binary data and are addressed by the *Y-register* and individually set or cleared by the SETR or RSTR instructions. The *O-lines* are the output of a 5- to 8-bit code converter that is itself a programmable logic array (PLA). All models have four binary input lines, referred to as the *K1*, *K2*, *K4*, and *K8* inputs that are read directly into the adder. Some models include a frequency divider input on the *K8* input line that can be programmed at manufacture to divide by 1, 2, 10 or 20. Some models include four additional inputs that can be latched under external control. These inputs are called the *L1*, *L2*, *L4*, and *L8*.

In addition to hardware configurations, the subroutine nesting level can differ among family members with only two options: one level in the same page or three levels in any page. Units that have high-voltage outputs (-35 volts) are also available to directly drive vacuum florescent displays.

There is also a TMS2100/2300 family, which is a derivative of the TMS1000 but with additional I/O capabilities, including, in all members[2]

1. 8-bit A/D
2. Zero crossing detector
3. Interval timer
4. Four levels of subroutine nesting

and, in some models,

5. Six to fifteen *R-lines*
6. Eight *O* lines
7. Eight to twelve *K-*, *J-*, and *R-lines*
8. Event counter

This family has an instruction that is a superset of the TMS1000 MCU.

5.2.2 TMS1000 Architecture

The architecture of the TMS1400 is shown in Fig. 5.2 and is representative of members of the TMS1000 family of MCUs. Note that program and data memory are distinct and have separate address and data lines, which makes this a Harvard architecture. Additionally, neither are addressed in the conventional manner but through a single address matrix with a page register specifying which block of 64 instructions is to be executed in the program memory and an *X-register* and *Y-register* specifying the data location. There is no external bus, and all inputs and outputs are handled through the previously explained *R-lines* and *O-lines* for outputs and *K-lines* and *L-lines* for input. The outputs are separately controlled through different busses and registers, but the *K-line* and *L-line* inputs are multiplexed through an externally controlled pin. Subroutines can be nested to one or three levels depending on the chip, and this limit is because the return addresses are stored in separate registers are integral to the chip rather than being implemented as a stack in RAM.

The control and timing for the CPU is implemented in hardware, and one instruction cycle consists of six oscillator pulses. All instructions are executed in one instruction cycle. The frequency of operation is determined by either a fixed external resistor or capacitor connected to the oscillator connections.

Instruction set.   There are two similar, yet distinct, instruction sets associated with the TMS1000 family. One set is for the TMS 10xx, TMS12xx, and TMS17xx MCUs and is shown in Table 5. 1, and another for the TMS11xx

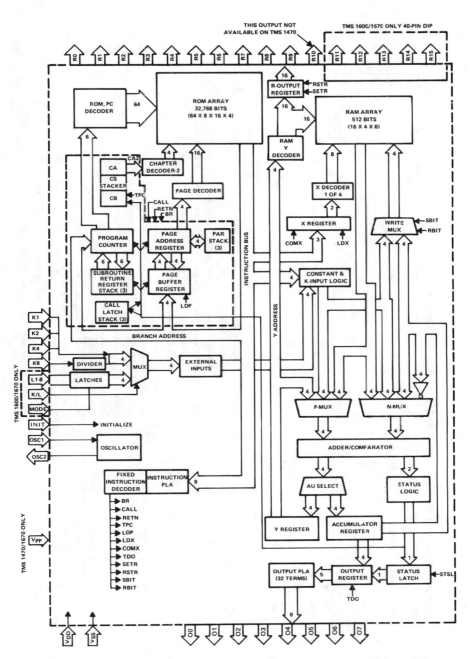

Figure 5.2 TMS 1400 series block diagram. *(Reprinted by permission of Texas Instruments)*

TABLE 5.1 TMS 1000/TMS1200/TMS1700 Standard Instruction Set

FUNCTION	MNEMONIC	STATUS EFFECTS C	STATUS EFFECTS N	DESCRIPTION
Register to Register	TAY			Transfer accumulator to Y register.
	TYA			Transfer Y register to accumulator.
	CLA			Clear accumulator.
Transfer Register to Memory	TAM			Transfer accumulator to memory.
	TAMIY			Transfer accumulator to memory and increment Y register.
	TAMZA			Transfer accumulator to memory and zero accumulator.
Memory to Register	TMY			Transfer memory to Y register.
	TMA			Transfer memory to accumulator.
	XMA			Exchange memory and accumulator.
Arithmetic	AMAAC	Y		Add memory to accumulator, results to accumulator. If carry, one to status.
	SAMAN	Y		Subtract accumulator from memory, results to accumulator. If no borrow, one to status.
	IMAC	Y		Increment memory and load into accumulator. If carry, one to status.
	DMAN	Y		Decrement memory and load into accumulator. If no borrow, one to status.
	IA			Increment accumulator, no status effect.
	IYC	Y		Increment Y register. If carry, one to status.
	DAN	Y		Decrement accumulator. If no borrow, one to status.
	DYN	Y		Decrement Y register. If no borrow, one to status.
	A6AAC	Y		Add 6 to accumulator, results to accumulator. If carry, one to status.
	A8AAC	Y		Add 8 to accumulator, results to accumulator. If carry, one to status.
	A10AAC	Y		Add 10 to accumulator, results to accumulator. If carry, one to status.
	CPAIZ	Y		Complement accumulator and increment. If then zero, one to status.
Arithmetic Compare	ALEM	Y		If accumulator less than or equal to memory, one to status.
	ALEC	Y		If accumulator less than or equal to a constant, one to status.
Logical	MNEZ	Y	Y	If memory not equal to zero, one to status.
Compare	YNEA	Y	Y	If Y register not equal to accumulator, one to status and status latch.
	YNEC	Y	Y	If Y register not equal to a constant, one to status.

FUNCTION	MNEMONIC	STATUS EFFECTS		DESCRIPTION
		C	N	
Bits in Memory	SBIT			Set memory bit.
	RBIT			Reset memory bit.
	TBIT1		Y	Test memory bit. If equal to one, one to status.
Constants	TCY			Transfer constant to Y register.
	TCMIY			Transfer constant to memory and increment Y.
Input	KNEZ		Y	If K inputs not equal to zero, one to status.
	TKA			Transfer K inputs to accumulator.
Output	SETR			Set R output addressed by Y.
	RSTR			Reset R output addressed by Y.
	TDO			Transfer data from accumulator and status latch to O outputs.
	CLO			Clear O-output register.
RAM 'X' Addressing	LDX			Load 'X' with a constant.
	COMX			Complement 'X'.
ROM Addressing	BR			Branch on status = one.
	CALL			Call subroutine on status = one.
	RETN			Return from subroutine.
	LDP			Load page buffer with constant.

NOTES: C-Y (Yes) means that if there is a carry out of the MSB, status output goes to the one state. If no carry is generated, status output goes to the zero state.

N-Y (Yes) means that if the bits compared are not equal, status output goes to the one state. If the bits are equal, status output goes to the zero state.

A zero in status remains through the next instruction cycle only. If the next instruction is a branch or call and status is a zero, then the branch or call is not executed successfully.

SOURCE: Reprinted by permission of Texas Instruments.

TABLE 5.2 TMS 1100/1300 Standard Instruction Set

FUNCTION	MNEMONIC	STATUS EFFECT C	STATUS EFFECT N	DESCRIPTION
Register-to-Register Transfer	TAY			Transfer accumulator to Y register
	TYA			Transfer Y register to accumulator
	CLA			Clear accumulator
Register to Memory	TAM			Transfer accumulator to memory
	TAMIYC	Y		Transfer accumulator to memory and increment Y register. If carry, one to status.
	TAMDYN	Y		Transfer accumulator to memory and decrement Y register. If no borrow, one to status.
	TAMZA			Transfer accumulator to memory and zero accumulator
Memory to Register	TMY			Transfer memory to Y register
	TMA			Transfer memory to accumulator
	XMA			Exchange memory and accumulator
Arithmetic	AMAAC	Y		Add memory to accumulator, results to accumulator. If carry, one to status.
	SAMAN	Y		Subtract accumulator from memory, results to accumulator. If no borrow, one to status.
	IMAC	Y		Increment memory and load into accumulator. If carry, one to status.
	DMAN	Y		Decrement memory and load into accumulator. If no borrow, one to status.
	IAC	Y		Increment accumulator. If carry, one to status.
	DAN	Y		Decrement accumulator. If no borrow, one to status.
	A2AAC	Y		Add 2 to accumulator. Results to accumulator. If carry, one to status.
	A3AAC	Y		Add 3 to accumulator. Results to accumulator. If carry, one to status.
	A4AAC	Y		Add 4 to accumulator. Results to accumulator. If carry, one to status.
	A5AAC	Y		Add 5 to accumulator. Results to accumulator. If carry, one to status.
	A6AAC	Y		Add 6 to accumulator. Results to accumulator. If carry, one to status.
	A7AAC	Y		Add 7 to accumulator. Results to accumulator. If carry, one to status.
	A8AAC	Y		Add 8 to accumulator. Results to accumulator. If carry, one to status.
	A9AAC	Y		Add 9 to accumulator. Results to accumulator. If carry, one to status.
	A10AAC	Y		Add 10 to accumulator. Results to accumulator. If carry, one to status.
	A11AAC	Y		Add 11 to accumulator. Results to accumulator. If carry, one to status.
	A12AAC	Y		Add 12 to accumulator. Results to accumulator. If carry, one to status.
	A13AAC	Y		Add 13 to accumulator. Results to accumulator. If carry, one to status.
	A14AAC	Y		Add 14 to accumulator. Results to accumulator. If carry, one to status.
	IYC	Y		Increment Y register. If carry, one to status.
	DYN	Y		Decrement Y register. If no borrow, one to status.
	CPAIZ	Y		Complement accumulator and increment. If then zero, one to status.

FUNCTION	MNEMONIC	STATUS EFFECT		DESCRIPTION
		C	N	
Arithmetic Compare	ALEM	Y		If accumulator less than or equal to memory, one to status.
Logical Compare	MNEA		Y	If memory is not equal to accumulator, one to status.
	MNEZ		Y	If memory not equal to zero, one to status.
	YNEA		Y	If Y register not equal to accumulator, one to status and status latch.
	YNEC		Y	If Y register not equal to a constant, one to status.
Bits in Memory	SBIT			Set memory bit
	RBIT			Reset memory bit
	TBIT1		Y	Test memory bit. If equal to one, one to status.
Constants	TCY			Transfer constant to Y register
	TCMIY			Transfer constant to memory and increment Y
Input	KNEZ		Y	If K inputs not equal to zero, one to status.
	TKA			Transfer K inputs to accumulator
Output	SETR			Set R output addressed by Y
	RSTR			Reset R output addressed by Y
	TDO			Transfer data from accumulator and status latch to O-outputs
RAM X Addressing	LDX			Load X with file address
	COMX			Complement the MSB of X
ROM Addressing	BR			Branch on status = one
	CALL			Call subroutine on status = one
	RETN			Return from subroutine
	LDP			Load page buffer with constant
	COMC			Complement chapter

NOTES: C-Y (Yes) means that if there is a carry out of the MSB, status output goes to the one state. If no carry is generated, status output goes to the zero state.

N-Y (Yes) means that if the bits compared are not equal, status output goes to the one state. If the bits are equal status output goes to the zero state.

A zero in status remains through the next instruction cycle only. If the next instruction is a branch or call and status is a zero, then the branch or call is not executed successfully.

SOURCE: Reprinted by permission of Texas Instruments.

and TMS13xx MCUs is shown in Table 5.2. An inspection of these two tables shows the compromises made in trying to keep the instruction set small and yet powerful enough to deal with a minimal MCU. The TMS1000 is an accumulator-based machine, and there are a number of instructions for transferring data among the registers, ALU, and memory. The arithmetic instructions include the normal add and subtract from the accumulator as well as increment and decrement. The instructions to add to the accumulator the numbers 2 through 14, each with its own instruction, are not common in other MCUs.

For branching, there are a number of instructions that affect the status bit, including testing memory for zero and comparing the *Y-register* to the accumulator. The test for a condition must be made with the instruction immediately following the instruction that sets the status, since the status is always set to its normal state, 1, after the execution of the next instruction. Individual bits in both the memory and the *R-lines* can be set or cleared, and individual memory bits and *K-inputs* can be tested.

For program control, conditional branches and subroutine calls are included. In order to handle the paged program memory, a load page buffer command is available.

On-chip peripherals.   There are essentially no on-chip peripherals in the TMS1000 series since it is designed for modest control applications, with the exception of those noted earlier for the TMS2100/TMS2300 series. The TMS1000 with the high-voltage output capability is capable of directly driving vacuum fluorescent displays. The input/output is particularly designed for reading keyboards and displaying information.

Other capabilities.   The TMS1000 family has none of the capabilities that have become standard in even the 8-bit MCUs, such as external memory, master-slave capability, power-down modes, and interrupts. The MCU is available in a variety of packages and fabrication processes which make it suitable for any environment, and a variety of power sources, including 9- and 15-volt power-supply models and low current CMOS versions.

5.2.3   Design and Application Tools

There is no high-level language support for the TMS1000 because none is required. There is insufficient memory to support even an interpreter.

5.3   NEC μPD7500 FAMILY OF 4-BIT MCUs

As with the Texas Instruments TMS1000 series of MCUs, NEC has developed its own 4-bit, single-chip CMOS microcontrollers for similar applications in high-volume, low-cost applications.[4] The μPD7500 series of

MCUs is entirely comprised of CMOS MCUs with common hardware and software so systems can be easily upgraded. Power supplies for these chips range from 2.0 volts (μPD7554A/7556A) to 6 volts in packages of 20, 24, 40, 42, 44, 64, or 80 pins.

5.3.1 μPD7500 Family Members

The entire family is most easily understood with reference to Fig. 5.3, which shows the basic 7500H EVA MCU that is used in the μPD7500 emulator. The range of program ROM among the various members of the family is from 1024 to 6144 bytes internal. Likewise, RAM can be selected from 64 to 256 nibbles. The number of I/O lines can vary from as few as 15 to as many as 35 individual lines, and one (the 7533), has an on-chip 4-channel, 8-bit A/D converter. Some models are designed to drive high-voltage vacuum fluorescent displays (FIP,* fluorescent indicator panel), and others are customized for directly driving liquid-crystal displays (LCDs) or light-emitting diode (LED) displays. Some models have a piggyback version that can be used for a final functional check of the program, preproduction testing, or small-volume production. This is particularly advantageous prior to mask programming thousands of MCUs for volume producing. All family members have an 8-bit timer to implement real-time clocks and, except for the 7506/7556/7566, all family members have an 8-bit serial I/O port.

---

*FIP is a registered trademark of NEC.

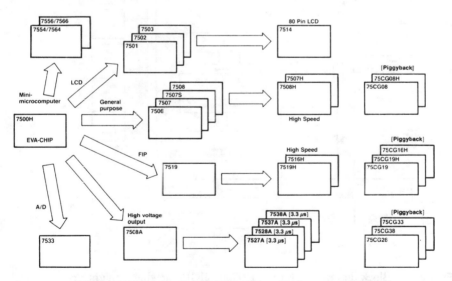

Figure 5.3   NEC μPD7500 family of MCUs. *(Provided courtesy of NEC Electronics Inc., © 1987 NEC Electronics Inc. All rights reserved.)*

5.3.2 μPD7500 Architecture

The architecture of the μPD7500 family of MCUs can best be understood by referring to Fig. 5.4, which is the block diagram of a μPD7502. There is a single 4-bit-wide bus that connects all of the major elements. A 12-bit program counter sequences through the program and is sufficient except for the largest ROM versions, which access the memory above 4096 bytes by setting bit 1 of the program status word (*BNK* flag) to 1 (μPD7516). Since multibyte instructions are allowed, the *PC* increments by the number of bytes in each instruction. Unlike the TMS1000, the μPD7500 maintains a stack pointer in RAM that allows for nesting of subroutines up to the limit of available RAM. Since the maximum RAM available is 256 nibbles, the stack pointer (SP) is only 8 bits wide. When an interrupt or subroutine call occurs, the SP is decremented first, so the *SP* should be initialized to the highest RAM address plus 1 through the use of the **TAMSP** instruction.

Restart and interrupt vectors as well as lookup tables for the LHLT and CALT instructions are contained in the 8-bit-wide program memory (ROM) as shown in Fig. 5.5. Data memory is static RAM with a maximum size of 256 × 4 bits. Since some of the RAM in some models is dedicated to LCD or FIP

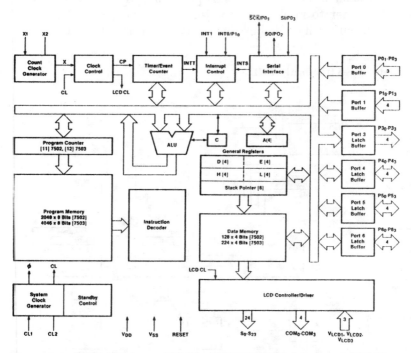

Figure 5.4 Block diagram of NEC μPD7502 MCU showing straightforward, single internal bus architecture. (*Provided courtesy of NEC Electronics Inc.*, © *1987 NEC Electronics Inc. All rights reserved.*)

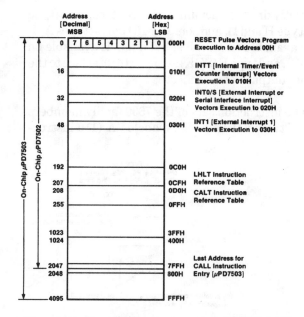

Figure 5.5 Program memory map of NEC 7500 series of MCUs. *(Provided courtesy of NEC Electronics Inc., © 1987 NEC Electronics Inc. All rights reserved.)*

display memory (00H to 01FH in the 7514 and to 03FH in the 7519/16), this area is not available to the stack. In the μPD7514, this leaves at least 192 × 4 for stack with each CALL instruction or interrupt requiring four 4-bit nibbles. PUSH instructions need only two nibbles since only 8-bit registers are stacked.

In addition to the Accumulator, all models contain two 4-bit registers, the *H* and *L*. Some models contain two more 4-bit, general-purpose registers labeled *D* and *E*. The registers can be used either alone or in register pairs including the *DE*, *DL*, and *HL* pairs, the leftmost lettered register containing the most significant bits. The *HL* pair can be used as a data pointer and some instructions perform automatic increment and decrement of the *L* register for ease of moving through data. As with the TMS1000, this is an accumulator based machine with a 4-bit accumulator. In addition to the 4-bit ALU, there is a 4-bit *program status word (PSW)* that contains the carry flag, the *BNK* bit for accessing memory higher than 4 kbytes, and two skip bits, *SK0* and *SK1*. These skip flags are set as a result of instruction execution.

The 7500 series of MCUs come with a variety of I/O interface circuits. There are nine different hardware configurations as shown in Fig. 5.6, some of which are suitable for directly driving vacuum fluorescent displays (type 6), some for driving LCDs (type H and I), and some for I/O ports (types A, B and D-F). Not all ports are available on all chips, but a mix is available on each. Type B (SI), type E (SO), and type F ($\overline{SCK}$) are used to interface to the serial I/O interface. All port outputs are latched.

μPD7500 instruction set.   The instruction sets of the 7500 family members actually consist of subsets of two instruction sets labeled A (110 instruc-

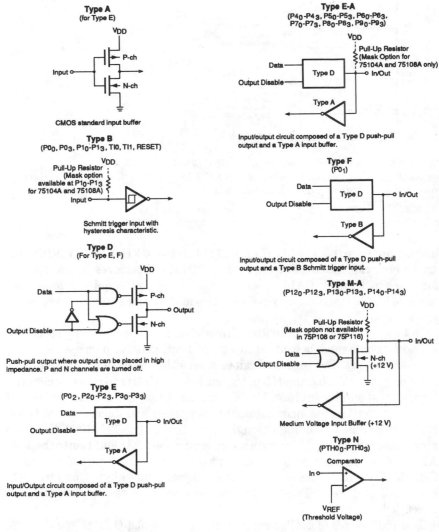

Figure 5.6   Nine types of interface circuits on NEC 7500 MCU. *(Provided courtesy of NEC Electronics Inc., © 1987 NEC Electronics Inc. All rights reserved.)*

tions) and B (70 instructions). The A instructions are available in the 2K or larger ROM versions, and the B instructions are available in the 1K ROM versions. There are four different A subsets and three different B subsets. The subsets are only selectable by choosing the appropriate device. For example, the μPD7502 uses the A3 instruction subset. The emulator chip that possesses both the A and B instruction sets is the 7500H/H-E. The A set is not just a superset of the B set, as each set contains some instructions that the other set does not. The minimum instruction execution time is 2.8 μsec.

As would be expected, there is a complete set of data movement instructions involving the accumulator and the general-purpose registers. A set of data exchange instructions exchange the data between the accumulator and the general-purpose registers as well as among the accumulator, H, and L registers and memory. The arithmetic operations are accumulator-based add and subtract. Logical instructions are XOR, AND, and inclusive OR as well as accumulator complement and single-bit rotate right and left. The general-purpose registers, as well as memory, can be directly incremented and/or decremented. Individual bits in RAM can also be directly set or cleared.

Branching can be accomplished by testing on individual bits of the accumulator, the carry bit, or memory. Branches can also occur as a result of testing the accumulator with memory, testing a memory location as well as a comparison of the general purpose registers with an immediate operand. Special instructions are included for controlling the on-chip serial I/O port, the timer/event counter, and the I/O ports.

The 7500 uses four methods for addressing operands in memory:

1. *Immediate*: The second byte of the instruction is the data.
2. *Direct*: The second byte of the instruction is the address of the operand.
3. *Register indirect*: The operand address is contained in the specified register pair.
4. *Stack indirect*: The stack pointer points to the data.

NEC 7500 on-chip peripherals. The timer/event counter of Fig. 5.7 consists of a single 8-bit up-counter that is supplied with a clock frequency derived from the MCU clock. The counter can be zeroed by software through the TIMER instruction and is automatically zeroed when there is a timer coincidence between the 8-bit modulo register and the 8-bit counter register. This coincidence generates $INTT$, a timer vectored interrupt, and also toggles the timer flip-flop, which is a clock source to the serial I/O.

An 8-bit serial interface is also on chip, as shown in Fig. 5.8. Although this is only a synchronous serial port without the usual control registers, it will provide bidirectional synchronous communications under direct software

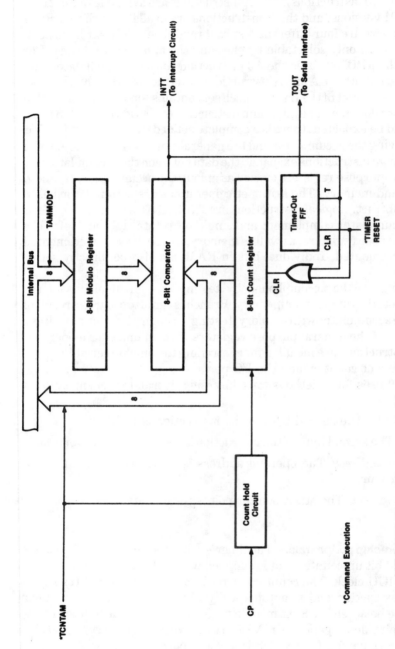

Figure 5.7 NEC 7502 Timer/Event counter block diagram. (*Provided courtesy of NEC Electronics Inc., © 1987 NEC Electronics Inc. All rights reserved.*)

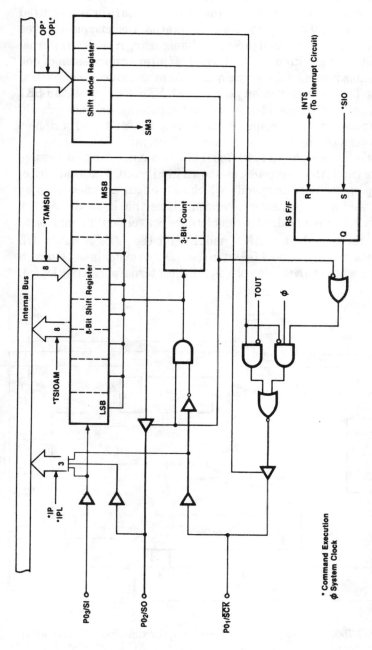

Figure 5.8 NEC 7500 on-chip serial interface. (*Provided courtesy of NEC Electronics Inc., © 1987 NEC Electronics Inc. All rights reserved.*)

or interrupt-driven control from/to external devices such as A/D converters, dot-matrix LCD controllers, and other MCUs. Data to be transmitted is loaded into the 8-bit shift register from the internal bus, and one of three clock sources is software selected to control the signaling rate. If an internal source is selected, it is automatically turned off after transmission of one byte. Reception is normally done by clocking from an external source and either waiting for an interrupt or polling the *INTS RQF* with the SKI instruction. The availability of a serial I/O capability on chip greatly minimizes the number of interconnections between the MCU and peripherals in cases where the required data rates will permit.

The third on-chip peripheral (μPD7533) is a 4-channel, 8-bit successive approximation (SA) A/D converter, as shown in Fig. 5.9. The 4-bit write-only *ADM-register* is used to specify which of the four channels is to be converted as well as to initiate conversion. Since this is an SA A/D, 9 machine clock cycles are required to complete the conversion. The analog to be converted is selected by bits *AN10* and *ANI1* in the *ADM-register*. A/D conversion is started when the *ADS* bit in the *ADM-register* is set. Conversion is complete when the *EOC* flag bit (bit 2 of the internal port A) is a logic 1.

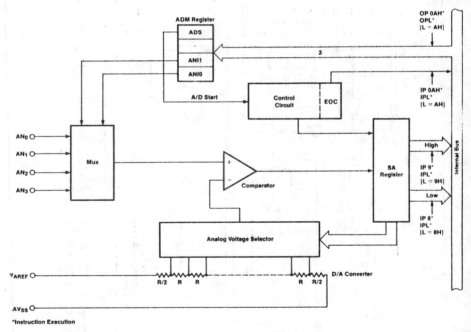

Figure 5.9 NEC 7500 4-channel, 8-bit successive approximation A/D converter. *(Provided courtesy of NEC Electronics Inc., © 1987 NEC Electronics Inc. All rights reserved.)*

Powerdown modes. All μPD7500 family members implement the STOP and HALT instructions, which put the MCU into a low-power operating mode. In the STOP mode the operation of the CPU ceases as well as all operations requiring the system clock. The only way the MCU can be brought out of this condition is through a hardware *RESET* or by interrupts. The HALT mode does not completely disable the MCU but rather disables the CPU while still allowing clocking of the timer/event counter. This reduces power consumption but not as much as with STOP, while at the same time allowing the occurrence of a timer interrupt to release the CPU from the HALT mode.

Interrupts. In the μPD7502 there are a maximum of four interrupts, two of which are individually brought out to external pins and are rising-edge triggered. The other two interrupts are internal, *INTS* and *INTT*, and are, respectively, the interrupt generated by the serial I/O interface and the timer/event counter interrupt. All of the interrupts are vectored through three dedicated vector addresses in program memory. If more than three interrupts are implemented in a single device, two interrupts share a single address. The interrupts can be enabled and disabled, but not selectively.

5.3.3   μPD7500 Design and Application Tools

NEC has a comprehensive array of development tools for their MCUs that can be hosted on PCs and other machines. The cross-assembler for the 7500 family will assemble code for all members of the family. No HLL is available, but again, that is understandable due to the limited resources of a 4-bit MCU. Hardware development is supported through two different methods. The EVAKITS are motherboards that accept plug-in daughter boards that emulate a specific MCU. The second option is SE boards which are in-circuit emulators that generally have more memory and more functionality than the MCU that they emulate. Up/down load software is supplied with both hardware alternatives, which allows for serial communications between the emulator and the host computer.

5.3.4   μPD75x Family Members

Members of the μPD75x family of 4-bit MCUs expand the capabilities of the μPD7500 in that they have higher clock speeds, more I/O ports, more peripherals, a larger instruction set (62, 107, 139, or 141 instructions), more ROM and RAM and reduced instruction execution time. Although the architecture provides for up to 64 Kbytes of ROM and 4 Knibbles of RAM, current version contain up to 32 Kbytes of ROM and 1 Knibbles of RAM.

Additionally, the number of 1-, 4-, and 8-bit registers has been increased as well as multiple banks of registers which are maintained in data memory. The CPU clock can be programmed so that the CPU will operate at high, medium, or low speeds. The number of interrupts and peripherals has also been increased. Versions are typically available in both one-time-programmable (OTP) as well as mask ROM and EPROM versions.

5.3.5 μPD75x Architecture

The architecture of the μPD75x family of MCUs can be understood through the example of Fig. 5.10, which is the block diagram of the μPD751xx general purpose MCU. There is a single bus that connects all of the major elements. There is a 12-, 13-, or 14-bit program counter whose size depends on the ROM (4K, 6K, 12K and 16 Kbytes) available to a particular family member. All locations in program ROM are usable except for 00H and 001H, which are reserved for the internal reset start address. The μPD751xx maintains a stack pointer in a dedicated 8-bit register, which allows for nesting of subroutines up to the limit of the reserved stack area which is the

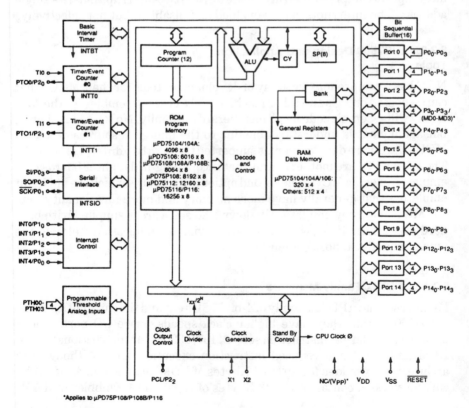

Figure 5.10   Block diagram of NEC μPD751xx. *(Courtesy NEC Electronics Inc., © 1987 NEC Electronics Inc. All rights reserved.)*

upper 224 locations of the 256 nibbles of bank-0 of RAM. The remaining 32 locations are the general purpose registers. Data memory consists of three memory banks labeled 0, 1, and 15. Memory bank-0 contains a set of general purpose registers, and memory bank-15 contains peripheral control registers. Switching among banks is controlled by setting bits in the Bank Select Register. The general register structure is shown in Fig. 5.11 and indicates four banks of eight 4-bit registers, four of which can be accessed as bytes. Although the RAM is organized as nibbles, data can be manipulated as bits, nibbles, or bytes. A 16-bit sequential buffer is located in the upper half of Bank 15 and is the only general purpose RAM in this area as the rest of Bank 15 is dedicated to peripheral control registers. This area can be bit, nibble, or byte manipulated and is used to control and send data to the onchip peripherals.

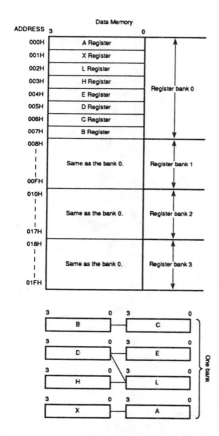

Figure 5.11 NEC μPD75x general purpose register configurations. *(Courtesy NEC Electronics Inc., © 1987 NEC Electronics Inc. All rights reserved.)*

Restart and interrupt vectors as well as the look-up table for the GETI instruction are contained in the 8-bit wide program memory as shown in Fig. 5.12. The GETI instruction is unique in that it allows for reducing the size of executable code by converting one 2-byte or two 1-byte instructions into a single 1-byte instruction. The five interrupt vectors are at the beginning of program memory and are followed by executable code which can be accessed by various call and branch instructions.

The μPD751x series of MCUs come with a variety of types of I/O. The μPD751x is the one of the most I/O populated of the μPD75x series with thirteen 4-bit parallel ports. All but two CMOS input ports are bidirectional.

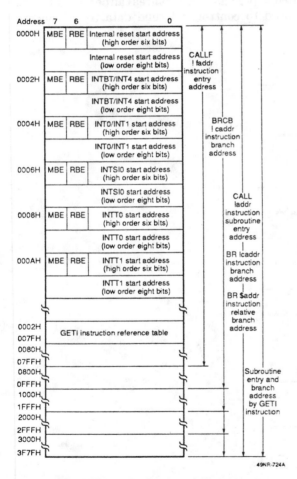

Figure 5.12   NEC μPD751xx program memory map. *(Courtesy NEC Electronics Inc., © 1987 NEC Electronics Inc. All rights reserved.)*

This makes for a total of 44 I/O lines. Thirty-two of the bidirectional lines are CMOS capable of sinking 15 mA directly. The other 12 are N-channel, open drain I/O lines with 12-volt maximum; they can be supplied with mask option pull-up resistors. These can also sink 15 mA for directly driving LEDs. Other members of the family have a lesser number of parallel I/O ports, but are nevertheless, very I/O oriented with 22, 28, 32, 34, 36, or 48 lines. One family member, the μPD755xx has 64 I/O lines. Some models have I/O ports with software selectable pull-up resistors.

μPD75x Instruction Set. The μPD75x instruction sets contain either 62, 107, 139, or 141 instructions. All sets contain bit manipulation instructions, 4-bit and 8-bit transfer instructions, table reference instructions, and 1-byte relative branches. All members of the μPD75x family except the μPD75402A have the GETI instruction explained above. MCUs with the 139 or 141 instruction set have the capability to perform 4-bit and 8-bit arithmetic, logical comparison, and increment/decrement in addition to nibble and byte transfers. The minimum instruction execution time is 0.95 μs and some newer family members operate at 0.67 μs.

The μPD75x has four methods for addressing operands in memory:

1. *Immediate:* The data is part of the instruction.

2. *Direct:* The second byte of the instruction is the address of the data to be manipulated. It can refer to either 1, 4, or 8 bits.

3. *Register indirect:* The address of the operand is contained in the specified register pair. Four-bit addressing of memory bank-0 can be done through the DE or DL registers. The HL register can address all memory banks and can be postincremented or postdecremented. The indirect reference can be either 1-, 4-, or 8-bit.

4. *Stack indirect:* The stack pointer points to the data in memory bank 0. In parts with 1 Knibblesof RAM, the stack can be located in memory banks 0, 1, 2, or 3.

All bits in general purpose RAM, I/O ports and some peripheral control registers (memory bank 15) can be bit manipulated. Bit manipulation is achieved using either direct addressing or a combination of direct and register indirect addressing.

μPD75x On-Chip Peripherals. The μPD-75x family contains 8-bit interval timers, 8-bit timer/event counters, 14-bit PWM timers, 14-bit watch timers (for time of day), 16-bit multifunction timers, I/O ports, 4-bit comparators, 8-bit A/D converters, LCD and FIP display controllers, EEPROMs, and 8-bit serial interfaces. The peripherals on an μPD-75x family member vary and depend upon which family member is used. Peripherals on the μPD-751x consist of a basic interval timer, two timer/event counters, a serial interface, and programmable threshold analog inputs. The two 8-bit timers are similar to the those of the 75xx, and differ from each other only in the clock selection to the count register. Timer 0 has divide by 16 prescaler and Timer 1 has a 4096 prescaler. The hardware comparator toggles a flip-flop and generates an interrupt each time the value of the counter matches that of the modulo register. The counter is zeroed at compare and continues to count up.

Also on chip is an 8-bit serial interface. Although this is only a synchronous serial port without the usual control registers, it will provide 3-wire bidirectional synchronous communications under direct software or interrupt driven control from/to external devices such as A/D converters, dot-matrix LCD controllers and other NEC MCUs. Data to be transmitted is loaded into the 8-bit shift register from the internal bus, and one of five clock sources is software selected to control the signaling rate.

Another peripheral is the comparator port which is useful for detecting when an analog signal exceeds a preset threshold without having to perform an A/D conversion and a subsequent comparison of the digitized value with the threshold. There are four inputs which are serially compared to a voltage from a resistor ladder. The ladder has 16 levels, and one of them is selected by control register PTHM. When the comparison begins, input PTHOx is compared to a D/A version of the stored valued through an analog comparator. The results of these four comparisons are stored as four bits in a single register which can be read by the program.

Two μPD75x family members have on-chip 8-channel, 8-bit successive approximation (SA) A/D converters: the NEC μPD75028 and μPD755xx. Other models have 8-bit counters, 14-bit watch timers, and a 16-bit multifunction timer which can be used as an 8-bit timer, a PWM output, a 16-bit-free-running timer, or a 16-bit counter for an integrating A/D converter. Like the 75xx, there is also a model with an FIP interface which can drive up to 24 segments, up to 16 digits, with eight dimming levels as well as provide key scan interrupt generation.

μPD-75x Powerdown Modes.   All NEC μPD75x family members implement the STOP and HALT instructions which can put the MCU into a low-power operating mode. In the STOP mode the operation of the CPU ceases as well as do all operations requiring the system clock. The MCU can be brought out of the STOP or HALT modes through a hardware RESET or by any interrupt request. The HALT mode does not completely disable the MCU but rather disables the CPU while still allowing all peripherals to continue normal operation. The HALT mode reduces power consumption but not as much as the STOP mode. A data retention mode may be entered after the Stop instruction by lowering the VDD to 2 volts. The contents of RAM and registers will be maintained. The MCU can be restarted with no loss of memory by first restoring VDD and then resetting or interrupting the MCU.

μPD75x Interrupts. Each family has a different number of interrupts ranging from 2 to 12 external interrupts, 2 to 7 internal interrupts, 3 to 7 vectored interrupts, and 2 to 12 edge-detecting interrupts. A typical interrupt structure is that of the μPD751x shown in Fig. 5.13. This device has seven external and four internal interrupt sources, some of which are mutually exclusive. Five interrupts are vectored, and five of the external sources are provided with edge-detection circuitry. The internal interrupts are generated by the basic interval timer, the serial interface, and Timers 0 and 1 compare. External interrupts are in response to rising and falling edges on input pin INT4, selectable edges on INT0 and INT1, rising edges on INT2 and INT3, and edge detection of input pins TI0 or TI1.

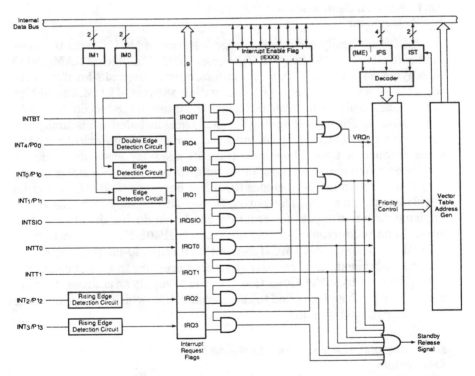

Figure 5.13 NEC μPD751xx interrupt controller block diagram. *(Courtesy NEC Electronics Inc., © 1987 NEC Electronics Inc. All rights reserved.)*

5.4 NATIONAL SEMICONDUCTOR COPS400

The National Semiconductor COPS400 (control-oriented processor system)[5,6] family exhibits many similarities to the 4-bit MCUs discussed previously in this chapter. There is a single instruction set that is common to all family members, and there is a single pinout configuration making it easy to transition from the simplest family members to the most sophisticated. The origins of the COPS400 family date to the hand-held calculators of 1970–75[5] which is the same time frame as the TMS1000. The instruction set is ROM efficient in that single instructions perform multiple functions, which can make programming difficult yet provide for a cost-effective solution to large volume control systems. These 4-bit MCUs with a control-oriented instruction set are intended for the high-volume consumer market, including consumer electronics, automotive and industrial control, toys/games, and telephones. As recently as 1989, 40 million COPS400 devices were shipped.[6]

5.4.1 National Semiconductor COPS400 Family Members

The COPS400 family comprises a large selection of MCUs, from the most limited (COP410) with 512 bytes of program ROM, 32 nibbles of RAM, 19 I/O pins, synchronous serial I/O, no interrupts, no counter, and 2-level stack, to the most powerful member (COP444C) with 2048 bytes of ROM, 128 nibbles of RAM, 23 I/O pins, synchronous serial I/O, 1 interrupt, 8-bit counter, and 3 levels of stack. The more than 60 different devices included in this family also share a common instruction set and a common pinout for devices with the same number of pins. These two features allow for easy upgrading or downsizing, depending on the application, with minimum retraining and/or retooling. The family also includes ROM-less devices intended for program development. This is an alternative to simulation that allows the direct implementation of a part for accurate emulation during development or low-volume production. For example, the COP401 MCU is functionally equivalent to the COP410 MCU except that it is in a 40-pin package rather than a 24-pin package. These extra pins are necessary to accommodate the multiplexed address/data lines that are used to supply an address stored in an external latch and supplied to an external ROM, which then returns the program data to the chip.

5.4.2 National Semiconductor COPS400 Architecture

Although all of the family members are based on the same architecture, it is best explained by first referring to the block diagram of a minimum implementation (Fig. 5.14), the COPS410L. As with the previous TMS1000

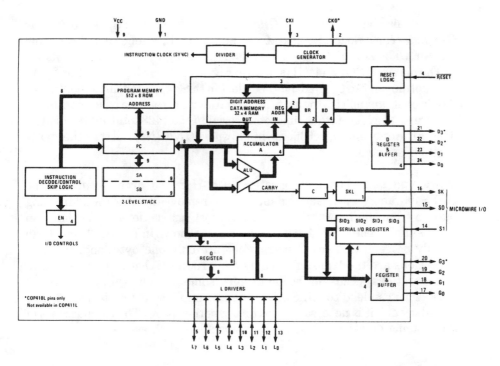

Figure 5.14 Block diagram of a minimal National Semiconductor COP400, the COP410L. *(Courtesy National Semiconductor)*

and NEC7500, the COPS400 family is accumulator based with a 4-bit accumulator which that can be loaded from RAM or the output of the 4-bit ALU. The COP410 has only two levels of subroutine nesting since the stack is implemented in dedicated registers. The data and program memory are separate, which yields a Harvard architecture, a feature that is somewhat compromised in the NS 8-bit COPS and abandoned in their 16-bit HPC series, which have the Von Neumann architecture. The program memory, directly addressed by the 9-bit incrementing PC, sends an 8-bit instruction to the instruction decoder. Because of certain paged jump instructions, it is at times necessary to view the program ROM as being organized into 8 pages of 64 bytes each. Almost all instructions are 1 byte and execute in one instruction cycle, but one cycle must be allowed for the execution of the skip instructions since the instruction is fetched even if not executed.

The data memory is more like a register bank than a RAM with 2 bits of the *B-register* used to address one of four register banks and with two more bits used to specify the 4-bit digit within that register. Data addressed in RAM can be transferred to/from, and exchanged with, the accumulator. RAM data can also be output to the *Q-latches* and loaded from the *L-ports*. The *G-register* can be loaded from the accumulator and is a general, bidirectional I/O port. The *D-register* is loaded through the digit part of the *B-register*.

The maximal COPS444C is shown in Fig. 5.15. It can be seen in this figure that nothing of the COP410 has been lost in this expanded implementation. Additional capabilities to this fully static device are a time base counter, an interrupt, 3 levels of stack, microbus serial I/O and 128 total digits of RAM. This is the only unit which has an in-only port *(IN0-IN3)*, all the other ports are bidirectional.

Instruction set. The incorporation of a Harvard architecture in the COPS400 family with its attendant dual bus removes much of the addressing burden from the instruction set[7] since most instructions do not have an address field. The COPS400 instruction set is shown in Table 5.3, and it can readily be seen that most instructions are only one byte long. Notable exceptions are the transfer of control instructions that also have special considerations because of the limitation of some of them to jumping within a page. This can lead to difficulties when a jump instruction is near a page boundary and it is necessary to jump to another page. The compactness of the instruction set can be seen in instructions such as

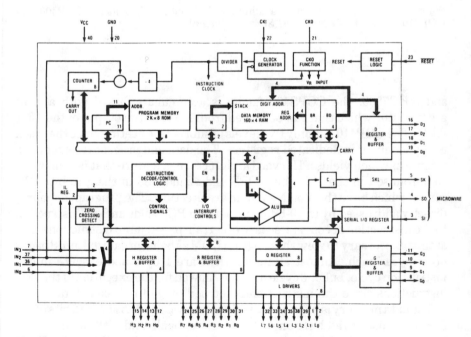

Figure 5.15 National Semiconductor COP444C MCU block diagram. *(Courtesy National Semiconductor)*

TABLE 5.3 National Semiconductor COP400 Instruction Set

Mnemonic Operand	Hex Code	Machine Language Code (Binary)	Data Flow	Skip Conditions	Description
ARITHMETIC/LOGIC INSTRUCTIONS					
ASC	30	$0011\|0000$	A + C + RAM(B) → A Carry → C	Carry	Add with Carry, Skip on Carry
ADD	31	$0011\|0001$	A + RAM(B) → A	None	Add RAM to A
ADT	4A	$0100\|1010$	A + 10_{10} → A	None	Add Ten to A
AISC y	5–	$0101\|$ y	A + y → A	Carry	Add Immediate, Skip on Carry (y ≠ 0)
CASC	10	$0001\|0000$	Ā + RAM(B) + C → A Carry → C	Carry	Complement and Add with Carry, Skip on Carry
CLRA	00	$0000\|0000$	0 → A	None	Clear A
COMP	40	$0100\|0000$	Ā → A	None	One's complement of A to A
NOP	44	$0100\|0100$	None	None	No Operation
OR	33 1A	$0011\|0011$ $0001\|1010$	A v M → A	None	OR RAM with A
RC	32	$0011\|0010$	"0" → C	None	Reset C
SC	22	$0010\|0010$	"1" → C	None	Set C
XOR	02	$0000\|0010$	A • RAM(B) → A	None	Exclusive-OR RAM with A
TRANSFER OF CONTROL INSTRUCTIONS					
JID	FF	$1111\|1111$	ROM ($PC_{10:8}$, A,M) → $PC_{7:0}$	None	Jump Indirect (Note 3)
JMP a	6–	$0110\|0\|a_{10:8}$ $a_{7:0}$	a → PC	None	Jump
JP a	––	$1\|a_{6:0}$ (pages 2,3 only) or $11\|a_{5:0}$ (all other pages)	a → $PC_{6:0}$ a → $PC_{5:0}$	None	Jump within Page (Note 4)
JSRP a	–––	$10\|a_{5:0}$	PC + 1 → RAM_N N + 1 → N 00010 → $PC_{10:8}$ a → $PC_{5:0}$	None	Jump to Subroutine Page (Note 5)
JSR a	6–	$0110\|1\|a_{10:8}$ $a_{7:0}$	PC + 1 → RAM_N N + 1 → N a → PC	None	Jump to Subroutine
RET	48	$0100\|1000$	N – 1 → N RAM_N → PC	None	Return from Subroutine
RETSK	49	$0100\|1001$	N – 1 → N RAM_N → PC	Always Skip on Return	Return from Subroutine then Skip

SOURCE: Courtesy National Semiconductor.

TABLE 5.3 National Semiconductor COP400 Instruction Set *(Continued)*

Mnemonic	Operand	Hex Code	Machine Language Code (Binary)	Data Flow	Skip Conditions	Description
MEMORY REFERENCE INSTRUCTIONS						
CAME		33	`0011|0011`	$A \to EN_{7:4}$	None	Copy A, RAM to EN
		1F	`0001|1111`	$RAM(B) \to EN_{3:0}$		
CAMQ		33	`0011|0011`	$A \to Q_{7:4}$	None	Copy A, RAM to Q
		3C	`0011|1100`	$RAM(B) \to Q_{3:0}$		
CAMT		33	`0011|0011`	$A \to T_{7:4}$	None	Copy A, RAM to T
		3F	`0011|1111`	$RAM(B) \to T_{3:0}$		
CEMA		33	`0011|0011`	$EN_{7:4} \to RAM(B)$	None	Copy EN to RAM, A
		0F	`0000|1111`	$EN_{3:0} \to A$		
CQMA		33	`0011|0011`	$Q_{7:4} \to RAM(B)$	None	Copy Q to RAM, A
		2C	`0010|1100`	$Q_{3:0} \to A$		
CTMA		33	`0011|0011`	$T_{7:4} \to RAM(B)$	None	Copy T to RAM, A
		2F	`0010|1111`	$T_{3:0} \to A$		
LD	r	-5	`00| r |0101` $r = 0{:}3$	$RAM(B) \to A$ $Br \oplus r \to Br$	None	Load RAM into A, Exclusive-OR Br with r
LDD	r,d	23	`00|10|0011`	$RAM(r,d) \to A$	None	Load A with RAM pointed to directly by r,d
		--	`0| r | d` $r = 0{:}7$			
LID		33	`0011|0011`	$ROM (PC_{10:8},A,M) \to M,A$	None	Load RAM, A Indirect
		19	`0001|1001`			
LQ D		BF	`1011|1111`	$ROM(PC_{10:8},A,M) \to Q$	None	Load Q Indirect (Note 3)
RMB	0	4C	`0100|1100`	$0 \to RAM(B)_0$	None	Reset RAM Bit
	1	45	`0100|0101`	$0 \to RAM(B)_1$		
	2	42	`0100|0010`	$0 \to RAM(B)_2$		
	3	43	`0100|0011`	$0 \to RAM(B)_3$		
SMB	0	4D	`0100|1101`	$1 \to RAM(B)_0$	None	Set RAM Bit
	1	47	`0100|0111`	$1 \to RAM(B)_1$		
	2	46	`0100|0110`	$1 \to RAM(B)_2$		
	3	4B	`0100|1011`	$1 \to RAM(B)_3$		
STII	y	7-	`0111| y`	$y \to RAM(B)$ $Bd + 1 \to Bd$	None	Store Memory Immediate and Increment Bd
X	r	-6	`00| r |0110` $r = 0{:}3$	$RAM(B) \leftrightarrow A$ $Br \oplus r \to Br$	None	Exchange RAM with A, Exclusive-OR Br with r
XAD	r,d	23	`0010|0011`	$RAM(r,d) \leftrightarrow A$	None	Exchange A with RAM pointed to directly by r,d
		--	`1| r | d` $r = 0{:}7$			
XDS	r	-7	`00| r |0111` $r = 0{:}3$	$RAM(B) \leftrightarrow A$ $Bd - 1 \to Bd$ $Br \oplus r \to Br$	Bd decrements past 0	Exchange RAM with A and Decrement Bd, Exclusive-OR Br with r
XIS	r	-4	`00| r |0100` $r = 0{:}3$	$RAM(B) \leftrightarrow A$ $Bd + 1 \to Bd$ $Br \oplus r \to Br$	Bd increments past 15	Exchange RAM with A and Increment Bd, Exclusive-OR Br with r

TABLE 5.3 National Semiconductor COP400 Instruction Set *(Continued)*

Mnemonic Operand	Hex Code	Machine Language Code (Binary)	Data Flow	Skip Conditions	Description
REGISTER REFERENCE INSTRUCTIONS					
CAB	50	0101|0000	A → Bd	None	Copy A to Bd
CBA	4E	0100|1110	Bd → A	None	Copy Bd to A
LBI r,d	--	00| r |(d − 1) r = 0:3,d = 0,9:15	r,d → B	Skip until not a LBI	Load B Immediate with r,d (Note 6)
		or			
	33	0011|0011			
	--	1| r | d r = 0:7,any d			
LEI y	33	0011|0011	y → EN$_{3:0}$	None	Load lower half of EN Immediate
	6–	0110| y			
XABR	12	0001|0010	A ↔ Br	None	Exchange A with Br
XAN	33	0011|0011	A ↔ N(0,0 → A$_3$.A$_2$)	None	Exchange A with N
	0B	0000|1011			
TEST INSTRUCTIONS					
SKC	20	0010|0000		C = "1"	Skip if C is True
SKE	21	0010|0001		A = RAM(B)	Skip if A Equals RAM
SKGZ	33	0011|0011		G$_{3:0}$ = 0	Skip if G is Zero (all 4 bits)
	21	0010|0001			
SKGBZ	33	0011|0011	1st byte		Skip if G Bit is Zero
0	01	0000|0001		G$_0$ = 0	
1	11	0001|0001	2nd byte	G$_1$ = 0	
2	03	0000|0011		G$_2$ = 0	
3	13	0001|0011		G$_3$ = 0	
SKMBZ 0	01	0000|0001		RAM(B)$_0$ = 0	Skip if RAM Bit is Zero
1	11	0001|0001		RAM(B)$_1$ = 0	
2	03	0000|0011		RAM(B)$_2$ = 0	
3	13	0001|0011		RAM(B)$_3$ = 0	
SKSZ	33	0011|0011		SIO = 0	Skip if SIO is Zero
	1C	0001|1100			
SKT	41	0100|0001		T counter carry has occurred since last test	Skip on Timer (Note 3)

TABLE 5.3   National Semiconductor COP400 Instruction Set *(Continued)*

Mnemonic Operand	Hex Code	Machine Language Code (Binary)	Data Flow	Skip Conditions	Description
INPUT/OUTPUT INSTRUCTIONS					
CAMR	33	0011|0011	$A \to R_{7:4}$	None	Output A,RAM to R Port
	3D	0011|1101	$RAM(B) \to R_{3:0}$		
ING	33	0011|0011	$G \to A$	None	Input G Port to A
	2A	0010|1010			
INH	33	0011|0011	$H \to A$	None	Input H Port to A
	2B	0010|1011			
ININ	33	0011|0011	$IN \to A$	None	Input IN Inputs to A (Note 2)
	28	0010|1000			
INIL	33	0011|0011	IL_3, CKO, IN_1Z, $IL_0 \to A$	None	Input IL Latches to A (Note 3)
	29	0010|1001			
INL	33	0011|0011	$L_{7:4} \to RAM(B)$	None	Input L Port to RAM,A
	2E	0010|1110	$L_{3:0} \to A$		
INR	33	0011|0011	$R_{7:4} \to RAM(B)$	None	Input R Port to RAM,A
	2D	0010|1101	$R_{3:0} \to A$		
OBD	33	0011|0011	$Bd \to D$	None	Output Bd to D Port
	3E	0011|1110			
OGI y	33	0011|0011	$y \to G$	None	Output to G Port Immediate
	5–	0101| y			
OMG	33	0011|0011	$RAM(B) \to G$	None	Output RAM to G Port
	3A	0011|1010			
OMH	33	0011|0011	$RAM(B) \to H$	None	Output RAM to H Port
	3B	0011|1011			
XAS	4F	0100|1111	$A \leftrightarrow SIO$, $C \to SKL$	None	Exchange A with SIO (Note 3)

Note 1: All subscripts for alphabetical symbols indicate bit numbers unless explicitly defined (e.g., Br and Bd are explicitly defined). Bits are numbered 0 to N where 0 signifies the least significant bit (low-order, right-most bit). For example, A_3 indicates the most significant (left-most) bit of the 4-bit A register.

Note 2: The ININ instruction is not available on the 24-pin COP442/COP342 since this device does not contain the IN inputs.

Note 3: For additional information on the operation of the XAS, JID, LQID, INIL, and SKT instructions, see below.

Note 4: The JP instruction allows a jump, while in subroutine pages 2 or 3, to any ROM location within the two-page boundary of pages 2 or 3. The JP instruction, otherwise, permits a jump to a ROM location within the current 64-word page. JP may not jump to the last word of a page.

Note 5: A JSRP transfers program control to subroutine page 2 (00010 is loaded into the upper 5 bits of P). A JSRP may not be used when in pages 2 or 3. JSRP may not jump to the last word in page 2.

Note 6: LBI is a single-byte instruction if d = 0, 9, 10, 11, 12, 13, 14, or 15. The machine code for the lower 4 bits equals the binary value of the "d" data *minus 1*, e.g., to load the lower four bits of B (Bd) with the value 9 (1001_2), the lower 4 bits of the LBI instruction equal 8 (1000_2). To load 0, the lower 4 bits of the LBI instruction should equal 15 (1111_2).

X   r

where r is a 4-bit operand field. The result of this single instruction is

RAM(B) ↔ A

and

Br ⊕ r → Br

This multiplicity of operations within a single instruction is not an isolated case but is common to many instructions. This coupling between operations can lead to exceedingly complex, tightly coupled programs. Of course, this coupling can also result in minimal code size.

One difficulty associated with Harvard architectures is the difficulty in fetching ROM-based data.[7] This difficulty has been alleviated in the COPS400 through the implementation of a LQID instruction. In this instruction, the *PC* and *Accumulator* are concatenated to form the address into program memory where a lookup table is located. An individual element is addressed within this table by a 4-bit offset specified by a RAM location. In the instruction set of Table 5.3, this whole process is briefly described as load the *Q-register* indirect, and symbolically shown as

ROM($PC_{10:8}$, A, M) → Q

In addition to these unusual instructions that result from the Harvard architecture and the desire to minimize ROM size, the COP400 includes the expected arithmetic and logical operations between the Accumulator and RAM. Subroutine calls can be nested two to four deep (depending on family member) and be located anyplace in memory. Most data movement instructions are combination instructions combine on the accumulator with movement to/from RAM. Individual bits can be directly manipulated RMB and SMB instructions that respectively clear or set the specified bit.

Many conditional operations can be performed through the use of skip instructions. Rather than jump to subroutines on detection of a condition, it is often sufficient either to execute or not execute a single instruction. There are six different skip instructions on the carry bit, RAM = *Accumulator*, *G-register* = 0, individual bits of *G-register* = zero, individual RAM bits = 0, *SIO* = 0, and skip on *Timer Overflow set*. Through these instructions, both internal and external single- and multiple-bit conditions can be directly sensed without resorting to masks or other multiple operation techniques.

The final type of instructions deals with I/O. Four-bit nibbles can be input/output through the *D-*, *G-*, and *H-registers* with some restrictions on the source. Eight bits can be input/output simultaneously to the *Accumulator* and a designated memory location through the *R-register* and *L-register*. The interrupt inputs can also be read into the *Accumulator*.

On-chip peripherals. The minimal implementation of the COP410 has only one on-chip peripheral, the synchronous serial I/O that National Semiconductor calls *MICROWIRE*. This is an example of a hardware driven, software-controlled peripheral in that the port or its clock is turned on/off by software, but the actual serial clocking of the data into a parallel register is under hardware control.

The most capable member of the COP400 family, the COP440, contains additional on-chip peripherals, including the following

1. *Zero-crossing detector circuitry with hysteresis:* Interrupt *IN1* can be programmed to respond to either a negative (1-0 transition) or zero crossing (0-1 or 1-0), with the type of zero crossing determined by a mask programmed option. The zero crossing can be either a logic (0-1 or 1-0) transition or a true voltage-level transition as detected by additional circuitry when the mask option is selected.

2. *Programmable time-base counter:* The *T-register* is an 8-bit up-counter that can be preloaded from both the accumulator and memory. The clock can be software selectable as being either from the an instruction clock divided by four source where it can be used as a time base or from an external input, *IN2*, in which case the *T-register* becomes an event counter. Upon overflow, an interrupt is generated and the register is cleared.

Notably missing from the list of on-chip peripherals is an A/D converter.

As with the NEC μPD700, the output pins are individually mask programmable at manufacture to provide a variety of hardware interfaces. The following options are available:

Standard: Enhancement mode device to ground, depletion mode device to V_{cc}.

Open-drain: Enhancement mode device to ground only, external pull-up.

Push-Pull: Enhancement mode device to ground, depletion mode device in parallel with an enhancement mode device to V_{cc} to allow for fast rise/fall times when driving capacitive loads.

Standard L, R: Same as standard, but may disabled on *L*-and *R-outputs* only at *RESET*.

LED Direct Drive: Enhancement-mode device to ground and V_{cc} together with depletion device to V_{cc}.

TRI-STATE push-pull: Enhancement mode device to ground and V_{cc} for *L*- and *R-outputs*.

Push-pull R: Same as push-pull, but may be disabled on *R-outputs*.

Additional depletion pull-up: Depletion load to V_{cc} for low source current pull-up.

Expansion capability.   Except for the ROM-less devices, there is no ability to expand the COP400 family's memory through parallel busses. External peripheral devices and memory may be connected through the synchronous serial I/O peripheral. Examples of these peripheral chips are A/D converters, frequency generators and counters, vacuum fluorescent display drivers, LCD/LED display controllers, low-power CMOS RAM, and PLL synthesizers. While this separation of peripherals from the chip may appear at first to indicate a limited MCU, there are occasions when this hardware partitioning is appropriate. For example, it may be more appropriate to put a display driver either on the display itself or in close proximity to it while the MCU itself is some distance away. In this case, the MICROWIRE reduces the number of interconnect lines required to pass data to the display.

Powerdown modes.   There are three low-power modes in the COP400 family with their availability dependent on the chip:

1. *Halt*: On CMOS COP400 MCUs without a timer, an additional instruction, **HALT**, is implemented that places the MCU in a low-power, nonoperating mode until pin *CK0* is forced low by an external signal.

2. *IT*: Idle until timer mode is implemented on those CMOS COP400s that have a timer/event counter. On these chips, the oscillator is stopped to the CPU to minimize power consumption, but not to the counter. When the timer overflows, the CPU comes out of its idle state and continues program execution.

3. *RAM keep-alive*: If care is taken in the way power is turned on/off to the MCU, a separate, low-voltage of 3.6 volts can keep the RAM data intact. The requirement is that the *#RESET* pin should go low before the V_{cc} drops below spec and should be returned to a logic one after V_{cc} is within spec.

With the application of these techniques and the correct choice of COP400 family member, standby power consumption can be as low as 100 μWatts.

NS COP400 Interrupts.   On the expanded MCUs, there are two addresses at which program execution continues depending on the source of the interrupt. If the interrupt is caused by *EN1*, and *EN4* (counter/timer) is not active, then the *PC* is stacked in RAM and execution continues at 0FFH, the interrupt being cleared. If *EN4* is active, then program execution continues at 0300H with the *PC* also having been stored on the stack. Interrupts may be enabled on exit from the interrupt service routines by executing a **CAME**

or LEI instruction immediately before the RET instruction. An interrupt status bit is stored along with the *PC* to indicate whether execution was interrupted during a skip instruction. This bit indicates if the next instruction to be executed on return is to be skipped.

5.4.3 Design and Application Tools

A full complement of evaluation boards, emulators, prototyping devices, and library software is available from National Semiconductor to aid in the design and development of COP400 based systems. In addition, software is available for both PCs and VAXs to provide a development environment without the need for an exclusive software development system. As with the other 4-bit MCUs, no HLL support is available.

5.5 OTHER 4-BIT MCUs

Four-bit MCUs are available from other manufacturers including Hitachi and Sharp. These devices are spinoffs of other consumer products and have the following advantages: extremely low-power operation, precise reference clock, large amount of I/O, small size, and built-in LCD driving capability at low cost.[8] A typical device from Hitachi is the HD44795, also known as the LCDIII. The device has a 2048 × 10-bit program memory and 160 × 4 bit RAM. There are 32 I/O pins, some of which are quasi-directional, meaning that if they are to be used as inputs, the appropriate output must be set high. The LCDIII also has two input interrupts and a timer interrupt. A standby mode is available in which power consumption is reduced to 21 μA. An LCD display with as many as 16 eight-segment digits can be directly driven without a special power supply.

A device similar to the LCDIII is the Sharp SM-4, which has a slightly larger ROM capacity (2268 bytes) at the expense of smaller RAM (96 nibbles). The other differences are that the use of the outputs for driving the LCDs is selectable in the Sharp whereas the outputs are dedicated in the LCDIII, and in the low-power standby mode, the Sharp LCD display keeps on working whereas the LCDIII blanks.

5.6 8-BIT MICROCONTROLLERS

Increasing the internal bus width from 4 to 8 bits increases both the cost and versatility of microcontrollers. On-chip ROM and RAM are still limited, but the use of a high-level language starts to become feasible if not for the final single-chip product, at least for the early prototyping and development stages on an expanded development system. Most chips are designed to be expandable, recognizing the occasional need for increased program store to allow for more sophisticated signal processing or complex control algorithms, or more RAM for data store-and-forward applications such as communications or network controllers. The architecture and features of several major manufacturers' families of microcontrollers are discussed here including Motorola and its 6801 and 6805 families as well as the 68HC11 MCU, Intel's MCS-51 families, and Texas Instruments' TMS370 family. The Motorola MC68HC11 is discussed in depth as prototypical of the capabilities of an 8-bit MCU.

Unlike the 4-bit MCUs previously studied, the number of 8-bit family members is much smaller and includes a limited number of alternatives. This is to be expected since the instruction sets are larger and more complete, and fewer compromises need to be made in optimizing silicon usage, at least in terms of the instructions. This does not mean that the instruction sets converge to a standard set across manufacturers, but that more options can be included in a single package rather than having many more individual packages. For example, many of the features of 4-bit MCUs that must be mask programmed are software programmed through the use of control registers that specify which pins are input, which are output, and the sense of edge triggering for latched external inputs or event counters. The greater variation is interfamily rather than intrafamily, and the selections for a particular application will probably be made across manufacturers at the 8-bit level rather than across members of a single family from a single manufacturer, as might be the case of the 4-bit MCUs.

5.7 MOTOROLA M6801 FAMILY

The Motorola M6801 family was first introduced to the market in 1978[9] and is built along the lines of the MC6800 microprocessor. In spite of its similarity to the M6800, its design and instruction set have been modified for control applications and single-chip operation while not compromising the ability to expand its operation with external RAM and ROM of up to 64 kbytes total. Unlike the M6800, the M6801 family members include on-chip I/O ports, an asynchronous serial communications interface, and a 16-bit timer as well as an expanded instruction set.

5.7.1 Motorola M6801 Family Members

There are few members of the M6801 family, since there are only three options for internal ROM size and even fewer for RAM. The options for ROM size are none, 2, or 4 kbytes. The MC6803 with no on-chip ROM is intended primarily as a development chip since it cannot operate in single-chip mode although in every other respect it is behaviorally equivalent to the 6801. The RAM options are also limited as being either 128 or 192 bytes on-chip. The number of I/O pins varies from 13 to 29, and the timer can be selected to have either three or six functions. All family members have an on-chip serial port. Unlike the serial ports of the 4-bit MCUs, transmission is not limited to synchronous signaling, but utilizes mark-space, non-return to zero (NRZ) asynchronous signaling with an on-chip baud rate generator to the CPU clock, and is software selectable from among four frequencies.[10]

5.7.2   M6801 Architecture

The block diagram of Fig. 5.16 shows the architecture of the M6801 family. There are separate internal address and data busses, but no distinction is made between program memory and data memory (i.e., it does not have a Harvard architecture like many of the 4-bit MCUs). There are three 8-bit parallel I/O ports and one 5-bit port. Two pins of I/O Port 3, *SC1* and *SC2*, are dedicated to I/O strobing when the M6801 is operated in *single-chip mode*

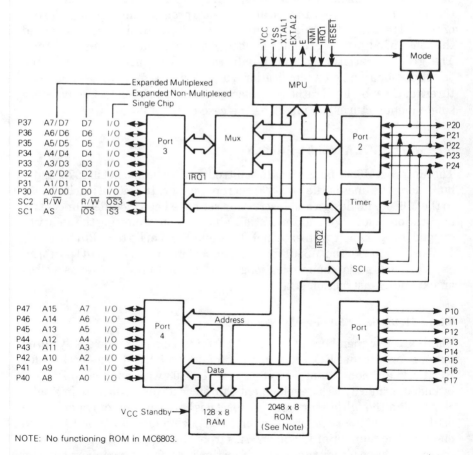

Figure 5.16   Motorola M6801 family block diagram. *(Courtesy Motorola, Inc.)*

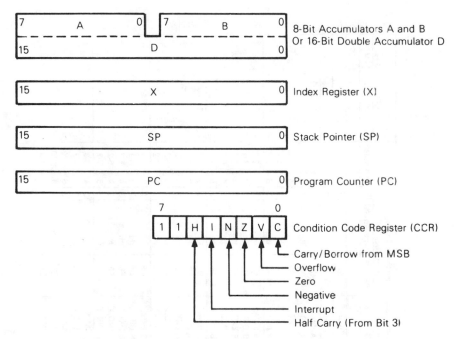

Figure 5.17 Register structure of Motorola M6801 family of MCUs. *(Courtesy Motorola, Inc.)*

(other modes are discussed later). The *SC1* input can be used to generate an interrupt and/or latch external data from another device, with selection of the operating mode being under software control by writing to the Port 3 Control/Status register. The *SC2* output is used to strobe external devices, indicating that there is valid data on the Port 3 I/O pins. The other 8-bit ports can likewise have their data directions specified on a single-bit basis by programming their respective data direction registers. Port 2, the five-line port, can likewise be programmed for input or output through its data direction register. Additionally, external access to the timer for input (event counting) or output are through two pins, and the serial communication device also uses three of these pins. In general, all pins are TTL compatible and can drive one TTL load. In the *expanded, nonmultiplexed mode* of operation, I/O Port 3 becomes the data bus, and Port 4 becomes the low order byte of the address (A0–A7). In the third operating mode, *expanded multiplexed mode*, Port 3 becomes a multiplexed address/data bus with the low-order byte of address needing to be latched externally. Port 4 becomes the high-order byte of address for external memory (A8–A15). Since I/O is memory mapped, the expanded, nonmultiplexed mode of operation allows for the direct addressing of 256 external I/O ports. The expanded multiplexed mode allows for the full 64-kbyte address space

TABLE 5.4 Availability of Instructions among the Motorola 6800, 6801, 6805 and 6811 Processors

Source Form(s)	Operation	Boolean Expression	Addressing Mode for Operand	Opcode	Operand(s)	Bytes	Cycle	Cycle by Cycle*	S	X	H	I	N	Z	V	C
ABA	Add Accumulators	A+B→A	INH	1B		1	2	2-1	-	-	↕	-	↕	↕	↕	↕
ABX	Add B to X	IX+00:B→IX	INH	3A		1	3	2-2	-	-	-	-	-	-	-	-
ABY	Add B to Y	IY+00:B→IY	INH	18 3A		2	4	2-4	-	-	-	-	-	-	-	-
ADCA (opr)	Add with Carry to A	A+M+C→A	A IMM	89	ii	2	2	3-1	-	-	↕	-	↕	↕	↕	↕
			A DIR	99	dd	2	3	4-1								
			A EXT	B9	hh ll	3	4	5-2								
			A IND,X	A9	ff	2	4	6-2								
			A IND,Y	18 A9	ff	3	5	7-2								
ADCB (opr)	Add with Carry to B	B+M+C→B	B IMM	C9	ii	2	2	3-1	-	-	↕	-	↕	↕	↕	↕
			B DIR	D9	dd	2	3	4-1								
			B EXT	F9	hh ll	3	4	5-2								
			B IND,X	E9	ff	2	4	6-2								
			B IND,Y	18 E9	ff	3	5	7-2								
ADDA (opr)	Add Memory to A	A+M→A	A IMM	8B	ii	2	2	3-1	-	-	↕	-	↕	↕	↕	↕
			A DIR	9B	dd	2	3	4-1								
			A EXT	BB	hh ll	3	4	5-2								
			A IND,X	AB	ff	2	4	6-2								
			A IND,Y	18 AB	ff	3	5	7-2								
ADDB (opr)	Add Memory to B	B+M→B	B IMM	CB	ii	2	2	3-1	-	-	↕	-	↕	↕	↕	↕
			B DIR	DB	dd	2	3	4-1								
			B EXT	FB	hh ll	3	4	5-2								
			B IND,X	EB	ff	2	4	6-2								
			B IND,Y	18 EB	ff	3	5	7-2								
ADDD (opr)	Add 16-Bit to D	D+M:M+1→D	IMM	C3	jj kk	3	4	3-3	-	-	-	-	↕	↕	↕	↕
			DIR	D3	dd	2	5	4-7								
			EXT	F3	hh ll	3	6	5-10								
			IND,X	E3	ff	2	6	6-10								
			IND,Y	18 E3	ff	3	7	7-8								

Processor availability columns (left margin): 68HC11, 6805, 6801, 6800

212

Source Form	Operation	Boolean/Arithmetic Operation	Addressing Mode	Opcode	Operand	~	#	Cycle Ref*	H	I	N	Z	V	C
ANDA (opr)	AND A with Memory	A•M → A	A IMM	84	ii	2	2	3-1	–	–	↕	↕	0	–
			A DIR	94	dd	3	2	4-1						
			A EXT	B4	hh ll	4	3	5-2						
			A IND,X	A4	ff	4	2	6-2						
			A IND,Y	18 A4	ff	5	3	7-2						
ANDB (opr)	AND B with Memory	B•M → B	B IMM	C4	ii	2	2	3-1	–	–	↕	↕	0	–
			B DIR	D4	dd	3	2	4-1						
			B EXT	F4	hh ll	4	3	5-2						
			B IND,X	E4	ff	4	2	6-2						
			B IND,Y	18 E4	ff	5	3	7-2						
ASL (opr)	Arithmetic Shift Left	$C \leftarrow [b7 \ldots b0] \leftarrow 0$	EXT	78	hh ll	6	3	5-8	–	–	↕	↕	↕	↕
			IND,X	68	ff	6	2	6-3						
			IND,Y	18 68	ff	7	3	7-3						
ASLA			A INH	48		2	1	2-1						
ASLB			B INH	58		2	1	2-1						
ASLD	Arithmetic Shift Left Double	$C \leftarrow [b15 \ldots b0] \leftarrow 0$	INH	05		3	1	2-2	–	–	↕	↕	↕	↕
ASR (opr)	Arithmetic Shift Right	$b7 \rightarrow [b7 \ldots b0] \rightarrow C$	EXT	77	hh ll	6	3	5-8	–	–	↕	↕	↕	↕
			IND,X	67	ff	6	2	6-3						
			IND,Y	18 67	ff	7	3	7-3						
ASRA			A INH	47		2	1	2-1						
ASRB			B INH	57		2	1	2-1						
BCC (rel)	Branch if Carry Clear	?C = 0	REL	24	rr	3	2	8-1	–	–	–	–	–	–
BCLR (opr) (msk)	Clear Bit(s)	$M \cdot (\overline{mm}) \rightarrow M$	DIR	15	dd mm	6	3	4-10	–	–	↕	↕	0	–
			IND,X	1D	ff mm	7	3	6-13						
			IND,Y	18 1D	ff mm	8	4	7-10						
BCS (rel)	Branch if Carry Set	?C = 1	REL	25	rr	3	2	8-1	–	–	–	–	–	–
BEQ (rel)	Branch if = Zero	?Z = 1	REL	27	rr	3	2	8-1	–	–	–	–	–	–

*Cycle-by-cycle number provides a reference to Tables 10-2 through 10-8 which detail cycle-by-cycle operation.

Example: Table 10-1 Cycle-by-Cycle column reference number 2-4 equals Table 10-2 line item 2-4.

TABLE 5.4 Availability of Instructions among the Motorola 6800, 6801, 6805 and 6811 Processors (Continued)

Source Form(s)	Operation	Boolean Expression	Addressing Mode for Operand	Machine Coding (Hexadecimal) Opcode	Operand(s)	Bytes	Cycle	Cycle by Cycle*	S	X	H	I	N	Z	V	C
BGE (rel)	Branch if ≥ Zero	?N⊕V=0	REL	2C	rr	2	3	8-1	-	-	-	-	-	-	-	-
BGT (rel)	Branch if > Zero	?Z+(N⊕V)=0	REL	2E	rr	2	3	8-1	-	-	-	-	-	-	-	-
BHI (rel)	Branch if Higher	?C+Z=0	REL	22	rr	2	3	8-1	-	-	-	-	-	-	-	-
BHS (rel)	Branch if Higher or Same	?C=0	REL	24	rr	2	3	8-1	-	-	-	-	-	-	-	-
BITA (opr)	Bit(s) Test A with Memory	A•M	A IMM	85	ii	2	2	3-1	-	-	-	-	↕	↕	0	-
			A DIR	95	dd	2	3	4-1								
			A EXT	B5	hh ll	3	4	5-2								
			A IND,X	A5	ff	2	4	6-2								
			A IND,Y	18 A5	ff	3	5	7-2								
BITB (opr)	Bit(s) Test B with Memory	B•M	B IMM	C5	ii	2	2	3-1	-	-	-	-	↕	↕	0	-
			B DIR	D5	dd	2	3	4-1								
			B EXT	F5	hh ll	3	4	5-2								
			B IND,X	E5	ff	2	4	6-2								
			B IND,Y	18 E5	ff	3	5	7-2								
BLE (rel)	Branch if ≤ Zero	?Z+(N⊕V)=1	REL	2F	rr	2	3	8-1	-	-	-	-	-	-	-	-
BLO (rel)	Branch if Lower	?C=1	REL	25	rr	2	3	8-1	-	-	-	-	-	-	-	-
BLS (rel)	Branch if Lower or Same	?C+Z=1	REL	23	rr	2	3	8-1	-	-	-	-	-	-	-	-
BLT (rel)	Branch If < Zero	?N⊕V=1	REL	2D	rr	2	3	8-1	-	-	-	-	-	-	-	-
BMI (rel)	Branch if Minus	?N=1	REL	2B	rr	2	3	8-1	-	-	-	-	-	-	-	-
BNE (rel)	Branch if Not = Zero	?Z=0	REL	26	rr	2	3	8-1	-	-	-	-	-	-	-	-
BPL (rel)	Branch if Plus	?N=0	REL	2A	rr	2	3	8-1	-	-	-	-	-	-	-	-
BRA (rel)	Branch Always	?1=1	REL	20	rr	2	3	8-1	-	-	-	-	-	-	-	-
BRCLR(opr) (msk) (rel)	Branch if Bit(s) Clear	?M•mm=0	DIR	13	dd mm rr	4	6	4-11	-	-	-	-	-	-	-	-
			IND,X	1F	ff mm rr	4	7	6-14								
			IND,Y	18 1F	ff mm rr	5	8	7-11								
BRN (rel)	Branch Never	?1=0	REL	21	rr	2	3	8-1	-	-	-	-	-	-	-	-

68HC11 / 6805 / 6801 / 6800

Mnemonic	Operation	Boolean Expression	Addressing Mode	Opcode	Operand	~		Ref	H	I	N	Z	V	C
BRSET (opr) (msk) (rel)	Branch if Bit(s) Set	?(M̄)•mm = 0	DIR IND,X IND,Y	12 1E 18 1E	dd mm rr ff mm rr ff mm rr	4 4 5	6 7 8	4-11 6-14 7-11	–	–	–	–	–	–
BSET (opr) (msk)	Set Bit(s)	M + mm → M	DIR IND,X IND,Y	14 1C 18 1C	dd mm ff mm ff mm	3 3 4	6 7 8	4-10 6-13 7-10	–	–	↕	↕	0	–
BSR (rel)	Branch to Subroutine	See Special Ops	REL	8D	rr	2	6	8-2	–	–	–	–	–	–
BVC (rel)	Branch if Overflow Clear	?V = 0	REL	28	rr	2	3	8-1	–	–	–	–	–	–
BVS (rel)	Branch if Overflow Set	?V = 1	REL	29	rr	2	3	8-1	–	–	–	–	–	–
CBA	Compare A to B	A - B	INH	11		1	2	2-1	–	–	↕	↕	↕	↕
CLC	Clear Carry Bit	0 → C	INH	0C		1	2	2-1	–	–	–	–	–	0
CLI	Clear Interrupt Mask	0 → I	INH	0E		1	2	2-1	–	0	–	–	–	–
CLR (opr)	Clear Memory Byte	0 → M	EXT IND,X IND,Y	7F 6F 18 6F	hh ll ff ff	3 2 3	6 6 7	5-8 6-3 7-3	–	–	0	1	0	0
CLRA	Clear Accumulator A	0 → A	A INH	4F		1	2	2-1	–	–	0	1	0	0
CLRB	Clear Accumulator B	0 → B	B INH	5F		1	2	2-1	–	–	0	1	0	0
CLV	Clear Overflow Flag	0 → V	INH	0A		1	2	2-1	–	–	–	–	0	–
CMPA (opr)	Compare A to Memory	A - M	A IMM A DIR A EXT A IND,X A IND,Y	81 91 B1 A1 18 A1	ii dd hh ll ff ff	2 2 3 2 3	2 3 4 4 5	3-1 4-1 5-2 6-2 7-2	–	–	↕	↕	↕	↕

* Cycle-by-cycle number provides a reference to Tables 10-2 through 10-8 which detail cycle-by-cycle operation.
Example: Table 10-1 Cycle-by-Cycle column reference number 2-4 equals Table 10-2 line item 2-4.

TABLE 5.4 Availability of Instructions among the Motorola 6800, 6801, 6805 and 6811 Processors (Continued)

6800	6801	6805	68HC11	Source Form(s)	Operation	Boolean Expression	Addressing Mode for Operand	Machine Coding (Hexadecimal) Opcode	Operand(s)	Bytes	Cycle	Cycle by Cycle*	S	X	H	I	N	Z	V	C
▨	▨			CMPB (opr)	Compare B to Memory	B − M	B IMM	C1	ii	2	2	3-1	-	-	-	-	↕	↕	↕	↕
							B DIR	D1	dd	3	3	4-1								
							B EXT	F1	hh ll	3	4	5-2								
							B IND,X	E1	ff	2	4	6-2								
							B IND,Y	18 E1	ff	3	5	7-2								
				COM (opr)	1's Complement Memory Byte	$FF − M → M	EXT	73	hh ll	3	6	5-8	-	-	-	-	↕	↕	0	1
							IND,X	63	ff	2	6	6-3								
							IND,Y	18 63	ff	3	7	7-3								
				COMA	1's Complement A	$FF − A → A	A INH	43		1	2	2-1	-	-	-	-	↕	↕	0	1
				COMB	1's Complement B	$FF − B → B	B INH	53		1	2	2-1	-	-	-	-	↕	↕	0	1
▨	▨	▨		CPD (opr)	Compare D to Memory 16-Bit	D − M:M + 1	IMM	1A 83	jj kk	4	5	3-5	-	-	-	-	↕	↕	↕	↕
							DIR	1A 93	dd	3	6	4-9								
							EXT	1A B3	hh ll	4	7	5-11								
							IND,X	1A A3	ff	3	7	6-11								
							IND,Y	CD A3	ff	3	7	7-8								
				CPX (opr)	Compare X to Memory 16-Bit	IX − M:M + 1	IMM	8C	jj kk	3	4	3-3	-	-	-	-	↕	↕	↕	↕
							DIR	9C	dd	2	5	4-7								
							EXT	BC	hh ll	3	6	5-10								
							IND,X	AC	ff	2	6	6-10								
							IND,Y	CD AC	ff	3	7	7-8								
▨	▨	▨		CPY (opr)	Compare Y to Memory 16-Bit	IY − M:M + 1	IMM	18 8C	jj kk	4	5	3-5	-	-	-	-	↕	↕	↕	↕
							DIR	18 9C	dd	3	6	4-9								
							EXT	18 BC	hh ll	4	7	5-11								
							IND,X	1A AC	ff	3	7	6-11								
							IND,Y	18 AC	ff	3	7	7-8								
		▨		DAA	Decimal Adjust A	Adjust Sum to BCD	INH	19		1	2	2-1	-	-	-	-	↕	↕	↕	↕

Mnemonic	Description	Operation	Address Mode	Opcode	Operand	~	#	Ref	H	I	N	Z	V	C
DEC (opr)	Decrement Memory Byte	M − 1 → M	EXT IND,X IND,Y	7A 6A 18 6A	hh ff ff	3 2 3	6 6 7	5-8 6-3 7-3	−	−	↕	↕	↕	−
DECA	Decrement Accumulator A	A − 1 → A	A INH	4A		1	2	2-1	−	−	↕	↕	↕	−
DECB	Decrement Accumulator B	B − 1 → B	B INH	5A		1	2	2-1	−	−	↕	↕	↕	−
DES	Decrement Stack Pointer	SP − 1 → SP	INH	34		1	3	2-3	−	−	−	−	−	−
DEX	Decrement Index Register X	IX − 1 → IX	INH	09		1	3	2-2	−	−	−	↕	−	−
DEY	Decrement Index Register Y	IY − 1 → IY	INH	18 09		2	4	2-4	−	−	−	↕	−	−
EORA (opr)	Exclusive OR A with Memory	A ⊕ M → A	A IMM A DIR A EXT A IND,X A IND,Y	88 98 B8 A8 18 A8	ii dd hh = ff ff	2 2 3 2 3	2 3 4 4 5	3-1 4-1 5-2 6-2 7-2	−	−	↕	↕	0	−
EORB (opr)	Exclusive OR B with Memory	B ⊕ M → B	B IMM B DIR B EXT B IND,X B IND,Y	C8 D8 F8 E8 18 E8	ii dd hh = ff ff	2 2 3 2 3	2 3 4 4 5	3-1 4-1 5-2 6-2 7-2	−	−	↕	↕	0	−
FDIV	Fractional Divide 16 by 16	D/IX → IX; r → D	INH	03		1	41	2-17	−	−	−	↕	↕	↕
IDIV	Integer Divide 16 by 16	D/IX → IX; r → D	INH	02		1	41	2-17	−	−	−	↕	0	↕
INC (opr)	Increment Memory Byte	M + 1 → M	EXT IND,X IND,Y	7C 6C 18 6C	hh = ff ff	3 2 3	6 6 7	5-8 6-3 7-3	−	−	↕	↕	↕	−
INCA	Increment Accumulator A	A + 1 → A	A INH	4C		1	2	2-1	−	−	↕	↕	↕	−
INCB	Increment Accumulator B	B + 1 → B	B INH	5C		1	2	2-1	−	−	↕	↕	↕	−
INS	Increment Stack Pointer	SP + 1 → SP	INH	31		1	3	2-3	−	−	−	−	−	−

*Cycle-by-cycle number provides a reference to Tables 10-2 through 10-8 which detail cycle-by-cycle operation.
Example: Table 10-1 Cycle-by-Cycle column reference number 2-4 equals Table 10-2 line item 2-4.

217

TABLE 5.4 Availability of Instructions among the Motorola 6800, 6801, 6805 and 6811 Processors (Continued)

6800	6801	6805	68HC11	Source Form(s)	Operation	Boolean Expression	Addressing Mode for Operand	Opcode	Operand(s)	Bytes	Cycle	Cycle by Cycle*	S	X	H	I	N	Z	V	C
				INX	Increment Index Register X	IX + 1 → IX	INH	08		1	3	2-2	-	-	-	-	-	-	-	-
				INY	Increment Index Register Y	IY + 1 → IY	INH	18 08		2	4	2-4	-	-	-	-	-	-	-	-
				JMP (opr)	Jump	See Special Ops	EXT	7E	hh ll	3	3	5-1	-	-	-	-	-	-	-	-
							IND,X	6E	ff	2	3	6-1								
							IND,Y	18 6E	ff	3	4	7-1								
				JSR (opr)	Jump to Subroutine	See Special Ops	DIR	9D	dd	2	5	4-8	-	-	-	-	-	-	-	-
							EXT	BD	hh ll	3	6	5-12								
							IND,X	AD	ff	2	6	6-12								
							IND,Y	18 AD	ff	3	7	7-9								
				LDAA (opr)	Load Accumulator A	M → A	A IMM	86	ii	2	2	3-1	-	-	-	-	↕	↕	0	-
							A DIR	96	dd	2	3	4-1								
							A EXT	B6	hh ll	3	4	5-2								
							A IND,X	A6	ff	2	4	6-2								
							A IND,Y	18 A6	ff	3	5	7-2								
				LDAB (opr)	Load Accumulator B	M → B	B IMM	C6	ii	2	2	3-1	-	-	-	-	↕	↕	0	-
							B DIR	D6	dd	2	3	4-1								
							B EXT	F6	hh ll	3	4	5-2								
							B IND,X	E6	ff	2	4	6-2								
							B IND,Y	18 E6	ff	3	5	7-2								
				LDD (opr)	Load Double Accumulator D	M → A, M + 1 → B	IMM	CC	jj kk	3	3	3-2	-	-	-	-	↕	↕	0	-
							DIR	DC	dd	2	4	4-3								
							EXT	FC	hh ll	3	5	5-4								
							IND,X	EC	ff	2	5	6-6								
							IND,Y	18 EC	ff	3	6	7-6								

Mnemonic	Operation	Boolean Expression / Diagram	Addressing Mode	Opcode	Operand	~	Cycle	Condition Codes
LDS (opr)	Load Stack Pointer	M:M + 1 → SP	IMM	8E	jj kk	3	3-2	- - - ↕ ↕ 0 -
			DIR	9E	dd	3	4-3	
			EXT	BE	hh ll	4	5-4	
			IND,X	AE	ff	5	6-6	
			IND,Y	18 AE	ff	6	7-6	
LDX (opr)	Load Index Register X	M:M + 1 → IX	IMM	CE	jj kk	3	3-2	- - - ↕ ↕ 0 -
			DIR	DE	dd	3	4-3	
			EXT	FE	hh ll	4	5-4	
			IND,X	EE	ff	5	6-6	
			IND,Y	CD EE	ff	6	7-6	
LDY (opr)	Load Index Register Y	M:M + 1 → IY	IMM	18 CE	jj kk	4	3-4	- - - ↕ ↕ 0 -
			DIR	18 DE	dd	4	4-5	
			EXT	18 FE	hh ll	5	5-6	
			IND,X	1A EE	ff	6	6-7	
			IND,Y	18 EE	ff	6	7-6	
LSL (opr)	Logical Shift Left	☐←─···─←☐ ← 0 C b7 b0	EXT	78	hh ll	6	5-8	- - - ↕ ↕ ↕ ↕
			IND,X	68	ff	6	6-3	
			IND,Y	18 68	ff	7	7-3	
LSLA			A INH	48		2	2-1	
LSLB			B INH	58		2	2-1	
LSLD	Logical Shift Left Double	☐←─···─←☐ ← 0 C b15 b0	INH	05		3	2-2	- - - ↕ ↕ ↕ ↕
LSR (opr)	Logical Shift Right	0 →☐─···─→☐→☐ b7 b0 C	EXT	74	hh ll	6	5-8	- - - 0 ↕ ↕ ↕
			IND,X	64	ff	6	6-3	
			IND,Y	18 64	ff	7	7-3	
LSRA			A INH	44		2	2-1	
LSRB			B INH	54		2	2-1	
LSRD	Logical Shift Right Double	0 →☐─···─→☐→☐ b15 b0 C	INH	04		3	2-2	- - - 0 ↕ ↕ ↕
MUL	Multiply 8 by 8	A x B → D	INH	3D		10	2-13	- - - - - 0 ↕

*Cycle-by-cycle number provides a reference to Tables 10-2 through 10-8 which detail cycle-by-cycle operation.
Example: Table 10-1 Cycle-by-Cycle column reference number 2-4 equals Table 10-2 line item 2-4.

219

TABLE 5.4 Availability of Instructions among the Motorola 6800, 6801, 6805 and 6811 Processors (Continued)

6800	6801	6805	68HC11	Source Form(s)	Operation	Boolean Expression	Addressing Mode for Operand	Opcode	Operand(s)	Bytes	Cycle	Cycle by Cycle*	S	X	H	I	N	Z	V	C
				NEG (opr)	2's Complement Memory Byte	0 − M → M	EXT / IND,X / IND,Y	70 / 60 / 18 60	hh ll / ff / ff	3 / 2 / 3	6 / 6 / 7	5-8 / 6-3 / 7-3	-	-	-	-	↕	↕	↕	↕
				NEGA	2's Complement A	0 − A → A	A INH	40		1	2	2-1	-	-	-	-	↕	↕	↕	↕
				NEGB	2's Complement B	0 − B → B	B INH	50		1	2	2-1	-	-	-	-	↕	↕	↕	↕
				NOP	No Operation	No Operation	INH	01		1	2	2-1	-	-	-	-	-	-	-	-
				ORAA (opr)	OR Accumulator A (Inclusive)	A + M → A	A IMM / A DIR / A EXT / A IND,X / A IND,Y	8A / 9A / BA / AA / 18 AA	ii / dd / hh ll / ff / ff	2 / 2 / 3 / 2 / 3	2 / 3 / 4 / 4 / 5	3-1 / 4-1 / 5-2 / 6-2 / 7-2	-	-	-	-	↕	↕	0	-
				ORAB (opr)	OR Accumulator B (Inclusive)	B + M → B	B IMM / B DIR / B EXT / B IND,X / B IND,Y	CA / DA / FA / EA / 18 EA	ii / dd / hh ll / ff / ff	2 / 2 / 3 / 2 / 3	2 / 3 / 4 / 4 / 5	3-1 / 4-1 / 5-2 / 6-2 / 7-2	-	-	-	-	↕	↕	0	-
				PSHA	Push A onto Stack	A→Stk, SP=SP−1	A INH	36		1	3	2-6	-	-	-	-	-	-	-	-
				PSHB	Push B onto Stack	B→Stk, SP=SP−1	B INH	37		1	3	2-6	-	-	-	-	-	-	-	-
				PSHX	Push X onto Stack (Lo First)	IX→Stk, SP=SP−2	INH	3C		1	4	2-7	-	-	-	-	-	-	-	-
				PSHY	Push Y onto Stack (Lo First)	IY→Stk, SP=SP−2	INH	18 3C		2	5	2-8	-	-	-	-	-	-	-	-
				PULA	Pull A from Stack	SP=SP+1, A→Stk	A INH	32		1	4	2-9	-	-	-	-	-	-	-	-
				PULB	Pull B from Stack	SP=SP+1, B→Stk	B INH	33		1	4	2-9	-	-	-	-	-	-	-	-
				PULX	Pull X from Stack (Hi First)	SP=SP+2, IX→Stk	INH	38		1	5	2-10	-	-	-	-	-	-	-	-
				PULY	Pull Y from Stack (Hi First)	SP=SP+2, IY→Stk	INH	18 38		2	6	2-11	-	-	-	-	-	-	-	-

(Machine Coding columns are grouped under "Machine Coding (Hexadecimal)": Opcode and Operand(s). The last eight columns are grouped under "Condition Codes".)

Mnemonic	Boolean/Arithmetic Operation	Addressing Mode	Opcode	Operand	#	~	Cycle-by-Cycle	H	I	N	Z	V	C
ROL (opr)	Rotate Left — C ← [b7 … b0] ← C	EXT	79	hh ll	3	6	5-8	–	–	↕	↕	↕	↕
		IND,X	69	ff	2	6	6-3						
		IND,Y	18 69	ff	3	7	7-3						
ROLA		A INH	49		1	2	2-1						
ROLB		B INH	59		1	2	2-1						
ROR (opr)	Rotate Right — C → [b7 … b0] → C	EXT	76	hh ll	3	6	5-8	–	–	↕	↕	↕	↕
		IND,X	66	ff	2	6	6-3						
		IND,Y	18 66	ff	3	7	7-3						
RORA		A INH	46		1	2	2-1						
RORB		B INH	56		1	2	2-1						
RTI	Return from Interrupt — See Special Ops	INH	3B		1	12	2-14	↕	↕	↕	↕	↕	↕
RTS	Return from Subroutine — See Special Ops	INH	39		1	5	2-12	–	–	–	–	–	–
SBA	Subtract B from A — A − B → A	INH	10		1	2	2-1	–	–	↕	↕	↕	↕
SBCA (opr)	Subtract with Carry from A — A − M − C → A	A IMM	82	ii	2	2	3-1	–	–	↕	↕	↕	↕
		A DIR	92	dd	2	3	4-1						
		A EXT	B2	hh ll	3	4	5-2						
		A IND,X	A2	ff	2	4	6-2						
		A IND,Y	18 A2	ff	3	5	7-2						
SBCB (opr)	Subtract with Carry from B — B − M − C → B	B IMM	C2	ii	2	2	3-1	–	–	↕	↕	↕	↕
		B DIR	D2	dd	2	3	4-1						
		B EXT	F2	hh ll	3	4	5-2						
		B IND,X	E2	ff	2	4	6-2						
		B IND,Y	18 E2	ff	3	5	7-2						
SEC	Set Carry — 1 → C	INH	0D		1	2	2-1	–	–	–	–	–	1
SEI	Set Interrupt Mask — 1 → I	INH	0F		1	2	2-1	–	1	–	–	–	–
SEV	Set Overflow Flag — 1 → V	INH	0B		1	2	2-1	–	–	–	–	1	–

* Cycle-by-cycle number provides a reference to Tables 10-2 through 10-8 which detail cycle-by-cycle operation.
Example: Table 10-1 Cycle-by-Cycle column reference number 2-4 equals Table 10-2 line item 2-4.

TABLE 5.4 Availability of Instructions among the Motorola 6800, 6801, 6805 and 6811 Processors (Continued)

68HC11	6805	6801	6800	Source Form(s)	Operation	Boolean Expression	Addressing Mode for Operand	Opcode	Operand(s)	Bytes	Cycle	Cycle by Cycle*	S	X	H	I	N	Z	V	C
				STAA (opr)	Store Accumulator A	A→M	A DIR	97	dd	2	3	4-2	-	-	-	-	↕	↕	0	-
							A EXT	B7	hh ll	3	4	5-3								
							A IND,X	A7	ff	2	4	6-5								
							A IND,Y	18 A7	ff	3	5	7-5								
				STAB (opr)	Store Accumulator B	B→M	B DIR	D7	dd	2	3	4-2	-	-	-	-	↕	↕	0	-
							B EXT	F7	hh ll	3	4	5-3								
							B IND,X	E7	ff	2	4	6-5								
							B IND,Y	18 E7	ff	3	5	7-5								
				STD (opr)	Store Accumulator D	A→M, B→M+1	DIR	DD	dd	2	4	4-4	-	-	-	-	↕	↕	0	-
							EXT	FD	hh ll	3	5	5-5								
							IND,X	ED	ff	2	5	6-8								
							IND,Y	18 ED	ff	3	6	7-7								
				STOP	Stop Internal Clocks		INH	CF		1	2	2-1	-	-	-	-	-	-	-	-
				STS (opr)	Store Stack Pointer	SP→M:M+1	DIR	9F	dd	2	4	4-4	-	-	-	-	↕	↕	0	-
							EXT	BF	hh ll	3	5	5-5								
							IND,X	AF	ff	2	5	6-8								
							IND,Y	18 AF	ff	3	6	7-7								
				STX (opr)	Store Index Register X	IX→M:M+1	DIR	DF	dd	2	4	4-4	-	-	-	-	↕	↕	0	-
							EXT	FF	hh ll	3	5	5-5								
							IND,X	EF	ff	2	5	6-8								
							IND,Y	CD EF	ff	3	6	7-7								
				STY (opr)	Store Index Register Y	IY→M:M+1	DIR	18 DF	dd	3	5	4-6	-	-	-	-	↕	↕	0	-
							EXT	18 FF	hh ll	4	6	5-7								
							IND,X	1A EF	ff	3	6	6-9								
							IND,Y	18 EF	ff	3	6	7-7								

Mnemonic	Operation	Boolean Expression	Addressing Mode for Operand	Opcode	Operand	~	#	Cycle Ref	S	X	H	I	N	Z	V	C
SUBA (opr)	Subtract Memory from A	A − M → A	A IMM	80	ii	2	2	3-1	–	–	–	–	↕	↕	↕	↕
			A DIR	90	dd	2	3	4-1								
			A EXT	B0	hh ll	3	4	5-2								
			A IND,X	A0	ff	2	4	6-2								
			A IND,Y	18 A0	ff	3	5	7-2								
SUBB (opr)	Subtract Memory from B	B − M → B	B IMM	C0	ii	2	2	3-1	–	–	–	–	↕	↕	↕	↕
			B DIR	D0	dd	2	3	4-1								
			B EXT	F0	hh ll	3	4	5-2								
			B IND,X	E0	ff	2	4	6-2								
			B IND,Y	18 E0	ff	3	5	7-2								
SUBD (opr)	Subtract Memory from D	D − M:M+1 → D	IMM	83	jj kk	3	4	3-3	–	–	–	–	↕	↕	↕	↕
			DIR	93	dd	2	5	4-7								
			EXT	B3	hh ll	3	6	5-10								
			IND,X	A3	ff	2	6	6-10								
			IND,Y	18 A3	ff	3	7	7-8								
SWI	Software Interrupt	See Special Ops	INH	3F		1	14	2-15	–	–	–	1	–	–	–	–
TAB	Transfer A to B	A → B	INH	16		1	2	2-1	–	–	–	–	↕	↕	0	–
TAP	Transfer A to CC Register	A → CCR	INH	06		1	2	2-1	↕	↓	↕	↕	↕	↕	↕	↕
TBA	Transfer B to A	B → A	INH	17		1	2	2-1	–	–	–	–	↕	↕	0	–
TEST	TEST (Only in Test Modes)	Address Bus Counts	INH	00		1	**	2-20	–	–	–	–	–	–	–	–
TPA	Transfer CC Register to A	CCR → A	INH	07		1	2	2-1	–	–	–	–	–	–	–	–
TST (opr)	Test for Zero or Minus	M − 0	EXT	7D	hh ll	3	6	5-9	–	–	–	–	↕	↕	0	0
			IND,X	6D	ff	2	6	6-4								
			IND,Y	18 6D	ff	3	7	7-4								
TSTA		A − 0	A INH	4D		1	2	2-1	–	–	–	–	↕	↕	0	0
TSTB		B − 0	B INH	5D		1	2	2-1	–	–	–	–	↕	↕	0	0
TSX	Transfer Stack Pointer to X	SP + 1 → IX	INH	30		1	3	2-3	–	–	–	–	–	–	–	–
TSY	Transfer Stack Pointer to Y	SP + 1 → IY	INH	18 30		2	4	2-5	–	–	–	–	–	–	–	–

*Cycle-by-cycle number provides a reference to Tables 10-2 through 10-8 which detail cycle-by-cycle operation.
Example: Table 10-1 Cycle by-Cycle column reference number 2-4 equals Table 10-2 line item 2-4.

TABLE 5.4 Availability of Instructions among the Motorola 6800, 6801, 6805 and 6811 Processors (Continued)

68HC11	6805	6801	6800	Source Form(s)	Operation	Boolean Expression	Addressing Mode for Operand	Machine Coding (Hexadecimal) Opcode	Operand(s)	Bytes	Cycle	Cycle by Cycle*	Condition Codes S	X	H	I	N	Z	V	C
▨				TXS	Transfer X to Stack Pointer	IX−1→SP	INH	35		1	3	2-2	-	-	-	-	-	-	-	-
▨				TYS	Transfer Y to Stack Pointer	IY−1→SP	INH	18 35		2	4	2-4	-	-	-	-	-	-	-	-
▨				WAI	Wait for Interrupt	Stack Regs & WAIT	INH	3E		1	***	2-16	-	-	-	-	-	-	-	-
				XGDX	Exchange D with X	IX→D, D→IX	INH	8F		1	3	2-2	-	-	-	-	-	-	-	-
				XGDY	Exchange D with Y	IY→D, D→IY	INH	18 8F		2	4	2-4	-	-	-	-	-	-	-	-

*Cycle-by-cycle number provides a reference to Tables 10-2 through 10-8 which detail cycle-by-cycle operation.
  Example: Table 10-1 Cycle-by-Cycle column reference number 2-4 equals Table 10-2 line item 2-4.

**Infinity or Until Reset Occurs

***12 Cycles are used beginning with the opcode fetch. A wait state is entered which remains in effect for an integer number of MPU E-clock cycles (n) until an interrupt is recognized. Finally, two additional cycles are used to fetch the appropriate interrupt vector (14 + n total).

dd = 8-Bit Direct Address ($0000 − $00FF) (High Byte Assumed to be $00)
ff = 8-Bit Positive Offset $00 (0) to $FF (255) (Is Added to Index)
hh = High Order Byte of 16-Bit Extended Address
ii = One Byte of Immediate Data
jj = High Order Byte of 16-Bit Immediate Data
kk = Low Order Byte of 16-Bit Immediate Data
ll = Low Order Byte of 16-Bit Extended Address
mm = 8-Bit Bit Mask (Set Bits to be Affected)
rr = Signed Relative Offset $80 (− 128) to $7F (+ 127)
      (Offset Relative to the Address Following the Machine Code Offset Byte)

SOURCE: Courtesy Motorola, Inc.

to be utilized for external memory or I/O with only the requirement for externally latching the low-order byte of address.

The register structure of the M6801, which is shown in Fig. 5.17, consists of both 8- and 16-bit registers in the familiar concatenation of two 8-bit registers (*A-register* and *B-register*) into a single 16-bit register *(D-register)*. This register structure is carried throughout the Motorola 8-bit microprocessors and MCUs with minor variations. For example, the MC68HC11, is be discussed later, has the identical registers with the addition of a second index register. A similar structure will also be seen in the Intel MCUs. The *condition code register* contains the same *CC* bits as the M6800 (*H, I, N, Z*, and *V*) with the addition of an *X-interrupt mask bit* and *stop disable bit*. In addition to the accumulators, there are three 16-bit registers: the *program counter*, the *stack pointer register*, and an *index register*. Limited arithmetic can be performed with these registers such as adding the contents of *B-register* to the *X-Register*.

Motorola M6801 instruction set.   The Motorola M6801 has a particularly rich set of addressing modes just as the M6800, and the addressing is completely orthogonal. There are seven addressing modes:

1. *Accumulator addressing*: Either *Accumulator A* or *Accumulator B* is specified within the 1-byte instruction.

2. *Immediate Addressing*: The data is part of the instruction.

3. *Direct addressing*: The second byte of the instruction is the address of the operand. This limits the addressing to the first 256 bytes of memory and is sometimes referred to as *base page addressing*.

4. *Extended addressing*: The 1-byte instruction is immediately followed by the 16-bit operand address which allows direct accessing of memory location.

5. *Indexed addressing*: The second byte of the instruction is added to the contents of the *Index register* without changing its contents to form the effective address of the operand.

6. *Implied addressing*: The 1-byte instruction refers to a particular 16-bit register.

7. *Relative addressing*: The second byte of the instruction is added to the preincremented PC allowing access to addresses -126 to $+129$ bytes from the current instruction.

Because of the memory mapped I/O, the availability of all of these addressing modes allows direct mathematical operations on the contents of I/O devices as well as memory without having to load the accumulator first.

Because of its similarity to the MC68HC11 instruction set, Table 5.4 will suffice for the 6800, 6801, 6805, and 6811 processors. The M6801 instruction is a superset of the M6800, with the additional instructions being **ABX**, **ADDD**, **ASLD**, **LDD**, **LSRD**, **MUL**, **PSHX**, **PULX**, **STD**, **SUBD**, **BRN**, and **CPX**. Although the **CPX** instruction is included in the M6800 set, internal processing has been modified in the 6801 to permit its use with any conditional branch instruction.

The M6801 instruction set has extensive branch capabilities to test and branch on the following conditions:

1. Branch if ≤ 0.
2. Branch if lower or same.
3. Branch if < 0.
4. Branch if minus.
5. Branch if = 0.
6. Branch if plus.
7. Branch always.
8. Branch never.
9. Branch if overflow clear.
10. Branch if overflow set.
11. Branch if carry clear.
12. Branch if carry set.
13. Branch if zero.

There is no need to specify to which addressing modes these branches apply since all addressing modes apply. In addition, operands can be tested for zero or minus without affecting the data or effecting a branch.

There are no single instructions that either set or clear individual bits; this must be done through the normal byte-wide logical processing. Single-bit processing is available in the 6805 and 6811. The normal complement of data movement instructions and arithmetic and logic operations are available. Of interest is the **SWI** (software interrupt), instruction since it behaves exactly as an interrupt but can be invoked at any time by the software. Instead of calling a subroutine, which does not stack the status of the MCU, the **SWI** saves the complete status of the machine as an interrupt would do before executing the code at the specified address. A return-from-interrupt (**RTI**) instruction must be used, rather than the normal return from subroutine instruction, to return from the **SWI** to unstack the status.

Programming the 68xx MCUs should take into account the following:[11]

1. Use base page (0-255) for frequently accessed data or expanded I/O since

it maximizes the use of direct addressing, which saves one byte per instruction over extended addressing. Since there are many instructions that directly address and operate on memory, the first 256 memory locations can be viewed almost as general-purpose registers.

2. Organize subroutines to lie near the calling line to maximize the usage of *PC*-relative addressing. This also uses one fewer bytes of instruction.

3. The *Index-register* can also be used as an offset to a data table for 1-byte addressing into memory if the data is not in base page or within the relative addressing range.

M6801 on-chip peripherals.   There are two peripherals on the M6801, a timer and serial communications interface. The 16-bit timer shown in Fig. 5.18 can independently measure an input waveform while at the same time generate an output signal. The timer consists of four hardware components:

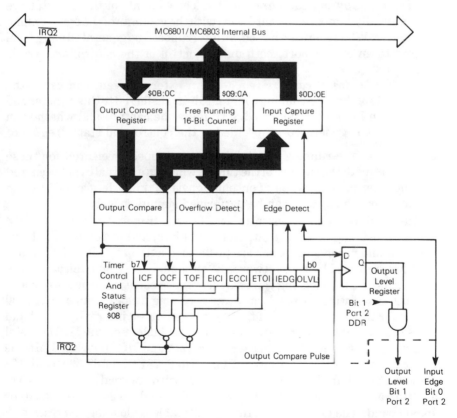

Figure 5.18 Input capture register and output compare register of 6801 MCU. *(Courtesy Motorola, Inc.)*

1. *Eight-bit control and status register*: Three bits are read only and indicate whether a correct input to the input capture register has been made, whether a match has been found between the counter and the output compare register, and when 0000H is in the free running counter. The other five bits are read/write and they specify
   a. The output level (0 or 1) that is to be set when a successful output compare occurs
   b. The type of edge (rising or falling) that causes the free-running counter to transfer to the input capture register
   c. Enable/disable the timer overflow interrupt
   d. Enable/disable the output compare interrupt
   e. Enable/disable the input capture interrupt.
2. *Sixteen-bit free-running counter*: This counter runs continuously at a software-selectable rate and can be read at any time by the software or transferred to the input capture register as the result of an external event.
3. *Sixteen-bit output compare register*: The contents of this register are continually compared by hardware with the contents of the free-running counter. When there is a match, an interrupt can be generated and a level sent to an output port, both under control of the control and status register.
4. *Sixteen-bit input capture register*: An external event can cause the contents of the free-running counter to be transferred to this register and cause an interrupt. Which edge triggers the capture and whether an interrupt is generated is specified in the control and status register.

Because the free-running counter is never stopped or cleared for these operations, both the capture of the time at which an external event occurred and the internal generation of output signals at times determined by the output compare register can be performed simultaneously.

The serial communications interface of Fig. 5.19 operates using the NRZ format consisting of a start bit, eight data bits, and a stop bit. Both the transmit and receive functions are performed automatically in hardware, once programmed by the software. Even though the communications are asynchronous, the clocking can be synchronous since there are provisions for both generating a clock locally and receiving a clock from an external source for the data transfer. If the clock is generated internally, four baud rates are available with the specific values dependent on the MCU crystal frequency. For example, with a 4 MHz crystal, an MCU clock of 1 MHz is generated that can produce either 244.1, 976.6, 7812.5, or 62,500 baud. If a standard baud is desired for communications with a normal terminal device, a 4.9152 MHz crystal can be chosen that will produce 300, 1200, 9600, or 76,800 baud. The desired baud rate is selectable by loading the *transmit/receive control and status register (TRCS)*, with the selected clock applied to both the transmitter and receiver.

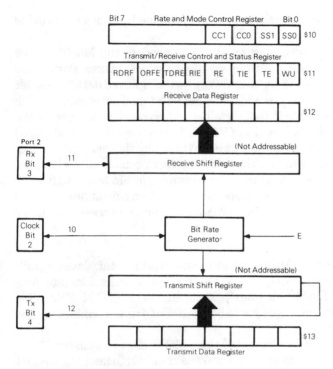

Figure 5.19 Serial communications interface in the Motorola M6801. *(Courtesy Motorola, Inc.)*

When the 6801 is used in a multiprocessor system, message reception overhead can be minimized through the use of a *wake-up* feature. In this method, which is implemented through software, an MCU that is receiving a serial message can disable its reception once it has determined that the message is not for it. It does this by setting the *WU-bit* in the *TRCS register*. This bit is reset by hardware with a sequence of 10 consecutive ones (marking condition), which indicates an idle data transmission line. The only proviso is that the transmitting 6801 ensure that each message it transmits is preceded by a preamble of 10 ones. This is easily implemented by setting the *TE-bit* in the *TRCS register*. In addition to wake-up, the 6801 automatically tests for overrun error (new byte received with old byte still in receive data register) or framing error (byte boundaries not synchronized to the bit counter).

M6801 expansion capability. The M6801 was designed to be expanded in two ways. The first, the expanded nonmultiplexed mode, reconfigures all or part of Port 4 as the low-order address lines (A0–A7) and Port 3 as the data bus for an additional 256 base-page locations. This effectively expands the

I/O capability of the 6801 since I/O is memory mapped. It does not provide any additional memory.

If the full range of addressing capability is needed, the M6801 can be operated in the expanded multiplexed mode, which makes Port 4 the high-order address lines (A8–A15) and Port 3 a multiplexed data/low-order address bus. The low-order bus address must be latched externally. Which mode the MCU will operate in is determined at hardware reset time by the MCU reading input bits 0, 1, and 2 of Port 2. This mode cannot be changed without resetting the MCU. Since there are three bits that select the mode, there are eight operating modes that offer different combinations of memory location and interrupt vectors. For example, in Mode 1 (001B), the interrupt vectors that are normally located in ROM on chip at hexadecimal addresses $FFF0 to $FFFF are fetched from externally addressed memory or switches.

Other M6801 capabilities.   Although there is no direct master/slave capability inherent in the M6801 MCU, the on-chip serial communications interface makes it easy to implement message passing among several M6801s. This allows for a variety of relationships to be maintained as determined by the controlling software.

Powerdown modes are implemented through the use of a standby V_{cc}. A maximum of 8 mA of current at 5 volts $\pm$ 5% is required to keep the first 64 bytes of memory (standby RAM) on the M6801 intact. In addition to the standby power supply, the *RAM enable bit (RAME)* of the *RAM control register* must be set to 1 before power is lost. The standby RAM is the only low power capability of the M6801.

M6801 interrupts.   Both maskable and nonmaskable interrupts are implemented in the M6801. A nonmaskable interrupt (NMI) is generated when the signal on the *NMI* pin makes a 1-0 transition. If this occurs, the registers and condition codes are stacked, and the instruction pointed to by the contents of locations $FFFC and $FFFD is executed. This input is typically used for emergency inputs that should not occur, such as power failure. The NMI ISR could load critical information in the standby RAM so it could continue where it left off when the power is restored. The other interrupt, IRQ, is maskable by a control bit in the condition code register. This is a level-sensitive input which can respond to slowly changing inputs. If the *I-bit* in the *CCR* is not set, the registers will be stacked after the completion of the current instruction, and the execution will transfer to the address that is stored in $FFF8 and $FFF9. Although there are only two interrupts, the interrupts are vectored, and the availability of operating modes that allow them to be off chip rather than in ROM allow for their values to be changed in real time.

5.7.3 Design and Application Tools for the M6801

In an earlier chapter, the INTROL cross-assembler and C cross-compiler were indicated as being one of a family of support software for MCUs, so high-language level support is available. In addition to the MC68HC11, INTROL has compilers for the M6801, as do other vendors. The front end of these cross-compilers is the same, with the only difference being the assembly language output that is unique to the target processor.

5.8 MOTOROLA M6805 FAMILY

The Motorola M6805 is not an enhanced M6801 on the way to the M6811, but rather a departure in the direction of even more austere, control-oriented, cost-conscious design for high-volume applications. Individual units are tailored with either A/D converters, phase-lock loops, or serial communications interfaces. The instruction set is neither a superset nor a subset of the M6801, but the register structure and enough of the instructions have been maintained that the heritage is evident. A significant deviation from the M6801 instruction set is the addition of powerful branch on bit-test instructions. Further supporting its intended application is the fact that it has only an 11-bit program counter allowing access to only 2048 bytes of code. The reduced address space also precludes any need for an expanded mode of operation.

5.81 M6805 Family Members

Both HMOS and CMOS technology are represented in the M6805 family members, allowing for the selection of parts that are optimized for speed or low power. The CMOS versions are fully static and can be operated at a clock rate down to DC. Packaging varies from 28 to 40 pins depending on the amount of I/O. Available RAM varies from a minimum of 64 to a maximum of 176 bytes. The amount of ROM varies from 1024 to 4096 bytes. All inputs and outputs are TTL compatible. Available on-chip peripherals, although not simultaneously available on a single model, include:[10]

1. Four-channel A/D
2. Standby RAM
3. Self-check capability
4. Two-serial ports
5. Phase-lock loop
6. Bootstrap ROM
7. UV EPROM

8. ROM-based self-programmer

9. Low-voltage inhibit.

The MC68705P3 version is able to program its own UV EPROM with a chip-resident ROM programmer program.[12] The only thing that is needed is an external ROM from which the 68705 is to copy its code and a 12-volt power supply.

5.8.2   M6805 Architecture

The register structure of the M6805 is a subset of the M6801 in that it has a single 8-bit accumulator *(A-register)*, a single 8-bit index register *(X-register)*, an 11-bit *program counter (PC)*, an 11-bit *stack pointer (SP)*, and a 5-bit *condition code register (CCR)*. Notably lacking is the 16-bit *D-register*. As was seen in the previous paragraph, various on-chip peripherals are available, but they do not materially affect the architecture of the M6805, which is (like the other Motorola products) non-Harvard architecture with memory-mapped I/O. It is a vectored interrupt machine with one external maskable interrupt and one internal maskable interrupt. There are no non-maskable interrupts other than a complete hardware reset.

The special function registers associated with the on-chip peripherals are all located beginning at $00 and hence are located in base page. RAM and even some ROM extends down into base page, and so the direct addressing mode can be used extensively in addressing variable storage and I/O ports as well as table lookup in ROM.

M6805 instruction set.   There is no change in the addressing modes available, and the instruction set is almost a subset of the M6800 instruction set. The exact instructions available can be seen in the same table that was used previously for the Motorola MCUs. Not surprisingly, all references to a *B-register* are absent, and there is no ability to test on the nonexistent *V-bit*. In spite of the need to do decimal arithmetic in some applications, the decimal adjust accumulator (DAA) instruction has been removed, although it was replaced with the ability to test and branch on the half-carry *(H-bit)* of the *CCR*. It is apparently more silicon efficient to implement DAA in software than in hardware.

A significant increase in power of the M6805's 59 basic instructions over those of the 6801 is the addition of bit test and branch as well as bit set and bit clear instructions. These bit-twiddling instructions are far more useful in small control applications than the computationally more powerful 8×8 multiply (which has been removed). With the exception of the data direction registers for the I/O Ports, any bit in base page RAM, ROM, or I/O can be set, cleared, or tested/branched. In addition to the ability to test individual memory bits, the M6805 is capable of testing and branching on carry set, carry clear, half-carry set, half-carry clear, interrupt set, interrupt clear, interrupt mask set, and interrupt mask clear.

Most stack pointer instructions are also absent. There is no way to alter the stack pointer short of resetting it to the top of RAM with the RSP instruction. The addressing capability of the *Index register* has been expanded to make up for the fact that it is only 1 byte wide and the address space of the 6805 is 11 bits. This has been resolved by allowing the index register to take a 0-, 1-, or 2-byte offset so the entire address space can be reached through indexing.

M6805 on-chip peripherals. On-chip peripherals for the M6805 family are discussed by type of peripheral rather than by individual chips. For this reason, some comments concerning addresses of registers may appear to overlap, but when applied to the appropriate peripherals on a single chip, there will be no conflict. This discussion of on-chip peripherals focuses on their capabilities and unique properties and is not a guide to chip selection. Three are discussed: timer, A/D converter, and phase-lock loop.

The M6805 timer of Fig. 5.20 is limited to an 8-bit preloadable down-counter. The clock input to this timer can be either from an external source or can be derived from the system clock and can be prescaled by a 1, 2, 4, 8, 16, 32, 64, or 128 divider. The actual divisor and source of the clock (internal or external) must be mask programmed at manufacture and is not software selectable. When the timer counts down to zero, an interrupt will be generated if it is not masked by its own timer mask bit or by the *I-bit* of the *CCR*. The interrupt will cause the MCU state registers and *CCR* to be stacked, and execution will transfer to the address stored at the timer

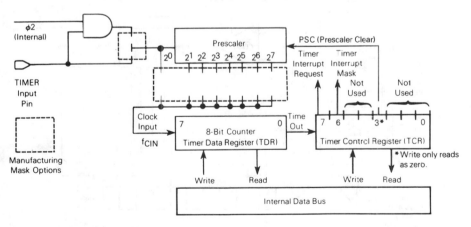

Figure 5.20 Motorola M6805 8-bit timer block diagram. *(Courtesy Motorola, Inc.)*

interrupt vector located at $7F8 and $7F9. It is not necessary to wait for an interrupt since the contents of the timer can be read at any time without affecting its state.

Four channels of A/D conversion are available through an on-chip successive-approximation, ratiometric converter. The M6805 A/D block diagram is shown in Fig. 5.21. The resolution of all four channels is 8 bits, with the input range from 0 to 5 Volts. Operating at a clock frequency of 1 MHz, the conversion time is 76 μs for the 30 cycles it takes to convert one channel. Only one channel can be converted at a time, and there is no way automatically to enable the conversion of all four channels sequentially. Sequential conversions must be done under software control. No interrupt is generated by the A/D conversion process, and even though other instructions can be executed while the conversion is taking place, the *CEND* bit of the *A/D Control Status Register (ADCSR)* must be polled to determine when conversion is complete. In the event of overflow (input voltage greater than V_{ref}) a value of $FF is returned with no indication of overflow.

For TV and radio applications, a phase-lock loop is often necessary to be used as a frequency synthesizer. The actual operation of a phase-lock loop (PLL) is outside the scope of this book and it is governed primarily by external components, consisting of a low-pass filter, voltage-controlled oscillator, and prescaler. Of importance in this discussion is that the on-chip PLL includes a 14-bit divider and a mask programmed reference frequency divider. There are also provisions for glitchless frequency switching by transferring 14-bit data from the 8- and 6-bit registers only when the data is written to the 8-bit register, assuming that the 6-bit register has already been loaded with the correct value.

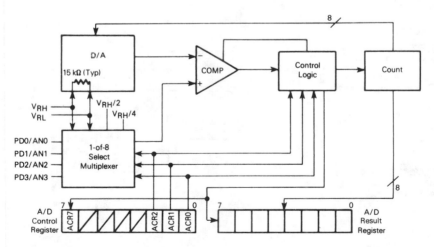

Figure 5.21 Motorola M6805 4-channel A/D converter block diagram. *(Courtesy Motorola, Inc.)*

Other capabilities of the M6805.   There is no capability to expand either RAM or ROM in the M6805. Nor is there the capability to operate easily in a master/slave relationship except by software and message passing through a parallel port. The M6805 does have other capabilities that are significant in large-volume, small-control applications. For example, on some non-CMOS family members, a low-voltage reset capability is available that restarts the program when the power supply voltage falls below a reliable operating level. A version is also available that has both ROM and UV EPROM on chip, where the ROM contains a bootstrap program for burning a copy of an external ROM into the EPROM. The ROM code is automatically initiated when one pin is raised to the required 12 volt programming voltage.

For a quick incoming inspection, a version with a self-check program already burned into ROM is available. This is initiated by applying 9 volts to the timer input and connecting the rest of the pins in a particular manner. If the chip passes its ROM, RAM, timer, interrupts, and I/O tests, an LED connected to Port C, output 3 will blink at an approximate 3 Hz. A standby mode is available for preserving the contents of the first 8 bytes of RAM and operates similar to the M6801 in that a register must be programmed to enable the standby mode and the standby voltage must be kept at greater than 4 volts.

M6805 interrupts.   Not only are the M6805 interrupts vectored, but they are also prioritized. There is one software interrupt, SWI, one overriding external interrupt, *RESET*, and a maskable external interrupt. Internally, the timer can generate an interrupt, but the A/D does not generate one and there is no need for the PLL to generate one. The priority of these interrupts, with *RESET* as the highest, is as follows:

1. *RESET* (nonmaskable)
2. SWI
3. External maskable interrupt
4. Internal maskable timer.

Associated with each of these interrupts is a unique address that contains the address of the appropriate interrupt-service routine.

5.9   MOTOROLA MC68HC11

The Motorola MC68HC11 is the most capable of the Motorola 8-bit MCUs in that it incorporates all of the capabilities of the M6801 and M6805 except for the PLL of the M6805. In this section, the MC68HC11 family is explored in detail through an examination of the architecture and characteristics of one of its variants, the MC68HC11A8. This variant is chosen for the following reasons:

1. It contains all of the on-chip peripherals of the family.

2. It is an EPROM version of the MC68HC11A1 which is the ROM-less version used on the M68HC11EVB evaluation board[13] available from Motorola at a modest cost for educational purposes.

3. It can operate in both a stand-alone or an expanded mode.

4. Its instruction set has been used for Chap. 3's examples of implementation of Petri tables in both C and assembly language.

General characteristics of the MC68HC11,[14] which are referred to here generically as the M6811, are as follows:

1. High-density CMOS technology (HCMOS)

2. Single voltage power supply operation

3. 2-MHz bus speed

4. Fully static operation (no minimum clock rate)

5. On-chip EEPROM for nonvolatile storage

6. Watchdog timer (COP)

7. Eight-bit arithmetic

8. Instruction set that is a superset of M6800/M6801

9. Sixteen-bit timer/event time capture

10. Eight-bit pulse accumulator

11. Synchronous and asynchronous hardware communications

12. Maskable and nonmaskable interrupts

13. 16 × 16 integer and fractional divide

14. Low-power operating modes under software control

15. Expandable to full address space

16. Memory-mapped I/O

17. Peripheral chip that emulates single-chip operation in expanded mode

18. Orthogonal addressing

19. Real-time interrupt.

Since the M6811 is fully static, it can operate down to zero clock frequency (DC). In this type of circuitry, power is consumed if the clock is running even if the CPU is not executing any instructions due to the charging and discharging of capacitors through small but finite resistances. The ability to operate at very low clock frequencies can reduce power consumption to a minimum for critical, low-power applications. Additionally, there are two modes of operation that can temporarily suspend operation causing the same effect: the wait mode and the stop mode.

The M6811 is an exceptionally versatile MCU that, when coupled with a low-cost evaluation board and BUFFALO monitor, allows one to study in detail a multiplicity of peripheral and I/O options in a single programming and development environment. Experience gained with this MCU is easily transferrable to a variety of other processors.

5.9.1   Motorola MC68HC11 Family Members

The primary differences among members of the M6811 family are the type and quantity of memory. Each major derivative has three configurations, labeled x8, x1, and x0, that use an identical die. The variants are made by semi-permanently disabling ROM in the x1 versions and disabling both ROM and EEPROM in the x0 versions. A no ROM version, labeled x2, is available for low volume applications that require less than 2 kbytes of EEPROM program memory or are intended to be operated only in the expanded mode with program memory off chip. An Ex series is available that has more on-chip memory, and an option for reconfiguring one of the output compares to make it an input capture register. A final, economy version, with no A/D and smaller memories, is designated with a Dx suffix. The amount of ROM varies across the family from 0 to 12 kbytes; RAM varies from 192 to 512 bytes; EPROM varies from 0 to 2 kbytes. The capabilities of the various versions are summarized in Table 5.5.

5.9.2   Architecture

Perhaps the single significant feature of the M6811 architecture is not that it has so many on-chip peripherals and options, but rather that it is user programmable on various levels. Earlier examples of MCUs that were discussed in this chapter have silicon-economical masked options to enable or disable certain features. Aside from its fundamental non-Harvard and memory-mapped I/O architecture, almost every aspect of the M6811 is software programmable at one level or another. Some semipermanent features must be programmed in nonvolatile EEPROM. These set up fundamental operating modes. Other features must be established within the first 64 clock cycles by the user program and remain fixed until a reset occurs. Still other features, such as the direction (input or output) of individual bits of some ports can be changed at any time by writing to the appropriate data direction registers. So, in many respects, the final architecture of the M6811 is determined by the end user and the software that is executed.

The major structural components of the M6811 are shown in Chap. 2, Fig. 2.10. It is an 8-bit MCU with a limited ability to perform 16-bit arithmetic.

TABLE 5.5 Characteristics of Motorola MC68HC11xx Family Members

Device Number	ROM	EEPROM	RAM	CONFIG[3]	Comments
MC68HC11A8	8K	512	256	$0F	Family Built Around this Device
MC68HC11A1	0	512	256	$0D	Same Die as 'A8 but ROM Disabled
MC68HC11A0	0	0	256	$0C	Same Die as 'A8 but ROM and EEPROM Disabled
XC68HC11B8	8K	512[1]	256	$0F	Early Experimental Version
XC68HC11B1	0	512[1]	256	$0D	'B8 with ROM Disabled
XC68HC11B0	0	0	256	$0C	'B8 with ROM and EEPROM Disabled
MC68HC11E9	12K	512	512	$0F	Four Input Captures and Bigger RAM and 12K ROM[4]
MC68HC11E1	0	512	512	$0D	'E9 with ROM Disabled[4]
M68HC11E0	0	0	512	$0C	'E9 with ROM and EEPROM Disabled[4]
MC68HC811E2	0	2K[2]	256	$FF	No ROM Part for Expanded Systems
MC68HC11D3	4K	0	192	N/A	Economy Version, No A-D and Smaller Memories[4]

NOTES:
1. The EEPROM on B Series parts requires an external 19-volt supply for programing and is not byte erasable.
2. This 2K EEPROM is relocatable to the top of any 4K memory page. Relocation is done with four bits in the CONFIG register.
3. CONFIG register values in this table reflect the value programmed prior to shipment from Motorola.
4. Available 1988.

SOURCE: Courtesy Motorola, Inc.

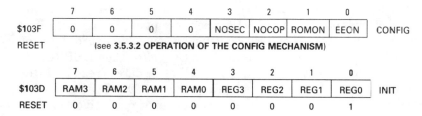

	7	6	5	4	3	2	1	0	
$103F	0	0	0	0	NOSEC	NOCOP	ROMON	EEON	CONFIG
RESET		(see **3.5.3.2 OPERATION OF THE CONFIG MECHANISM**)							

	7	6	5	4	3	2	1	0	
$103D	RAM3	RAM2	RAM1	RAM0	REG3	REG2	REG1	REG0	INIT
RESET	0	0	0	0	0	0	0	1	

Figure 5.22 CONFIG AND INIT registers of the M68HC11 MCU.

As with the other Motorola products, the M6811 is a non-Harvard architecture with no separate I/O addressing scheme. In addition to the expected on-chip ROM and RAM, there is an EEPROM that enhances the capability of the MCU since it is able to be erased and programmed under software control and implements nonvolatile storage. Some of the less often changed architectural features are under the control of the *system configuration register (CONFIG)*, which is implemented in EEPROM shown in Fig. 5.22, but is not part of the EEPROM memory. Through the programming of the *CONFIG* register, the following options can be enabled/disabled:

Security mode: This mode is available only if specified at the mask level. If it is, then the security mode can be enabled/disabled by clearing/setting the *NOSEC bit* (bit 3) of *CONFIG*. If enabled, the contents of EEPROM, RAM, and *CONFIG* are erased before the bootstrap mode (described later) is entered and the expanded mode is disabled.

COP system: The watchdog timer can be enabled or disabled by setting *NOCOP* (bit 2) of *CONFIG* to 0/1.

On-chip ROM: The *ROMON* (bit 1) bit of *CONFIG* determines whether the internal ROM is enabled or disabled. In the single-chip mode, this bit has no effect.

On-chip EEPROM: The *EEON* (bit 0) bit of *CONFIG* enables/disables the on-chip EEPROM when set to 1/0.

These bits specify the fundamental operating modes for the M6811 and therefore are not usually changed. If one wants to change the *CONFIG* register, then it must be erased by erasing all of EEPROM and reprogramming at its register address of $103F.

Another hardware option for selecting the operating architecture of the M6811 is through the use of two mode select pins. These two pins (MODA, MODB) select one of four operating modes:

1. *Special bootstrap mode* (00): If this mode is enabled at reset, a special 192 byte ROM is enabled that executes a program to load, through the serial

communications interface, a 256-byte program into base page. If the internal clock is operating at 2 MHz, upon receipt of $FF at either 7812 baud or 1200 baud, the bootstrap loader will receive the next 256 bytes at that rate. If the security bit is enabled in *CONFIG*, then EEPROM is cleared and RAM is overwritten with $FF before data is read in. After the program is loaded, execution is transferred to $0000. Other options are available in response to first characters being received other than $FF.

2. *Single-chip* (01): In this mode, no external data or address bus is enabled and all communications with other devices are made through the I/O ports and serial interfaces.

3. *Special test* (10): This is primarily a factory test mode that disables some protection features, allows access to other registers, and changes the location of interrupt vectors. It is not recommended for end-user use.

4. *Expanded multiplexed* (11): I/O Ports B and C are sacrificed to high-order address and a multiplexed low-order address/data bus, respectively, to access the complete 64-kbyte address space. The low-order address must be latched externally in the same manner as was done with the M6801 expanded mode.

Unlike the M6801, which has eight operating modes that also select memory mapping, the M6811 has only these four operating modes. This does not mean that memory and I/O control registers cannot be mapped to various locations within the address space of the machine. On the contrary, the *RAM and I/O Mapping register (INIT)* can be programmed to move them to various locations depending on operating mode. Because of their special-purpose nature, Special Test and Bootstrap modes are not discussed and attention is focused on the Single-chip and Expanded Multiplexed Mode of operation. The *INIT* register is shown in Fig. 5.22. Four bits of this register *(INIT4—INIT7)* control the location of the start of RAM. The default location for RAM is $0000, but if this register is loaded within the first 64 clock cycles, these four bits become the high-order four bits of the starting address of RAM. That is, RAM can be moved to any of sixteen 4-kilobyte boundaries. The other four bits of the *INIT* register have the same timing constraint but control the starting address of the 64 control registers. By default, these start at $1000 but can also be moved in four Kbyte increments just like RAM.

The programming register structure is shown in Chap. 2, Fig. 2.11. It is an expansion of the M6800/M6801 set in that the M6800 did not treat the A and B registers as a single 16-bit register whereas the M6801 and the M6811 do. The M6800/M6801 each had one index register whereas the M6811 has two. In terms of the condition code registers, the M6800 and M6801 are identical with the H, I, N, Z, and V bits. The M6811 has these same condition codes with the addition of a stop disable bit and an *X-interrupt mask bit,* accounting for all 8 bits of the *CCR*. In other words, programming the M6811 is a little bit easier, but in the same vein, as the earlier Motorola MCUs. The addition of a second index register allows for maintaining a source pointer and destination pointer simultaneously, which is useful for transferring data.

In addition to the programming registers, there are 64 data and control registers that, as mentioned previously, are by default located starting at $1000. These registers are listed in Table 5.6. These registers are not detailed here but are referred to by *mnemonic* in the ensuing discussion of particular on-chip peripherals.

Instruction set. While the instruction sets of the M6800/M6801 have been mentioned, no details have been given except for the addressing modes available. Because the instruction set is orthogonal, all of these modes are available to all instructions. Since the M6811 instruction is a proper superset of the M6800/M6801 MCUs, it is able to execute code written for these other machines.

The M6811 instruction set can be subdivided into five areas[15,16] for the sake of discussing general properties. These areas and representative instructions are as follows:

1. *Move instructions:* LDAA, LDAB, LDD, LDX, LDY, LDS, STAA, STAB, STD, STX, STY, STS, TAB, TBA, TAP, TPA, TXS, TSX, TSY, TYX, PSHA, PULA, PSHB, PULB, PSHX, PULX, PSHY, PULY, CLRA, CLRB, CLR, TSTA, TSTB, TST.

2. *Arithmetic instructions:* ADDA, ADDB, ADCA, ADCB, ABA, ABX, ABY, SUBA, SUBB, SBCA, SBCB, SBA, CMPA, CMPB, CBA, INCA, INCB, INC, DECA, DECB, DEC, NEGA, NEGB, NEG, ASLA, ALSB, ASL, ASRA, ASRB, ASR, LSRA, LSRB, LSR, LSLA, LSLB, LSL, ADDD, SUBD, CPD, CPX, CPY, INX, NY, INS, DEX, DEY, DES, MUL, FDIV, IDIV, ASLD, LSRD, LSLD, DAA.

3. *Logic instructions:* EORA, EORB, ORAA, ORAB, ANDA, ANDB, BITA, BITB, COMA, COMB, COM, SEC, SEI, SEV, CLC, CLI, CLV, BSET, BCLR.

4. *Edit instructions:* ASLA, ASLB, ASL, ASRA, AASRB, ASR, LSLA, LSLB, LSL, LSRA, LSRB, LSR, LSLD, ASLD, LSRD, ROLA, ROLB, ROL, RORA, RORB, ROR.

5. *Control instructions:* JMP, BRA, BRN, NOP, SKIP2, BEQ, BNE, BMI, BPL, BCS, BCC, BVS, BVC, BGT, BGE, BEQ, BLE, BLT, BHI, BHS, BEQ, BLS, BLO, BRSET, BRCLR, JSR, BSR, RTS, RTI, SWI, STOP, WAI.

Looking at the move instructions, and with reference to Table 5.4, one can see that all 8- and 16-bit registers can be loaded and stored. Data can be transferred between registers as well as the *CCR* (TAP: transfer *A* to *CCR*) that cannot be loaded directly. Sixteen-bit transfers can also be effected

TABLE 5.6   Motorola MC68HC11 Data and Control Registers

	Bit 7	Bit 6	Bit 5	Bit 4	Bit 3	Bit 2	Bit 1	Bit 0		
$1000	Bit 7	—	—	—	—	—	—	Bit 0	PORTA	I/O Port A
$1001									Reserved	
$1002	STAF	STAI	CWOM	HNDS	OIN	PLS	EGA	INVB	PIOC	Parallel I/O Control Register
$1003	Bit 7	—	—	—	—	—	—	Bit 0	PORTC	I/O Port C
$1004	Bit 7	—	—	—	—	—	—	Bit 0	PORTB	Output Port B
$1005	Bit 7	—	—	—	—	—	—	Bit 0	PORTCL	Alternate Latched Port C
$1006									Reserved	
$1007	Bit 7	—	—	—	—	—	—	Bit 0	DDRC	Data Direction for Port C
$1008			Bit 5	—	—	—	—	Bit 0	PORTD	I/O Port D
$1009			Bit 5	—	—	—	—	Bit 0	DDRD	Data Direction for Port D
$100A	Bit 7	—	—	—	—	—	—	Bit 0	PORTE	Input Port E
$100B	FOC1	FOC2	FOC3	FOC4	FOC5				CFORC	Compare Force Register
$100C	OC1M7	OC1M6	OC1M5	OC1M4	OC1M3				OC1M	OC1 Action Mask Register
$100D	OC1D7	OC1D6	OC1D5	OC1D4	OC1D3				OC1D	OC1 Action Data Register
$100E	Bit 15	—	—	—	—	—	—	Bit 8	TCNT	Timer Counter Register
$100F	Bit 7	—	—	—	—	—	—	Bit 0		
$1010	Bit 15	—	—	—	—	—	—	Bit 8	TIC1	Input Capture 1 Register
$1011	Bit 7	—	—	—	—	—	—	Bit 0		
$1012	Bit 15	—	—	—	—	—	—	Bit 8	TIC2	Input Capture 2 Register
$1013	Bit 7	—	—	—	—	—	—	Bit 0		
$1014	Bit 15	—	—	—	—	—	—	Bit 8	TIC3	Input Capture 3 Register
$1015	Bit 7	—	—	—	—	—	—	Bit 0		
$1016	Bit 15	—	—	—	—	—	—	Bit 8	TOC1	Output Compare 1 Register
$1017	Bit 7	—	—	—	—	—	—	Bit 0		
$1018	Bit 15	—	—	—	—	—	—	Bit 8	TOC2	Output Compare 2 Register
$1019	Bit 7	—	—	—	—	—	—	Bit 0		
$101A	Bit 15	—	—	—	—	—	—	Bit 8	TOC3	Output Compare 3 Register
$101B	Bit 7	—	—	—	—	—	—	Bit 0		
$101C	Bit 15	—	—	—	—	—	—	Bit 8	TOC4	Output Compare 4 Register
$101D	Bit 7	—	—	—	—	—	—	Bit 0		
$101E	Bit 15	—	—	—	—	—	—	Bit 8	TOC5	Output Compare 5 Register
$101F	Bit 7	—	—	—	—	—	—	Bit 0		

TABLE 5.6   Motorola MC68HC11 Data and Control Registers (Continued)

	Bit 7	Bit 6	Bit 5	Bit 4	Bit 3	Bit 2	Bit 1	Bit 0		
$1020	OM2	OL2	OM3	OL3	OM4	OL4	OM5	OL5	TCTL1	Timer Control Register 1
$1021			EDG1B	EDG1A	EDG2B	EDG2A	EDG3B	EDG3A	TCTL2	Timer Control Register 2
$1022	OC1I	OC2I	OC3I	OC4I	OC5I	IC1I	IC2I	IC3I	TMSK1	Timer Interrupt Mask Register 1
$1023	OC1F	OC2F	OC3F	OC4F	OC5F	IC1F	IC2F	IC3F	TFLG1	Timer Interrupt Flag Register 1
$1024	TOI	RTII	PAOVI	PAII			PR1	PR0	TMSK2	Timer Interrupt Mask Register 2
$1025	TOF	RTIF	PAOVF	PAIF					TFLG2	Timer Interrupt Flag Register 2
$1026	DDRA7	PAEN	PAMOD	PEDGE			RTR1	RTR0	PACTL	Pulse Accumulator Control Register
$1027	Bit 7	—	—	—	—	—	—	Bit 0	PACNT	Pulse Accumulator Count Register
$1028	SPIE	SPE	DWOM	MSTR	CPOL	CPHA	SPR1	SPR0	SPCR	SPI Control Register
$1029	SPIF	WCOL		MODF					SPSR	SPI Status Register
$102A	Bit 7	—	—	—	—	—	—	Bit 0	SPDR	SPI Data Register
$102B	TCLR		SCP1	SCP0	RCKB	SCR2	SCR1	SCR0	BAUD	SCI Baud Rate Control
$102C	R8	T8		M	WAKE				SCCR1	SCI Control Register 1
$102D	TIE	TCIE	RIE	ILIE	TE	RE	RWU	SBK	SCCR2	SCI Control Register 2
$102E	TDRE	TC	RDRF	IDLE	OR	NF	FE		SCSR	SCI Status Register
$102F	Bit 7	—	—	—	—	—	—	Bit 0	SCDR	SCI Data (Read RDR, Write TDR)
$1030	CCF		SCAN	MULT	CD	CC	CB	CA	ADCTL	A/D Control Register
$1031	Bit 7	—	—	—	—	—	—	Bit 0	ADR1	A/D Result Register 1
$1032	Bit 7	—	—	—	—	—	—	Bit 0	ADR2	A/D Result Register 2
$1033	Bit 7	—	—	—	—	—	—	Bit 0	ADR3	A/D Result Register 3
$1034	Bit 7	—	—	—	—	—	—	Bit 0	ADR4	A/D Result Register 4
$1035 Thru $1038									Reserved	
$1039	ADPU	CSEL	IRQE	DLY	CME		CR1	CR0	OPTION	System Configuration Options
$103A	Bit 7	—	—	—	—	—	—	Bit 0	COPRST	Arm/Reset COP Timer Circuitry
$103B	ODD	EVEN		BYTE	ROW	ERASE	EELAT	EEPGM	PPROG	EEPROM Programming Control Register
$103C	RBOOT	SMOD	MDA	IRV	PSEL3	PSEL2	PSEL1	PSEL0	HPRIO	Highest Priority I-Bit Int and Misc
$103D	RAM3	RAM2	RAM1	RAM0	REG3	REG2	REG1	REG0	INIT	RAM and I/O Mapping Register
$103E	TILOP		OCCR	CBYP	DISR	FCM	FCOP	TCON	TEST1	Factory TEST Control Register
$103F	—	—	—	—	NOSEC	NOCOP	ROMON	EEON	CONFIG	COP, ROM, and EEPROM Enables

SOURCE: Courtesy Motorola, Inc.

between the stack pointer and the two index registers. The 8-bit accumulators as well as the index registers can be pushed to and pulled from the stack for temporary storage, but the 16-bit *D-register* cannot. The accumulators can be cleared (and set to all ones by **CMPA** if needed) as well as arbitrary memory locations, with the restriction that there is no base page addressing mode for **CLR**. Both accumulators and arbitrary memory locations (no base page addressing) can be tested for zero or minus, setting the appropriate *CCR* bits, but not branching.

The arithmetic instructions allow for the expected addition and subtraction with or without carry/borrow as well as the incrementing/decrementing of all registers and arbitrary memory locations (no base page addressing). Sixteen-bit addition is also available between the *D-register* and memory. Equality of the accumulators with memory can be tested in all addressing modes. The memory-mapped I/O and extensive addressing modes of the M6811 allow for easy manipulation of ports as well as variables. Although the accumulators can be loaded directly from an input port, arithmetic can often be performed while loading an accumulator by preloading the accumulator with one operand and then **ADDA**ing or **SUBA**ing a port to the accumulator.

The logic instructions allow logical operations between the accumulators and operands addressed by all addressing modes including base page. Three *CCR* bits can be directly set or cleared (carry, overflow, and interrupt). Individual bits of both accumulators and memory locations can be set through the use of an immediate operand. In this manner, individual bits in I/O ports or their control registers can be set/cleared.

Edit instructions allow both 8- and 16-bit arithmetic and logical shifts of the accumulator, the difference between these instructions being whether the carry bit is involved in the data movement. Except for the direct addressing mode, arithmetic and logical shifts can be performed on memory as well as the registers.

The final group of instructions is the control instructions which affect the order of program execution. (An extensive list of conditions in which branching can occur was given in Sec. 5.7.2.) Conditional branches can be made both to distinct memory locations or to subroutines. In addition to these versatile branch instructions, unconditional jumps to subroutines can be made through the **JSR/RTS** pair of instructions. Interrupt handling is also included in this list of control instructions with the inclusion of the return from interrupt (**RTI**) instruction and the software interrupt (**SWI**) instruction. The **SWI** instruction causes the registers to be stacked just as if a normal hardware interrupt had occurred with the address of the **SWI** ISR being specified in a vector location at $FFF6 and $FFF7. To unstack the registers after the **SWI** ISR, the **RTI** instruction must be executed. The last two instructions are used for low-power operation since they both reduce the amount of circuitry is being clocked. If the stop bit in the *CCR* is set, and the **STOP** instruction is executed, then all system clocks are halted, which results in a minimum-power-consumption mode.[17] The only way that the M6811 can be reactivated is through an external reset, $XIRQ$, or an unmasked IRQ. If the x-bit of the CCR is set, masking $XIRQ$ interrupts, execution will continue with the instruction that follows the STOP instruction. The other low-power mode instruction is the Wait (WAI) instruction.[14] When WAI is executed, it prepares for an interrupt by stacking all registers. In this mode, and depending on the status of the various timers and A/D converter control registers, the clocking of the CPU is minimized, thereby reducing current consumption. The MCU is taken out of this wait state to the appropriate ISR by the occurrence of any unmasked interrupt.

MC68HC11 on-chip peripherals. The I/O registers themselves must be considered as peripherals because of their ability to strobe external registers (indicating that data is valid at the output port) and, on inputs, their ability to use one input as a latching signal to effect the capture of data from an external device to an input port of the M6811. Ignoring the expanded mode of operation, Port C and Port D are bidirectional parallel I/O in that the direction of data flow (in or out) is determined by the contents of each port's *data direction register* (*DDRC* and *DDRD*). These registers are independent of the port registers themselves and are, by default, initialized to inputs. The other ports, except for Port A, bit 7, have fixed data directions. (For a detailed listing of the appropriate control and data registers, refer to Table 5.6.) Port C can be used as a strobed input by clearing the appropriate bit *(HNDS)* in the *parallel I/O control register (PIOC)*. When this is done, pin STRA of the A Port is used by an external device to latch data coming into the M6811 on the edge determined by the *EGA* bit of the *PIOC register*. Likewise, Port B can be used as a strobed output port to latch data into an external latch. When Port B is written to, the STRB output is pulsed or set to a level as determined by the *PLS* and *INVB bits* of the *PIOC register*. Port E serves dual duty as either the input to the A/D converters (8 channels) or as binary inputs.

There are two serial interfaces built into the M6811, one asynchronous and one synchronous. The asynchronous one is full duplex in NRZ format consisting of one start bit, eight or nine data bits, and one stop bit. While similar to the SCI of the M6801, the M6811's SCI has the additional capability of 32 software-selectable baud rates, eight or nine data bits, and interrupt-driven operation. There is also a wake-up feature identical to the M6801 as well as framing error, overrun, and noise detection. All of these options are selectable through the *serial communications control registers 1 and 2* (*SCCR1* and *SCCR2*) and the *serial communications status register (SCSR)*, which provides inputs to the interrupt logic circuitry. Both the transmit and receive baud rate are the same and can only be generated internally (unlike the M6801), but there are two different prescalers that allow for baud rates from 75 to 131,072 baud.

The second serial interface is designed to communicate synchronously with peripherals and consists of a full-duplex three-wire interface. This coexists with the SCI and is called the *serial peripheral interface (SPI)*; both interfaces can operate simultaneously as they have independent hardware. The SPI can either supply the clock (master) from one of four software programmable rates up to 1.05 MHz or use an external clock (slave) of up to 2.1 MHz. Both the polarity of the clock and its phase are also software selectable. Since both the SCI and SPI share pins with Port D, if they are used, Port D cannot be used as a general I/O port. Software control of the SPI is established through the *serial peripheral control register (SPCR)* and the *serial peripheral status register (SPSR)*. There is also provision for write collision detection to detect when an attempt is made to write to the SPI while a data transfer is taking place. Fully interrupt-driven control of the SPI is also an option. The advantage of the SPI over the SCI is that synchronous communications allows for the elimination of both the start and stop bits as well as higher clock rates. These two combine effectively to establish a significantly higher information rate.

Eight channels of analog-to-digital conversion are available on (some) M6811 MCUs. These eight inputs are time-division multiplexed to a single 8-bit, successive-approximation A/D converter. There is no sample and hold (S/H) circuitry, so time-varying input signals can cause some errors in converting data.[18] External references for both the high and low endpoints are allowed to improve accuracy, but they must both be between V_{SS} and V_{DD} and have a range of at least 2.5 Vs. The conversion accuracy is ±1 LSB, including quantization error. Provided that the MCU clock rate is at least 750 kHz, each conversion is complete in 32 clock cycles, which equates to 16 μs at an MCU clock rate of 2 MHz. The result of the conversion is $00 for a minimum voltage level and $FF for a maximum level, with no indication of overflow in the event of signals larger than V_{RH}. There are actually 16 inputs to the A/D that can be specified, with eight being external from Port E, four for internal reference and test, and four channels reserved for future use. There are two single-channel conversion modes under control of the *SCAN* bit of the *A/D control/status register (ADCTL)*. Either one channel can be converted four successive times with the results being stored in the four *A/D result registers (ADR1—ADR4)* and the conversion sequence halted, or the conversion can continue, with the registers being filled in a round-robin fashion with the most recent data replacing earlier measurements. In multichannel operation, as specified by the *MULT* bit of the *ADCTL register*, the *SCAN* bit causes four channels to be converted, with each result being stored in data registers *ADR1* through *ADR4*, respectively. After the fourth conversion, the activity is halted until the *ADCTL* register is written to again. If the *SCAN bit* = 1, conversions continue after the fourth with each channel replacing its respective data in *ADR1—ADR4*. No interrupt is generated by the conversion process, and the end of conversion must be sensed by polling the *CCF bit* of the *ADCTL register*.

Similar to the counter/timer of the M6801, the M6811 has a single counter that runs continuously and cannot be set or reset except by an external reset. The counter is supplied with a clock from the MCU clock that is prescaled by software selectable values that must be programmed during the first 64 clock cycles and that cannot be changed until an external reset occurs. Given a nominal 2-MHz MCU clock, the counter can be programmed with a clock that will make it overflow every 32.77 msec at its fastest down to overflowing once every 524.3 ms at its slowest. The 16-bit counter can be read at any time by reading the *TCNT* register. The *timer interrupt flag register 2 (TFLG2)* can be programmed to generate an interrupt on each counter overflow. To determine the time of an external event without having to poll it, input capture registers (which are not affected by reset) can be used to latch the value of the free-running counter when a pin experiences a specified transition. This action is the same as the M6801, except that there are three timer input capture registers rather than the M6801's one. As with the M6801, the output compare registers can be programmed to generate an interrupt. The M6801 has one output compare register and the M6811 has five. These registers are not required to generate interrupts and their status can be read without an interrupt being generated. The output compare registers have the additional capability that they can be programmed to generate output levels on specified pins of Port A when a compare occurs. In addition to the input capture registers and the output compare registers, there is a *real-time interrupt (RTI)* that can interrupt the MCU at one of four software-selectable rates. At the nominal 2 MHz, an RTI can be generated at 4.1 ms, 8.19 ms, 16.38 ms, or 32.77 ms. Finally, there is an 8-bit-wide, read/write pulse accumulator that can be configured either to count the number of inputs on Port A, bit 7 (at a maximum rate of half the MCU clock rate) or the number of MCU clocks divided by 64 that occur while the same pin is activated. The mode of operation of the pulse accumulator and the edges that are used are programmed in the *pulse accumulator control register (PACTL)*.

Although they are not exactly peripherals, there are other on-chip capabilities that enhance the operation of the M6811 in real control problems, including a watchdog timer and clock monitor. The *NOCOP* bit of the nonvolatile *CONFIG* register has already been discussed. If this bit is clear, the watchdog timer, or *computer-operating-properly (COP)* system will be enabled. The COP functions as other watchdog timers in that a COP reset sequence must be executed periodically before the timer times out, or a reset is initiated that will restart the MCU. The maximum time between these COP reset sequences is set by 2 bits in the *OPTION* register and for the nominal 2-MHz MCU clock can be set at 16.384 ms, 65.536 ms, 262.14 ms, or 1.049 s. The other monitor function that can be enabled is the clock monitor. This is done by setting the *CME* bit in the *OPTION* register. If the MCU clock is missing for 5 to 100 µs a reset will occur. Particular consideration should be given if a slow clock is used or if the STOP instruction is executed since the clock monitor may time out. The reason for the clock monitor in addition to the watchdog timer is that the COP is activated by the MCU clock, and if it is absent, the COP may not detect the system malfunction.

MC68HC11 expansion capability.   The availability of an expanded multiplexed mode of operation has already been discussed. In addition, a peripheral chip is available, the MC68HC24, that allows the M6811 to operate in expanded mode while at the same time preserving the same I/O port structure as if it were operating in single-chip mode. This capability allows for easy development of software in expanded memory systems, such as proposed in the program development sequence of the previous chapters, followed by compression of the software to fit the space allotted.

MC68HC11 interrupts.   The M6811 has a strict hierarchy of interrupts that is complex because of the number of on-chip peripherals. Interrupts are vectored through dedicated locations in memory. At reset, all interrupts are masked including the two external interrupts, IRQ and $XIRQ$. The IRQ is always a maskable interrupt, but the $XIRQ$ can be made nonmaskable by clearing the X bit of the CCR by a TAP instruction. Once enabled, the $XIRQ$ interrupt is nonmaskable. Although there is a strict interrupt hierarchy, any one interrupt that is maskable by the I bit of the CCR can be selected to have the highest priority. This is done by disabling the I interrupts and writing to the low-order 4 bits of the $HPRIO$ register. The other bits of this register are latched at reset and are determined by the values on the MODA and MODB pins.

5.10   INTEL MCS-51 FAMILY

The first Intel 8-bit MCU was the MCS-48, which was introduced in 1976.[19] The availability of HMOS technology allowed for the development of a higher-performance MCU, resulting in the MCS-51 family of 8-bit MCUs. As seen in this discussion, there are architectural features of the MCS-51 that are distinctly different from the Motorola MCUs discussed previously.

5.10.1   Intel MCS-51 Family Members

The members of the MCS-51 family are listed in Table 5.7, which shows that on-chip ROM can vary from 0 to 16 kbytes and RAM from 128 to 256 bytes. The number of 8-bit I/O ports ranges from a minimum of four 8-bit ports to seven, for a total of 32 to 56 I/O pins. Either two or three 16-bit timers are also available on all models as well as a UART. One family member has an eight-channel A/D, and there are four models with on-chip DMA. All family members have a ROMless version, and most are available in an EPROM

TABLE 5.7 Characteristics of MCS-51 Family Members

Device	ROMless Version	EPROM Version	ROM Bytes	RAM Bytes	8-Bit I/O Ports	16-Bit Timer/ Counters	Programmable Counter Array (PCA)	UART	Serial Expansion Port (SEP)	Global Serial Channel (GSC)	DMA Channels	A/D Channels	Interrupt Sources/ Vectors	Power Down and Idle Modes
8051	8031	—	4K	128	4	2		✓					6/5	
8051AH	8031AH	8751H 8751BH	4K	128	4	2		✓					6/5	
8052AH	8032AH	8752BH	8K	256	4	3		✓					8/6	
80C51BH	80C31BH	87C51	4K	128	4	2		✓					6/5	✓
83C51FA	80C51FA	87C51FA	8K	256	4	3	✓	✓					14/7	✓
83C51FB	80C51FA	87C51FB	16K	256	4	3	✓	✓					14/7	✓
83C51GA	80C51GA	87C51GA	4K	128	4	2		✓	✓			8	8/7	✓
83C152JA	80C152JA	—	8K	256	5	2		✓		✓	2		19/11	✓
—	80C152JB	—	—	256	7	2		✓		✓	2		19/11	✓
83C152JC	80C152JC	—	8K	256	5	2		✓		✓	2		19/11	✓
—	80C152JD	—	—	256	7	2		✓		✓	2		19/11	✓
83C451	80C451	—	4K	128	7	2		✓					6/5	✓
83C452	80C452	87C452P	8K	256	5	2		✓					9/8	✓

SOURCE: Courtesy Motorola, Inc.

version. A low-power mode of operation is available on most models. General features of the MCS-51 family are as follows:[19]

1. Single voltage power supply and HMOS technology

2. Four register banks

3. Expandable to utilize complete address space

4. Direct bit manipulation

5. Binary or decimal arithmetic

6. Hardware multiply and divide

7. Single 40-pin package with same pinout and electrical characteristics.

5.10.2 MCS-51 Architecture

Unlike the Motorola MCUs, the MCS-51 is decidedly Harvard architecture, as are the TI 4-bit MCUs. The separate *RAM address register (RAR)* and *program address register (PCH/PCL)* do not share a common bus, as shown in Fig. 5.23. Not only are the busses distinct, but the addressing also allows for 64 kbytes of program and 64 kbytes of RAM when operating in expanded mode. Computationally, this is an 8-bit, accumulator-based machine with a limited number of computational registers available to the programmer. In addition to the accumulator, an 8-bit *B-register* is available as a general-purpose register when it is not being used for the hardware multiply/divide operations. The 8-bit *program status word (PSW)* implements the following:

0. *P*: Even parity of *Accumulator*

1. - -: User-defined flag

2. *OV*: *Overflow*

3. *RS0*: *Register-bank-select* bit

4. *RS1*: *Register-bank-select* bit

5. *FO*: User general-purpose flag

6. *AC*: Auxiliary carry for BCD operations

7. *CY*: Carry.

The register-bank-select bits refer to four 8-byte regions at the bottom of memory. Certain instructions can access these registers in RAM directly with 1-byte instructions, since 3 bits of the instruction specify the desired register. While the instruction specifies which register is being referred to within a bank, the ambiguity in specifying the bank is resolved by bits *RS0/1* of the *PSW*. These bits must be set to the appropriate bank before execution of the instruction. This method of addressing saves one byte per instruction

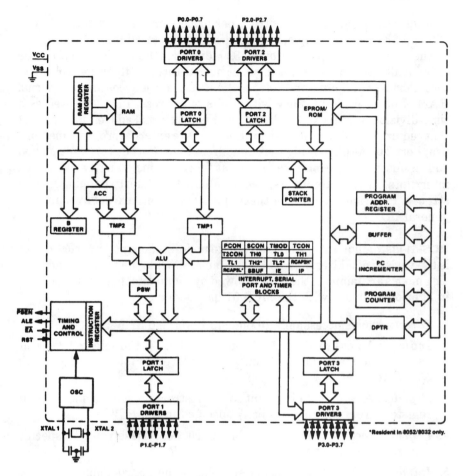

Figure 5.23 Block Diagram of Intel MCS-51 family of MCUs. *(Courtesy of Intel Corporation)*

less the overhead of loading the *PSW* with the bank number. An 8-bit stack pointer can be initialized to point to any location in on-chip RAM. The *data pointer register (DPTR)* consists of two concatenated 8-bit registers *(DPH/DPL)* that can be manipulated individually or as a 16-bit pair. All of these registers are implemented in the *Special Function Register (SFR)* address space.

A familiarity with Intel microprocessors causes one to expect separate I/O control lines and separate I/O address space. Unlike their microprocessors and like the Motorola MCUs, the MCS-51 uses memory-mapped I/O through a set of special function registers that are implemented in the address space immediately above the 128 bytes of RAM. All access to the four I/O ports, the CPU registers, interrupt-control registers, the timer/

counter, UART, and power control is performed through registers between $80 and $FF. For example, the *Accumulator* is at address $E0.

The four 8-bit I/O ports provide 32 bidirectional data lines that can be individually programmed for input or output through their control registers. Certain ports have multiple functions such as the accessing of external RAM. For external memory with 8 bits of address, Port 0 is used as a time-division multiplexed address/data bus. External addressing of RAM that requires 16-bit addresses and all accesses to external program memory use, Port 2 to output the high-order address byte. Port 0 is for multiplexed address/data. Port 3 pins have special functions including two external interrupts, two counter inputs, two serial data lines, and two timing control strobes. Pins that are used for these special functions are not available for general I/O.

MCS-51 instruction set.   Fundamental to understanding an instruction set is the variety of addressing modes available.[20] While the following addressing modes are familiar, they are followed by comments that interpret them in light of their use with the MCS-51 instruction set.

1. *Direct*: Allows access only to on-chip RAM and *SFRs*.

2. *Indirect*: Access to all RAM through R_0 or R_1 of the selected register bank or the stack pointer. Sixteen-bit addresses must use the 16-bit *DPTR* register.

3. *Register*: A 3-bit code inherent in the instruction selects one of eight registers from the register bank specified in the *PSW*.

4. *Register Specific*: Inherent in the instruction, these refer to specific registers such as the accumulator or *DPTR*.

5. *Immediate*: The data immediately follows the instruction.

6. *Indexed*: Only program memory can be accessed with indexed addressing and it is read-only. This mode is intended for lookup tables in ROM. Either the *DPTR* or *PC* can be used.

The complete list of MCS-51 instructions is shown in Table 5.8. As a contrast to the orthogonality of addressing the Motorola MCUs, Table 5.9 shows some of the logical instructions that cannot operate on all locations. In fairness, though, most addressing modes are available in most instructions. It is interesting to note that all on-chip RAM and *SFR* can be moved to/from the stack without going through the *Accumulator*. All addressing modes are available for moving data directly from one memory location to another, again without going through the *Accumulator*. There is extensive bit set/clear capability as well as *PC*-relative branching (jumping) based on individual bits in both 128 bits of RAM as well as 128 bits of the *SFRs*. Not only can bits be set or cleared, but they can also be operated on with the

TABLE 5.8   Instruction Set of Intel MCS-51 Family of MCUs

ARITHMETIC OPERATIONS

Mnemonic		Description	Byte	Cyc
ADD	A,Rn	Add register to Accumulator	1	1
ADD	A,direct	Add direct byte to Accumulator	2	1
ADD	A,@Ri	Add indirect RAM to Accumulator	1	1
ADD	A,#data	Add immediate data to Accumulator	2	1
ADDC	A,Rn	Add register to Accumulator with Carry	1	1
ADDC	A,direct	Add direct byte to A with Carry flag	2	1
ADDC	A,@Ri	Add indirect RAM to A with Carry flag	1	1
ADDC	A,#data	Add immediate data to A with Carry flag	2	1
SUBB	A,Rn	Subtract register from A with Borrow	1	1
SUBB	A,direct	Subtract direct byte from A with Borrow	2	1
SUBB	A,@Ri	Subtract indirect RAM from A w. Borrow	1	1
SUBB	A,#data	Subtract immed. data from A w. Borrow	2	1
INC	A	Increment Accumulator	1	1
INC	Rn	Increment register	1	1
INC	direct	Increment direct byte	2	1
INC	@Ri	Increment indirect RAM	1	1
DEC	A	Decrement Accumulator	1	1
DEC	Rn	Decrement register	1	1
DEC	direct	Decrement direct byte	2	1
DEC	@Ri	Decrement indirect RAM	1	1
INC	DPTR	Increment Data Pointer	1	2
MUL	AB	Multiply A & B	1	4
DIV	AB	Divide A by B	1	4
DA	A	Decimal Adjust Accumulator	1	1

LOGICAL OPERATIONS

Mnemonic		Destination	Byte	Cyc
ANL	A,Rn	AND register to Accumulator	1	1
ANL	A,direct	AND direct byte to Accumulator	2	1
ANL	A,@Ri	AND indirect RAM to Accumulator	1	1
ANL	A,#data	AND immediate data to Accumulator	2	1
ANL	direct,A	AND Accumulator to direct byte	2	1
ANL	direct,#data	AND immediate data to direct byte	3	2
ORL	A,Rn	OR register to Accumulator	1	1
ORL	A,direct	OR direct byte to Accumulator	2	1
ORL	A,@Ri	OR indirect RAM to Accumulator	1	1
ORL	A,#data	OR immediate data to Accumulator	2	1
ORL	direct,A	OR Accumulator to direct byte	2	1
ORL	direct,#data	OR immediate data to direct byte	3	2
XRL	A,Rn	Exclusive-OR register to Accumulator	1	1
XRL	A,direct	Exclusive-OR direct byte to Accumulator	2	1
XRL	A,@Ri	Exclusive-OR indirect RAM to A	1	1
XRL	A,#data	Exclusive-OR immediate data to A	2	1
XRL	direct,A	Exclusive-OR Accumulator to direct byte	2	1
XRL	direct,#data	Exclusive-OR immediate data to direct	3	2
CLR	A	Clear Accumulator	1	1
CPL	A	Complement Accumulator	1	1
RL	A	Rotate Accumulator Left	1	1
RLC	A	Rotate A Left through the Carry flag	1	1
RR	A	Rotate Accumulator Right	1	1
RRC	A	Rotate A Right through Carry flag	1	1
SWAP	A	Swap nibbles within the Accumulator	1	1

DATA TRANSFER

Mnemonic		Description	Byte	Cyc
MOV	A,Rn	Move register to Accumulator	1	1
MOV	A,direct	Move direct byte to Accumulator	2	1
MOV	A,@Ri	Move indirect RAM to Accumulator	1	1
MOV	A,#data	Move immediate data to Accumulator	2	1
MOV	Rn,A	Move Accumulator to register	1	1
MOV	Rn,direct	Move direct byte to register	2	2
MOV	Rn,#data	Move immediate data to register	2	1
MOV	direct,A	Move Accumulator to direct byte	2	1
MOV	direct,Rn	Move register to direct byte	2	2
MOV	direct,direct	Move direct byte to direct	3	2
MOV	direct,@Ri	Move indirect RAM to direct byte	2	2
MOV	direct,#data	Move immediate data to direct byte	3	2
MOV	@Ri,A	Move Accumulator to indirect RAM	1	1
MOV	@Ri,direct	Move direct byte to indirect RAM	2	2
MOV	@Ri,#data	Move immediate data to indirect RAM	2	1
MOV	DPTR,#data16	Load Data Pointer with a 16-bit constant	3	2

SOURCE: Courtesy Intel.

TABLE 5.8 Instruction Set of Intel MCS-51 Family of MCUs (Continued)

DATA TRANSFER (cont.)

Mnemonic		Description	Byte	Cyc
MOVC	A,@A+DPTR	Move Code byte relative to DPTR to A	1	2
MOVC	A,@A+PC	Move Code byte relative to PC to A	1	2
MOVX	A,@Ri	Move External RAM (8-bit addr) to A	1	2
MOVX	A,@DPTR	Move External RAM (16-bit addr) to A	1	2
MOVX	@Ri,A	Move A to External RAM (8-bit addr)	1	2
MOVX	@DPTR,A	Move A to External RAM (16-bit addr)	1	2
PUSH	direct	Push direct byte onto stack	2	2
POP	direct	Pop direct byte from stack	2	2
XCH	A,Rn	Exchange register with Accumulator	1	1
XCH	A,direct	Exchange direct byte with Accumulator	2	1
XCH	A,@Ri	Exchange indirect RAM with A	1	1
XCHD	A,@Ri	Exchange low-order Digit ind. RAM w A	1	1

BOOLEAN VARIABLE MANIPULATION

Mnemonic		Description	Byte	Cyc
CLR	C	Clear Carry flag	1	1
CLR	bit	Clear direct bit	2	1
SETB	C	Set Carry flag	1	1
SETB	bit	Set direct Bit	2	1
CPL	C	Complement Carry flag	1	1
CPL	bit	Complement direct bit	2	1
ANL	C,bit	AND direct bit to Carry flag	2	2
ANL	C, bit	AND complement of direct bit to Carry	2	2
ORL	C,bit	OR direct bit to Carry flag	2	2
ORL	C, bit	OR complement of direct bit to Carry	2	2
MOV	C,bit	Move direct bit to Carry flag	2	1
MOV	bit,C	Move Carry flag to direct bit	2	2

PROGRAM AND MACHINE CONTROL

Mnemonic		Description	Byte	Cyc
ACALL	addr11	Absolute Subroutine Call	2	2
LCALL	addr16	Long Subroutine Call	3	2
RET		Return from subroutine	1	2
RETI		Return from interrupt	1	2
AJMP	addr11	Absolute Jump	2	2
LJMP	addr16	Long Jump	3	2
SJMP	rel	Short Jump (relative addr)	2	2
JMP	@A+DPTR	Jump indirect relative to the DPTR	1	2
JZ	rel	Jump if Accumulator is Zero	2	2
JNZ	rel	Jump if Accumulator is Not Zero	2	2
JC	rel	Jump if Carry flag is set	2	2
JNC	rel	Jump if No Carry flag	2	2
JB	bit,rel	Jump if direct Bit set	3	2
JNB	bit,rel	Jump if direct Bit Not set	3	2
JBC	bit,rel	Jump if direct Bit is set & Clear bit	3	2
CJNE	A,direct,rel	Compare direct to A & Jump if Not Equal	3	2
CJNE	A,#data,rel	Comp. immed. to A & Jump if Not Equal	3	2
CJNE	Rn,#data,rel	Comp. immed. to reg. & Jump if Not Equal	3	2
CJNE	@Ri,#data,rel	Comp. immed. to ind. & Jump if Not Equal	3	2
DJNZ	Rn,rel	Decrement register & Jump if Not Zero	2	2
DJNZ	direct,rel	Decrement direct & Jump if Not Zero	3	2
NOP		No operation	1	1

Notes on data addressing modes:
Rn   Working register R0-R7
direct   128 internal RAM locations, any I O port, control or status register
@Ri   Indirect internal RAM location addressed by register R0 or R1
#data   8-bit constant included in instruction
#data16   16-bit constant included as bytes 2 & 3 of instruction
bit   128 software flags, any I O pin, control or status bit

Notes on program addressing modes:
addr16   Destination address for LCALL & LJMP may be anywhere within the 64-Kilobyte program memory address space.
addr11   Destination address for ACALL & AJMP will be within the same 2-Kilobyte page of program memory as the first byte of the following instruction.
rel   SJMP and all conditional jumps include an 8-bit offset byte. Range is +127 -128 bytes relative to first byte of the following instruction.

TABLE 5.9   MCS-51 Logical Instructions and Their Associated Addressing Modes

Mnemonic	Operation	Addressing Modes				Execution Time (μs)
		Dir	Ind	Reg	Imm	
ANL A, <byte>	A = A .AND. <byte>	X	X	X	X	1
ANL <byte>,A	<byte> = <byte> .AND. A	X				1
ANL <byte>,#data	<byte> = <byte> .AND. #data	X				2
ORL A, <byte>	A = A .OR. <byte>	X	X	X	X	1
ORL <byte>,A	<byte> = <byte> .OR. A	X				1
ORL <byte>,#data	<byte> = <byte> .OR. #data	X				2
XRL A, <byte>	A = A .XOR. <byte>	X	X	X	X	1
XRL <byte>,A	<byte> = <byte> .XOR. A	X				1
XRL <byte>,#data	<byte> = <byte> .XOR. #data	X				2
CRL A	A = 00H	Accumulator only				1
CPL A	A = .NOT. A	Accumulator only				1
RL A	Rotate ACC Left 1 bit	Accumulator only				1
RLC A	Rotate Left through Carry	Accumulator only				1
RR A	Rotate ACC Right 1 bit	Accumulator only				1
RRC A	Rotate Right through Carry	Accumulator only				1
SWAP A	Swap Nibbles in A	Accumulator only				1

SOURCE: Courtesy Intel.

logical OR and AND instructions, a feature not found in the Motorola families.

MCS-51 on-chip peripherals.   In contrast to Motorola's concept of a single 16-bit timer with multiple compare and capture registers, the MCS-51 has two 16-bit timers that can be operated in a variety of modes. These timers can also be used as counters that increment when there is a 1-0 transition of the corresponding external input pin, T_0 or T_1 (plus T_3 in the 8052). Timers 0 and 1 have additional operating modes:

1. *Mode 0*: The timer acts as an 8-bit counter with a divide-by-32 prescaler. The timer can counter either the internal clock or an external input with the clock being gated to the timer by a software-setable bit or an external pin. An interrupt is generated on overflow.

2. *Mode 1*: Same as Mode 0 but with a 16-bit counter.

3. *Mode 2*: The timer is set for 8-bits with automatic reload of an initial count from a preloaded register.

4. *Mode 3*: Timer 1 holds its count. Timer 0 operates as two separate 8-bit counters with different clock input logic. Both 8-bit counters will generate an interrupt.

Complete timer control is embodied in two timer/counter registers that are part of the *SFR* registers.

A full-duplex asynchronous serial interface is implemented on chip. The receiver is double buffered so that it can receive a second character while the first one is being held in an intermediate register until the software fetches it. This double buffering reduces the time criticality of servicing a serial interface and reduces the chances of an overrun error. The Motorola SCIs are not double buffered. As with the timer, there are four modes of operation for the UART:

1. *Mode 0*: Half-duplex operation with the serial data being received and transmitted through the RXD input. The shift clock is output through the TXD line. The baud rate is fixed at one-twelfth of the oscillator frequency. Eight bits only are transmitted, thus emulating the SCI with its lack of start/stop bits.

2. *Mode 1*: Normal full-duplex NRZ operation with 10 bits being transmitted. The baud rate is variable.

3. *Mode 2*: Normal full-duplex NRZ operation with 11 bits, including a ninth programmable data bit. The baud rate is programmable at either 1/32 or 1/64 the MCU clock rate.

4. *Mode 3*: Same as Mode 2 but with a variable baud rate.

Variable baud rates are determined by the timer 1 overflow rate. In the 8052, the third timer can be used to establish a different transmit baud rate than receive baud rate. On some models, an additional half-duplex synchronous serial interface is provided. There is no wake-up mode for network considerations.

One family member, the 8xC51GA, contains an A/D converter. It is an 8-channel, 8-bit converter that includes a sample/hold and conversion speed control circuitry. The sample/hold is important for the capture and correct conversion of rapidly changing analog signals as well as the minimization of external noise. The sample/hold circuit capacitor can be taken out of the circuit under program control. The nominal conversion time is 22 µs with an accuracy of ±1 LSB. Four of the inputs are tied to input port 1, and these will be lost as general-purpose inputs. Unlike the polling required by the M6811, the completion of an A/D conversion causes the *A/D maskable interrupt flag (AIF)* to be set. Analog channels are individually selected and started, and there is no automatic or cyclic mode for the conversion of multiple analog channels.

Additional MCS-51 capabilities.   Just as in the Motorola M6811, the MCS-51 implements both a watchdog timer and an oscillator failure detector that will reset the machine in the event of a detectable malfunction. Both program memory and RAM can be extended at the expense of one or more general purpose ports. Additional I/O can be easily implemented in the expanded mode through the use of memory-mapped I/O. A powerdown mode is under software control, and the only restart mechanism is through an external hardware reset. There are both internal and external interrupts implemented in the MCS-51. Two are external and the rest internal. All interrupts are maskable, each individually and all simultaneously under the control of a single bit. Internal interrupts are generated by the timer, the serial interface and the A/D converter. Each interrupt can be assigned one of two priorities: high or low. High-priority interrupts will interrupt low-priority ones, but among a single priority level, the order of servicing is determined by a fixed polling sequence.

5.11 TEXAS INSTRUMENTS TMS370 FAMILY

By now one should have an intuitive understanding of the basic capabilities of 8-bit MCUs, and the TMS370 family from Texas Instruments fits well within those capabilities. This section does not detail the TMS370 except to point out notable exceptions from previously discussed capabilities or unique methods of implementing peripherals. The TMS370 family consists of three major models: TMS370Cx1x, TMS370Cx3x, and TMS370Cx5x.[12] The particular characteristics of the three families are listed in Table 5.10, and particular details of all of the models in Table 5.11. In general, on-chip ROM varies from 4 to 16 kbytes, RAM from 128 to 512 bytes, and nonvolatile EEPROM data memory from 256 to 512 bytes.

5.11.1 TMS370 Architecture

Architecturally, the TMS370 is a mixture of concepts implemented in other machines. The block diagram of Fig. 5.24 shows the programmers model and, Fig. 5.25 is the block diagram of the most capable member of the TMS370 family, the TMS370Cx5x. The register structure consists of only three dedicated registers, the 16-bit PC; the 8-bit, incrementing SP, and the *Status register*. All other registers (128 to 256) exist in the base page of RAM, with access to these registers accomplished in a single instruction cycle, allowing them to be considered registers rather than RAM. Other parts of RAM require a two-cycle access. Two of the registers, the *A-register* and the *B-register*, are the first two addresses in RAM memory and have special properties in that they can be accessed inherently by some instructions. This makes them the equivalent of two accumulators. The status register includes two interrupt-enable bits for the two-level, prioritized, software-controllable, vectored interrupt structure. Since there are many registers, there is no need to save them, and the only registers that are stacked are the PC and the *Status register* upon interrupt. Since the stack pointer increments when data is pushed onto the stack, it can roll over from

TABLE 5.10 TMS370 Family Feature Summary

Feature	TMS370Cx1x	TMS370Cx3x	TMS370Cx5x
Oscillator Frequency Range	2 - 20 MHz	2 - 20 MHz	2 - 20 MHz
Voltage	5 V ±10%	5 V ±10%	5 V ±10%
Operating temperature	-40°C to 85°C	-40°C to 85°C	-40°C to 85°C
Program Memory	4 Kbyte ROM / EEPROM	8 Kbyte ROM / EPROM	0–16 Kbyte ROM / EPROM / EEPROM
Static RAM	128 bytes	256 bytes	256 / 512 bytes
Data EEPROM	0 / 256 bytes	0 / 256 bytes	0 / 256 / 512 bytes
Modules			
SPI	Yes	–	Yes
Timer 1	Yes	–	Yes
Watchdog Timer	Yes	See Note.	Yes
Timer 2	–	–	Yes
PACT	–	Yes	–
SCI	–	See Note.	Yes
A/D Converter	–	Yes	Yes
Maximum Digital I/O			
Bidirectional	21	14	46
Input Only	1	13	9
Output Only	0	9	0
External Memory Bus Expansion	No	No	Yes
Interrupts/Reset			
External	4	4	4
Vectors total	6	23	10
Sources total	13	25	23
Pin Count	28	44	68
Packaging	DIP / PLCC	PLCC / CLCC	PLCC / CLCC

Note: This function is included in the PACT module.

SOURCE: Reprinted by permission of Texas Instruments.

TABLE 5.11 Detail Characteristics of TMS370 family members

Device	Program Memory (Bytes)			Data Memory (Bytes)		Off-Chip Memory Expansion (Bytes)	Serial Interface Modules ◄	Timer Modules ◊	A/D Channels	I/O Pins	No. of Pins /Package
	ROM	EPROM	EEPROM	EEPROM	RAM						
ROM											
370C010	4K	—	—	256	128	None	SPI	T1	—	22	28 DIP/PLCC
370C050	4K	—	—	256	256	112K	SPI/SCI	T1/T2	8	55	68 PLCC
370C032	8K	—	—	256	256	None	PACT-SCI	PACT	8	36	44 PLCC
370C052	8K	—	—	256	256	112K	SPI/SCI	T1/T2	8	55	68 PLCC
370C056	16K	—	—	512	512	112K	SPI/SCI	T1/T2	8	55	68 PLCC
370C310	4K	—	—	—	128	None	SPI	T1	—	22	28 DIP/PLCC
370C350	4K	—	—	—	256	112K	SPI/SCI	T1/T2	8	55	68 PLCC
370C332	8K	—	—	—	256	None	PACT-SCI	PACT	8	36	44 PLCC
370C352	8K	—	—	—	256	112K	SPI/SCI	T1/T2	8	55	68 PLCC
370C356	16K	—	—	—	512	112K	SPI/SCI	T1/T2	8	55	68 PLCC
ROM-less§											
370C150	—	—	—	—	256	112K	SPI/SCI	T1/T2	8	55	68 PLCC
370C250	—	—	—	256	256	112K	SPI/SCI	T1/T2	8	55	68 PLCC
370C156	—	—	—	—	512	112K	SPI/SCI	T1/T2	8	55	68 PLCC
370C256	—	—	—	512	512	112K	SPI/SCI	T1/T2	8	55	68 PLCC
FFE											
370C810	—	—	4K	256	128	None	SPI	T1	—	22	28 DIP/PLCC
370C850	—	—	4K	256	256	112K	SPI/SCI	T1/T2	8	55	68 PLCC
370C732	—	8K	—	256	256	None	PACT-SCI	PACT	8	36	44 CLCC¶
370C756	—	16K	—	512	512	112K	SPI/SCI	T1/T2	8	55	68 CLCC¶

◊ Timer 1 module has a Watchdog timer included which can be programmed to a general purpose 16 bit timer.
PACT module has a Watchdog timer included.

◄ PACT module has a mini SCI port.

§ In ROM-less (microprocessor) mode all Address, Data, and Control lines are fixed as their function.

¶ For OTP (PLCC) availability information, contact local TI sales office or distributor.

SOURCE: Reprinted by permission of Texas Instruments.

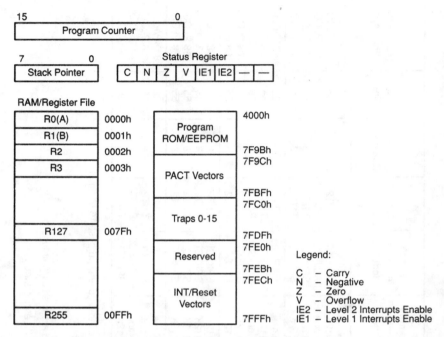

Figure 5.24 Programmers model of the register structure of the Texas Instrument TMS370 family of MCUs. *(Reprinted by permission of Texas Instruments)*

0FFH to 00H if care is not taken to monitor stack size in software. This is important since there is no hardware indication of this rollover. If the 128 byte RAM is selected, the *SP* increments to 07FH and then stops, with no rollover or warning. Additional data pushed on the stack is put into nonexistent memory.

As with the other machines, peripherals are memory mapped, with all registers accessed through a peripheral file of 256 addresses. Some of the operating characteristics of peripherals can be initialized by software at startup while operating in a privileged mode, and then the configuration can be fixed by setting a privilege bit. If external memory or peripherals are added, a wait pin can be used to halt the computer until the peripheral or memory responds. This allows for a reduction of power consumption while operating with slow devices or for the adding of slow memory without having to slow down the inherent operation of CPU.

TMS370 instruction set. Fourteen addressing modes are available in the TMS370. In addition to implied, register, immediate, direct, indexed, and *PC*-relative, the TMS370 implements the following:

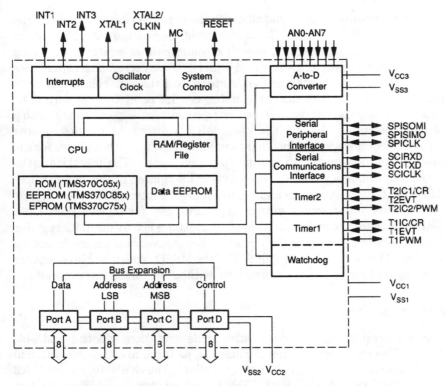

Figure 5.25 Block Diagram of the TI TMS370Cx5x MCU. *(Reprinted by permission of Texas Instruments)*

1. Peripheral
2. Stack-pointer relative
3. Indirect (both absolute and relative)
4. Offset indirect (both absolute and relative).

Stack-pointer-relative addressing is particularly useful for passing parameters from a calling to a called function. This method of parameter passing on the stack was first shown in the examples in Chap. 4, in which assembly language and HLL were linked using the INTROL compiler. With this instruction, the *SP* itself does not need to be manipulated, but all of the operands that are passed can be accessed by adding offsets to it. Binary-coded decimal (BCD) arithmetic is available as well as integer multiply and divide. Nibble swapping is supported in the *A*- and *B-Registers*. Software interrupts are supported by the TRAP instruction and extensive conditional jumping on the normal conditions as well as single bits of the registers.

TMS370 on-chip peripherals and other capabilities.   Peripherals available on chip in the TMS370 family include a serial peripheral interface, serial communications interface, a complex counter/timer system, watchdog timer, clock monitor, and an eight channel A/D converter. The capabilities of the timer/counter system are similar to those of the Motorola M6811, with the exception that the master counter can be reset by software. The successive approximation A/D converter consists of eight channels with 8 bits of resolution. Unlike the other MCUs discussed previously, the TMS370 contains a single internal sample/hold circuitry that allows for the capture and subsequent conversion of dynamic signals. The limitation on the maximum frequency that can be converted is related to the source impedance. The TMS370 requirement is to enable the internal sampler for 1 μs for each kilohm of source output impedance but for no less than 1 μs. Although there are eight channels of input, there is only one data register that necessitates reading the result of one conversion before initiating another. This is in contrast to the A/D of the M6811, which is able to capture as many as four channels sequentially without software intervention.

5.12   SUMMARY

This chapter presented representative implementations of both 4- and 8-bit MCUs. The coverage was not exhaustive, as there are also the National Semiconductor COP800 (conceptually similar to the COP400), the NEC K2 series uD782xx and NEC UPD-7514, as well as other 4- and 8-bit MCUs. Through the examples, which included both Harvard and Von Neumann architectures, orthogonal and nonorthogonal instruction sets, multieffect instructions and single-effect instructions, mask programmed options and software-configurable MCUs, the basic implementations of architectural alternatives were shown. No single criterion for MCU selection can be given since there are many variables. The ultimate luxury of choice is in the design of MCU systems for large-volume production. In these cases, the cost of nonrecurring engineering can be offset more easily by the profit per system. For small production systems or one-off designs, more capable MCUs that are software configurable and locally programmable, with HLL support as well as an adequate development system, are more cost effective. Even though single-chip costs are higher, the nonrecurring costs will be minimized. A detailed analysis for either alternative can only be completed with accurate estimates of expected volume, production costs, and NRE costs.

REFERENCES

1. J. Murakami, "Development of Microcontroller: F²MC," *Fujitsu Scientific and Technical Journal*, 24(4):328–34, winter 1988.
2. C. Moller, "Texas Instrument Microcomputers," in *Single-chip Microcomputers*, P.F. Lister, ed. New York: McGraw-Hill, 1984.

3. "TMS1000 Family Microcomputer," Texas Instruments, document numbers MP056, June 1982.

4. "Microcomputer Products 1987 Data Book," vol. 1, NEC Electronics Inc., Mountain View, CA, NECEL-000175, 1986.

5. R. M. Hohol, "The COPS 400 Microcontrollers from National Semiconductor," in *Single-chip Microcomputers*, P. F. Lister, ed. New York: McGraw-Hill, 1984.

6. "Microcontroller Databook," National Semiconductor, Santa Clara, CA, 1989.

7. L. A. Distaso, "A ROM Efficient Instruction Set," *WESCON '82 Conference Record*, Anaheim, CA, session 5A/1, pp. 1–5, 14–16, September 1982.

8. J. Brodie, "A Comparative Guide to Low-power 4-bit Microcontrollers," *New Electronics*, pp. 34–36, August 13, 1985.

9. G. J. Livey, "The Motorola M6801 and M6805 Families," *Single-chip Microcomputers*, P. F. Lister, ed. New York: McGraw-Hill, 1984.

10. 8-bit Single-Chip Microcomputer Data Book, Hitachi America, Ltd., Semiconductor and IC Division, San Jose, CA, publication #U71, July 1985.

11. T. Dollhoff, "P Software: How to Optimize Timing & Memory Usage," *Digital Design*, pp. 56–70, November 1976.

12. T. Henry, "68705 Microcontroller," *Radio-Electronics*, 60(9):82–86, September, 1989.

13. *M68HC11EVB Evaluation Board User's Manual*. Motorola Semiconductor Technical Data Publication #M68HC11EVB/D1, Phoenix, AZ, 1st ed., September 1986.

14. *HCMOS Single-Chip Microcontrollers*. Motorola Semiconductor Technical Data Publication #ADI1207R2, Phoenix, AZ, 1988.

15. G. J. Lipovski, *Single- and Multiple-Chip Microcomputer Interfacing*. Englewood Cliffs, NJ: Prentice-Hall, 1988.

16. J. B. Peatman, *Design with Microcontrollers*. New York: McGraw-Hill, 1988.

17. *M68HC11 HCMOS Single-Chip Microcomputer Programmer's Reference Manual*. Motorola Semiconductor Technical Data Publication #M68HC11PM/AD, Phoenix, AZ, 1st ed., September 1985.

18. R. Wilson, "Microcontrollers with On-chip A-Ds Raise Speed, Accuracy Issues," *Computer Design*, pp. 26–29, August 15, 1988.

19. J. Wharton, "An Introduction to the Intel MCS-51 Single-chip Microcomputer Family," AP-69, in *Embedded Control Applications*, Intel Publication #270645-001, Santa Clara, CA, 1989.

20. *8-Bit Embedded Controller Handbook*. Intel publication #270645-001, Santa Clara, CA, 1989.

21. *TMS370 Family Data Manual*. Texas Instruments, #SPNS014A, 1990.

16-Bit Microcontrollers

Numerous embedded controller applications require a better precision and higher speed than the 8-bit microcontrollers, but not as precise and fast as the 32-bit units. The 16-bit microcontrollers were created to fill the need for intermediate performance at an acceptable cost. Two 16-bit embedded controller products are described in this chapter: Intel MCS-96 microcontroller family and Motorola MC68332.

INTEL MCS-96 MICROCONTROLLER FAMILY

Intel created a multichip, versatile, 16-bit microcontroller family denoted as MCS-96.[1,3] The top product of this family as of 1991 is the 80C196KC.[2,3] For the sake of brevity, it is referred to as 196KC in this chapter. The 196KC is a CHMOS technology, 16-MHz, 68-pin (or optionally 80-pin) chip. It is about 33 percent faster than its predecessor in the MCS-96 family, the 12 MHz 80C196KB. Other differences between the two are discussed later in this chapter. All of the MCS-96 processors share a common instruction set and architecture. As stated by Intel,[2] the MCS-96 family implements a register-to-register architecture. However, the MCS-96 CPU registers take up locations in the on-chip main memory space. The MCS-96 implements a fast, on-chip, memory-to-memory architecture. In assembly language, memory locations designated as registers can be represented symbolically by register names such as AX, BX, CX, and others. The names used are the same as those in the Intel 80x86 family.[4] Typical applications of the MCS-96 family embedded processors are closed-loop control and midrange digital signal processing. MCS-96 processors are also used in modems, motor controls, printers, engine controls, photocopiers, antilock brakes, disk drives, and medical instrumentation.

A block diagram of the 196KC is shown in Fig. 6.1.[2] The major components of the 196KC CPU are the register file and the register arithmetic logic unit (RALU). The register file is part of the on-chip portion of the main memory. Communication with the circuitry outside of the chip is done through either the special function registers (SFRs) or the memory controller (Fig. 6.1). The 196KC CPU is 16 bit wide and connects to the interrupt controller and the memory controller by a 16-bit bus. In addition, there is an 8-bit bus that transfers instruction bytes from the memory controller to the CPU. An extension of the 16-bit bus connects the CPU to the peripheral devices. The CPU implements microcoded control.

A more detailed block diagram of the RALU and memory controller is shown in Fig. 6.2.[2] The RALU contains a 17-bit ALU (16 + sign extension), the program status word (PSW), the program counter (PC), a loop counter, and 3 temporary registers. All of the registers are 16, or 17-bits (16 + sign extension) wide. Some of the registers have the ability to perform simple operations (such as increment) to reduce the load on the ALU. The PC has a separate incrementer. However, other PC modifications are handled by the ALU. Two of the temporary registers have their own shift logic. They are

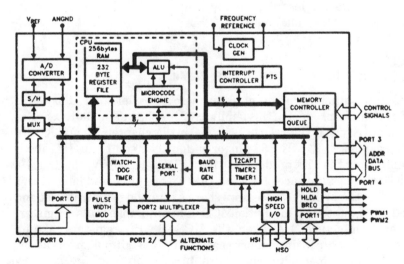

Figure 6.1   80C196KC block diagram. *(Courtesy Intel Corporation)*

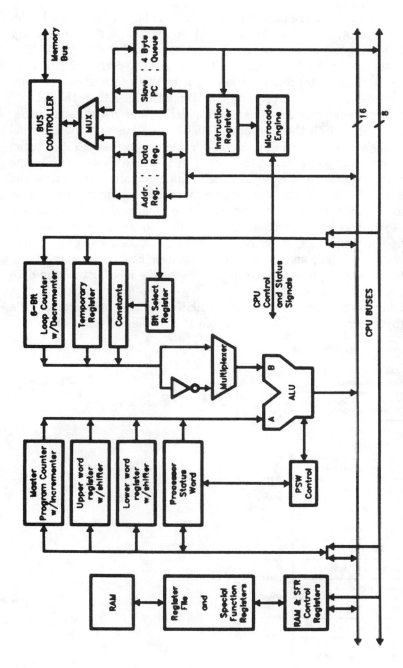

Figure 6.2 RALU and memory controller block diagram. *(Courtesy Intel Corporation)*

267

used for operations that require logical shifts, including multiply and divide. Repetitive shifts are counted by the 6-bit loop counter. A third temporary register stores the second operand of two operand instructions. This includes the multiplier during multiplications and the divisor during divisions. To perform subtractions, the output of this register can be complemented before being placed into the B input (Fig. 6.2) of the ALU.

The memory controller contains a bus controller, a 4-byte prefetch queue, and a slave PC. Both the internal ROM/EPROM bus and the external memory bus are driven by the bus controller. Memory access requests to the bus controller can come from either the RALU or the queue, with queue accesses having priority. Requests from the queue are always for data at the address in the slave PC. By having program fetches from memory referenced to the slave PC, the processor saves time as addresses seldom have to be sent to the memory controller. If the address sequence changes because of a jump, interrupt, call, or return, the slave PC is loaded with a new value, the queue is flushed, and processing continues. Execution speed is increased by using a queue since it usually keeps the next instruction byte available.

The addressable memory space on the 196KC (its main memory) consists of 64 kbytes, most of which is available to the user for code and data.

EXTERNAL MEMORY OR I/O	0FFFFH
	6000H
INTERNAL ROM/EPROM OR EXTERNAL MEMORY	
	2080H
RESERVED	
	205EH
PTS VECTORS	
	2040H
UPPER INTERRUPT VECTORS	
	2030H
ROM/EPROM SECURITY KEY	
	2020H
RESERVED	
	2019H
CHIP CONFIGURATION BYTE	
	2018H
RESERVED	
	2014H
LOWER INTERRUPT VECTORS	
	2000H
PORT 3 AND PORT 4	
	1FFEH
EXTERNAL MEMORY	
	200H
ADDITIONAL RAM	
	100H
REGISTER FILE AND EXTERNAL PROGRAM MEMORY	
	0

Figure 6.3   80C196KC memory map.
(Courtesy Intel Corporation)

Locations that have special purposes are 0000H (hexadecimal) through 01FFH and 1FFEH through 2080H. All other locations can be used for either code or data storage or for memory mapped peripherals. A memory map is shown in Fig. 6.3.[2] The first 512 locations 00H through 1FFH contain the register file, special function registers (SFRs), and 256 bytes of additional RAM, which can also be used as registers. If an attempt to execute instructions from locations 000H through 1FFH is made, the instructions will be fetched from external memory. This section of external memory is reserved for use by Intel development tools.

The on-chip RAM from location 018H (24 decimal) to 0FFH (255 decimal) is the *register file*. It contains 232 bytes of RAM that can be accessed as bytes (8 bits), words (16 bits), or doublewords (32 bits). The extra 256 bytes of on-chip RAM from 100H to 1FFH can be accessed as registers by the RALU with vertical register windowing (discussed later in this chapter.) Locations 18H and 19H contain the stack pointer (SP). These are not SFRs and may be used as standard RAM if stack operations are not being performed. The SP must be initialized by the user program and can point anywhere in the unrestricted memory space. Stack operations cause it to build down, so the SP should be initialized to 2 bytes above the highest stack location. The stack must be word aligned.

Locations 00H through 17H (a total of 24 bytes) are the I/O control registers or SFRs. All of the peripheral devices on the 196KC (except Ports 3 and 4, which are shared with the data and address bus, as shown in Fig. 6.1) are controlled through the SFRs. There are four arrangements for SFR assignment, called *Hwindows*, as shown in Fig. 6.4.[2] Switching between windows is controlled by the window select register (WSR), to be discussed later in this chapter. Hwindow 0 is a superset of the SFR space on an earlier member of the MCS-96 family, the 8096BH[3] and identical to Hwindow 0 of 196KB.[3] It has 24 registers, some of which have different functions when read than when written. This is why Hwindow 0 was counted as two arrangements. Hwindow 1 contains the additional SFRs needed to support the added functionality of the 196KC, compared to earlier members of the MCS-96 family. These SFRs support the peripheral transaction server (PTS), the two new PWMs, Timer 2, and the new functions of the A/D converter. All SFRs are read/write in this window. In Hwindow 15, the operation of the SFRs is changed, so those that were read-only in Hwindow 0 are write-only and vice versa. The only major exception is that Timer 2 is read/write in Hwindow 0, and T2CAPTURE is read/write in Hwindow 15 (see Fig. 6.4). A brief description of the SFR registers is given in Fig. 6.5.[2]

The 87C196KC version of the 196KC contains an on-chip 16-kbyte EPROM, and the 83C196KC contains a 16-kbyte ROM. The internal EPROM or ROM is mapped into memory locations 2080H to 5FFFH. Memory locations 2000H to 207FH are dedicated to interrupt vectors, peripheral transaction server (PTS) vectors, a chip configuration register,

	HWINDOW 0 when Read		HWINDOW 0 when Written		HWINDOW 1 Read/Write		HWINDOW 15
19H	SP (HI)	19H	SP (HI)	19H	SP (HI)	19H	SP (HI)
18H	SP (LO)	18H	SP (LO)	18H	SP (LO)	18H	SP (LO)
17H	IOS2	17H	PWM0_CONTROL	17H	PWM1_CONTROL	17H	
16H	IOS1	16H	IOC1	16H	PWM2_CONTROL	16H	
15H	IOS0	15H	IOC0	15H	RESERVED	15H	
14H	WSR	14H	WSR	14H		14H	WSR
13H	INT_MASK1	13H	INT_MASK1	13H	INT_MASK1	13H	INT_MASK1
12H	INT_PEND1	12H	INT_PEND1	12H	INT_PEND1	12H	INT_PEND1
11H	SP_STAT	11H	SP_CON	11H	RESERVED	11H	
10H	PORT2	10H	PORT2	10H	RESERVED	10H	RESERVED
0FH	PORT1	0FH	PORT1	0FH	RESERVED	0FH	RESERVED
0EH	PORT0	0EH	BAUD RATE	0EH	RESERVED	0EH	RESERVED
0DH	TIMER2 (HI)	0DH	TIMER2 (HI)	0DH	RESERVED	0DH	T2CAPTURE (HI)
0CH	TIMER2 (LO)	0CH	TIMER2 (LO)	0CH	IOC3*	0CH	T2CAPTURE (LO)
0BH	TIMER1 (HI)	0BH	IOC2	0BH	RESERVED	0BH	
0AH	TIMER1 (LO)	0AH	WATCHDOG	0AH	RESERVED	0AH	
09H	INT_PEND	09H	INT_PEND	09H	INT_PEND	09H	INT_PEND
08H	INT_MASK	08H	INT_MASK	08H	INT_MASK	08H	INT_MASK
07H	SBUF (RX)	07H	SBUF (TX)	07H	PTSSRV (HI)	07H	
06H	HSI_STATUS	06H	HSO_COMMAND	06H	PTSSRV (LO)	06H	
05H	HSI_TIME (HI)	05H	HSO_TIME (HI)	05H	PTSSEL(HI)	05H	
04H	HSI_TIME (LO)	04H	HSO_TIME (LO)	04H	PTSSEL(LO)	04H	
03H	AD_RESULT (HI)	03H	HSI_MODE	03H	AD_TIME	03H	
02H	AD_RESULT (LO)	02H	AD_COMMAND	02H	RESERVED	02H	
01H	ZERO_REG (HI)	01H	ZERO_REG (HI)	01H	ZERO_REG (HI)	01H	ZERO_REG (HI)
00H	ZERO_REG (LO)	00H	ZERO_REG (LO)	00H	ZERO_REG (LO)	00H	ZERO_REG (LO)
	HWINDOW 0 when Read		HWINDOW 0 when Written		HWINDOW 1 Read/Write		HWINDOW 15

NOTE:
*This was previously called T2CONTROL or T2CNTC.

Figure 6.4 Multiple register windows. *(Courtesy Intel Corporation)*

and a security key (see Fig. 6.3). Instruction and data accesses to the internal ROM or EPROM occur only if pin EA# is tied high.

The MCS-96 architecture supports the following *unsigned* data types:

8-bit byte

16-bit word, aligned on even byte address boundaries (little-endian)

32-bit doubleword, aligned at an address divisible by 4 (little-endian)

and the following *signed* data types:

8-bit short integer

16-bit integer

32-bit long integer.

Integers are aligned as words, and long integers are aligned as double-

Register	Description
R0	Zero Register - Always reads as a zero, useful for a base when indexing and as a constant for calculations and compares.
AD_RESULT	A/D Result Hi/Low - Low and high order results of the A/D converter
AD_COMMAND	A/D Command Register - Controls the A/D
HSI_MODE	HSI Mode Register - Sets the mode of the High Speed Input unit.
HSI_TIME	HSI Time Hi/Lo - Contains the time at which the High Speed Input unit was triggered.
HSO_TIME	HSO Time Hi/Lo - Sets the time or count for the High Speed Output to execute the command in the Command Register.
HSO_COMMAND	HSO Command Register - Determines what will happen at the time loaded into the HSO Time registers.
HSI_STATUS	HSI Status Registers - Indicates which HSI pins were detected at the time in the HSI Time registers and the current state of the pins.
SBUF(TX)	Transmit buffer for the serial port, holds contents to be outputted.
SBUF(RX)	Receive buffer for the serial port, holds the byte just received by the serial port.
INT_MASK	Interrupt Mask Register - Enables or disables the individual interrupts.
INT_PEND	Interrupt Pending Register - Indicates that an interrupt signal has occurred on one of the sources and has not been serviced. (also INT_PENDING)
WATCHDOG	Watchdog Timer Register - Written periodically to hold off automatic reset every 64K state times.
TIMER1	Timer 1 Hi/Lo - Timer1 high and low bytes.
TIMER2	Timer 2 Hi/Lo - Timer2 high and low bytes.
IOPORT0	Port 0 Register - Levels on pins of Port 0.
BAUD_RATE	Register which determines the baud rate, this register is loaded sequentially.
IOPORT1	Port 1 Register - Used to read or write to Port 1.
IOPORT2	Port 2 Register - Used to read or write to Port 2.
SP_STAT	Serial Port Status - Indicates the status of the serial port.
SP_CON	Serial Port Control - Used to set the mode of the serial port.
IOS0	I/O Status Register 0 - Contains information on the HSO status.
IOS1	I/O Status Register 1 - Contains information on the status of the timers and of the HSI.
IOC0	I/O Control Register 0 - Controls alternate functions of HSI pins, Timer 2 reset sources and Timer 2 clock sources.
IOC1	I/O Control Register 1 - Controls alternate functions of Port 2 pins, timer interrupts and HSI interrupts.
PWM_CONTROL	Pulse Width Modulation Control Register - Sets the duration of the PWM pulse.
INT_PEND1	Interrupt Pending register for the 8 new interrupt vectors (also INT_PENDING1)
INT_MASK1	Interrupt Mask register for the 8 new interrupt vectors
IOC2	I/O Control Register 2
IOS2	I/O Status Register 2 - Contains information on HSO events
WSR	Window Select Register - Selects register window
AD_TIME	Determines A/D Conversion Time
IOC3	New 80C196KC features (T2 internal clocking, PWMs) (Previously T2CONTROL or T2CNTC)
PTSSEL	Individually enables PTS channels
PTSSRV	End-of-PTS Interrupt Pending Flags

Figure 6.5 Special function register description. *(Courtesy Intel Corporation)*

words. In addition, the MCS-96 architecture supports testing of any bit within the register file. Bit 0 is always the least significant bit.

There are six basic *addressing modes*:

1. *Register direct*: The operand is within the 256-byte on-chip register file. Registers are addressed in a symbolic manner, similar to the 80x86 Intel family.[4] Examples include

```
ADD    AX,BX,CX    ;AX<---BX+CX
MUL    AX,BX       ;AX<---AX*BX
INCB   CL          ;CL<---CL+1
```

where CL is the lower byte of CX, and B at the end of INCB denotes a byte operation.

2. *Indirect*: The address of the operand in memory is in a word variable (16-bit) in the register file. Examples include

```
LD     AX,[BX]     ;AX<---MEM WORD(BX)
ADDB   AL,BL,[CX]  ;AL<---BL+MEM BYTE(CX)
POP    [AX];                   MEM WORD(AX)<---MEM WORD(SP),SP<---SP+2
```

3. *Indirect with autoincrement*: Addressing same as in indirect, except that the word variable containing the indirect address is incremented after it is used to address the operand. The increment is 1 for a byte and 2 for a word operation. Examples include

```
LD     AX,[BX]+    ;AX<---MEM WORD(BX), BX<---BX+2
ADDB   AL,BL,[CX]+ ;AL<---BL+MEM BYTE(CX), CX<---CX+1
PUSH   [AX]+       ;SP<---SP-2,MEM WORD(SP)<---MEM WORD(AX),AX<---AX+2
```

4. *Immediate*: The operand (signed or unsigned, 8- or 16-bit) is a part of the instruction. Examples include

```
ADD    AX,#340     ;AX<---AX+340
PUSH   #1234H      ;SP<---SP-2, MEM WORD(SP)<---1234H
DIVS   AX,#10      ;AL<---AX/10, AH<---AX mod 10 (remainder)
```

5. *Short indexed*: A word variable in the register file serves as an indirect address register. An 8-bit variable is sign extended and added to the register content to form an address of the operand. Examples include

```
LD     AX,12[BX]   ;AX<---MEM WORD(BX+12),
MULB   AX,BL,3[CX] ;AX<---BL*MEM BYTE(CX+3)
```

6. *Long-indexed*: Same as short-indexed, except a 16-bit variable is added to the word register content. Examples include

```
AND    AX,BX,TABLE[CX]   ;AX<---BX AND MEM WORD(TABLE+CX)
ADDB   AL,BL,LOOKUP[CX]  ;AL<---BL+MEM BYTE(LOOKUP+CX)
```

where TABLE and LOOKUP are symbolic 16-bit variables. The user can use symbolic variables. Their addresses and addressing modes are established by the assembler.

The first two bytes in the register file (or rather in the memory space) are fixed at zero by the 196KC hardware (see Fig. 6.4). This is the same practice used in a number of RISC-type systems for the first register in the CPU register file r0.[5] This feature provides a constant zero for calculations and a capability to synthesize a direct addressing mode into any location in memory, as demonstrated in the following examples:

```
ADD   AX,1234[0]    ;AX<---AX+MEM WORD(1234)
POP 5678[0];          MEM WORD(5678)<---MEM WORD(SP),SP<---SP+2
```

The stack pointer can be mentioned explicitly in a number of operations, as in the following examples:

```
PUSH   [SP]      ;Duplicate TOP OF STACK
LD     AX,2[SP]  ;AX<---NEXT TO TOP
```

The 196KC RALU accesses the register file directly using 8-bit addressing, making 256 bytes available to the RALU at one time. Register windowing was implemented so the RALU could have access to more than 256 bytes of registers simply by switching windows. There are two types of windows: *horizontal* and *vertical*. *Switching between windows is controlled by the 8-bit window select register* (WSR). The seven least significant bits of the WSR control the windowing, and the most significant bit, called HLDEN, enables the HOLD#/HLDA# pin function. Horizontal windows (see Fig. 6.4) contain the extra SFRs for the 196KC. Switching to the horizontal windows option maps the 24 bytes of the H window into the lowest 24 bytes of the register file. There is no other way to access a horizontal window. To switch between windows, one has to store the Hwindow number into the four least significant bits of WSR. Bits 4 to 6 of the WSR must be zero when accessing horizontal windows.

Vertical windows (Vwindows) can be used to map sections of the 512 bytes of RAM from 000H to 1FFH into the upper section of the register file. The Vwindows reside directly in the 196KC addressing space. Therefore, locations 100H to 1FFH can be addressed directly with 16-bit addressing using an indexed or indirect addressing mode, or as registers using Vwindows. Vertical windowing allows 32-, 64-, or 128-byte windows from 000H to 1FFH to be mapped onto the top 32-, 64-, or 128-byte block of the register file, as illustrated in Fig. 6.6.[2] Switching between Vwindows is done by setting bit 6, 5, or 4 in the WSR and writing the number of the Vwindow into WSR's four least significant bits. A number of examples for window setting on the WSR is illustrated in Figs. 6.7 and 6.8.[2] The example in Fig. 6.8 maps

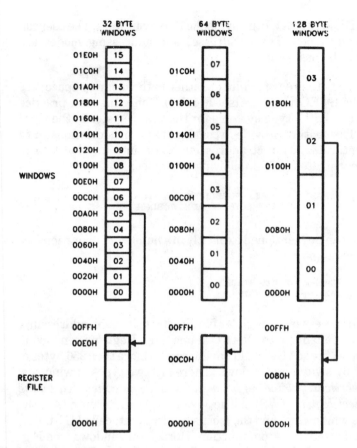

Figure 6.6 Vertical windows. *(Courtesy Intel Corporation)*

Forms A 9 Bit Address

	7							0	
WSR	HLDEN	W6	W5	W4	W3	W2	W1	W0	14H

```
WWW WWWW = 000 XXXX: Select Horizontal Window (3 Possible)
         = 100 XXXX: Select 32-Byte Window (16 Possible)
         = 010 0XXX: Select 64-Byte Window (8 Possible)
         = 001 00XX: Select 128-Byte Window (4 Possible)
```

```
32-Byte Window Addresses :  W3 W2 W1 W0 A4 A3 A2 A1 A0
64-Byte Window Addresses :  W2 W1 W0 A5 A4 A3 A2 A1 A0
128-Byte Window Addresses : W1 W0 A6 A5 A4 A3 A2 A1 A0
```

Figure 6.7 Accessing a Vwindow. *(Courtesy Intel Corporation)*

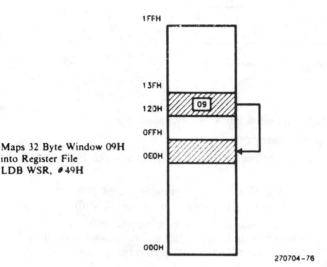

Maps 32 Byte Window 09H
into Register File
LDB WSR, #49H

270704-76

```
ldb WSR_IMAGE, WSR
ldb WSR, #49H          ;switch windows
add 40H, 0FOH          ;adds the value at 130H to the
                       ;value at 40H
add 40H, 0FOH[0]       ;adds the value at 0FOH to the
                       ;value at 40H
add 40H, 130H [0]      ;adds the value at 130H to the
                       ;value at 40H
ldb WSR, WSR_IMAGE     ;switch back to previous window
```

Figure 6.8   VWindow example. *(Courtesy Intel Corporation)*

a 32-byte block from 120H to 13FH into the upper part of the register file
from 0E0H to 0FFH. Now any access to locations 0E0H to 0FFH using a
register direct mode will actually access the memory at 120H to 13FH.
However, the two location intervals can still be accessed directly with 16-bit
addressing. Vwindowing provides for fast context switching of register
sets. For example, an interrupt service routine could have its own set of local
registers in a Vwindow, and pass results to a main routine through global
registers in the register file.

The 196KC program status word (PSW) is a collection (16-bit) of flags that
retain information concerning the state of the user's program. The PSW's
lower 8 bits individually mask the lowest 8 sources of interrupt. A logical 1 in
these bit positions enables the corresponding interrupt. Bit 9 (I) in the PSW
is the global interrupt disable (when I = 0). The upper byte of the PSW,
illustrated in Fig. 6.9, contains the status flags

Z   Zero flag, set (Z = 1) if the result is zero.

N   Negative flag, set (N = 1) if the result is negative.

V   Overflow flag, set (V = 1) if there is an overflow.

	7	6	5	4	3	2	1	0
PSW:	Z	N	V	VT	C	PSE	I	ST

Figure 6.9 PSW register. *(Courtesy Intel Corporation)*

VT   Overflow trap flag, set when V is set. Facilitates testing.

C   Carry flag, set on carry out.

PSE   Peripheral transaction server enable flag, enables the PTS when set.

I   Interrupt disable flag, disables all interrupts except NMI, TRAP, and unimplemented opcode when cleared (I = 0).

ST   Sticky bit flag, set to indicate that during a right shift a 1 has been shifted first into the C flag and then been shifted out. It can be used along with the C flag to control rounding after a right shift.

The 196KC instruction set is summarized in Table 6.1.[2] The following six instructions are new on the 196KC and were not available on its predecessor 80C196KB or any other members of the MCS-96 family: DPTS, EPTS, BMOVI, XCH, XCHB, TIJMP (see Table 6.1).

The 196KC contains six major peripherals:

1. *Pulse width modulation (PWM) outputs*: There are three PWM outputs (only one on 80C196KB). Digital-to-analog (D/A) conversion can be done with any of the PWM outputs. A block diagram of the PWM subsystem is shown in Fig. 6.10.[2] The 8-bit counter is incremented every state time. When it equals 0, the PWM output is set to 1. When the counter matches the value in the corresponding PWM register, the output is switched low. When the counter overflows, the outputs are once again switched high.

2. *Two 16-bit timers*: Timer 1 and timer 2 serve as time bases for the high-speed I/O units.

3. *High-speed inputs (HSI) unit*: The HSI unit, illustrated in Fig. 6.11, can capture the value of timer 1 when an event takes place on one of four input pins HSI.0 to HSI.3. Four types of events can trigger a capture: rising edges only, falling edges only, rising or falling edges, or every eighth rising edge. Each HSI pin can be independently programmed to look for any of these conditions. When events occur, the timer 1 value gets stored in the 7 × 20 FIFO queue (see Fig. 6.11) along with four status bits that indicate the input line(s) that caused the event. The next event ready to be unloaded from the FIFO is placed in the HSI holding register. Data is taken off the FIFO by reading the HSI status register, followed by reading the HSI time register. When the time register is read, the next FIFO location is loaded into the holding register.

TABLE 6.1 Instruction Summary

Mnemonic	Operands	Operation (Note 1)	Z	N	C	V	VT	ST	Notes
ADD/ADDB	2	D ← D + A	✓	✓	✓	✓	↑	–	
ADD/ADDB	3	D ← B + A	✓	✓	✓	✓	↑	–	
ADDC/ADDCB	2	D ← D + A + C	↓	✓	✓	✓	↑	–	
SUB/SUBB	2	D ← D – A	✓	✓	✓	✓	↑	–	
SUB/SUBB	3	D ← B – A	✓	✓	✓	✓	↑	–	
SUBC/SUBCB	2	D ← D – A + C – 1	↓	✓	✓	✓	↑	–	
CMP/CMPB	2	D – A	✓	✓	✓	✓	↑	–	
MUL/MULU	2	D,D + 2 ← D × A	–	–	–	–	–	–	2
MUL/MULU	3	D,D + 2 ← B × A	–	–	–	–	–	–	2
MULB/MULUB	2	D,D + 1 ← D × A	–	–	–	–	–	–	3
MULB/MULUB	3	D,D + 1 ← B × A	–	–	–	–	–	–	3
DIVU	2	D ← (D,D + 2) /A,D + 2 ← remainder	–	–	–	✓	↑	–	2
DIVUB	2	D ← (D,D + 1) /A,D + 1 ← remainder	–	–	–	✓	↑	–	3
DIV	2	D ← (D,D + 2) /A,D + 2 ← remainder	–	–	–	✓	↑	–	
DIVB	2	D ← (D,D + 1) /A,D + 1 ← remainder	–	–	–	✓	↑	–	
AND/ANDB	2	D ← D AND A	✓	✓	0	0	–	–	
AND/ANDB	3	D ← B AND A	✓	✓	0	0	–	–	
OR/ORB	2	D ← D OR A	✓	✓	0	0	–	–	
XOR/XORB	2	D ← D (excl. or) A	✓	✓	0	0	–	–	
LD/LDB	2	D ← A	–	–	–	–	–	–	
ST/STB	2	A ← D	–	–	–	–	–	–	
XCH/XCHB	2	D ← A, A ← D	–	–	–	–	–	–	
LDBSE	2	D ← A; D + 1 ← SIGN(A)	–	–	–	–	–	–	3,4
LDBZE	2	D ← A; D + 1 ← 0	–	–	–	–	–	–	3,4
PUSH	1	SP ← SP – 2; (SP) ← A	–	–	–	–	–	–	
POP	1	A ← (SP); SP + 2	–	–	–	–	–	–	
PUSHF	0	SP ← SP – 2; (SP) ← PSW; PSW ← 0000H; I ← 0	0	0	0	0	0	0	
POPF	0	PSW ← (SP); SP ← SP + 2; I ← ✓	✓	✓	✓	✓	✓	✓	
SJMP	1	PC ← PC + 11-bit offset	–	–	–	–	–	–	5
LJMP	1	PC ← PC + 16-bit offset	–	–	–	–	–	–	5
BR[indirect]	1	PC ← (A)	–	–	–	–	–	–	
TIJMP	3	PC ← [A] + 2 * ([B] AND C)	–	–	–	–	–	–	
SCALL	1	SP ← SP – 2; (SP) ← PC; PC ← PC + 11-bit offset	–	–	–	–	–	–	5
LCALL	1	SP ← SP – 2; (SP) ← PC; PC ← PC + 16-bit offset	–	–	–	–	–	–	5

SOURCE: Courtesy Intel Corporation.

TABLE 6.1   Instruction Summary *(Continued)*

Mnemonic	Operands	Operation (Note 1)	Z	N	C	V	VT	ST	Notes
				Flags					
RET	0	PC ← (SP); SP ← SP + 2	–	–	–	–	–	–	
J (conditional)	1	PC ← PC + 8-bit offset (if taken)	–	–	–	–	–	–	5
JC	1	Jump if C = 1	–	–	–	–	–	–	5
JNC	1	jump if C = 0	–	–	–	–	–	–	5
JE	1	jump if Z = 1	–	–	–	–	–	–	5
JNE	1	Jump if Z = 0	–	–	–	–	–	–	5
JGE	1	Jump if N = 0	–	–	–	–	–	–	5
JLT	1	Jump if N = 1	–	–	–	–	–	–	5
JGT	1	Jump if N = 0 and Z = 0	–	–	–	–	–	–	5
JLE	1	Jump if N = 1 or Z = 1	–	–	–	–	–	–	5
JH	1	Jump if C = 1 and Z = 0	–	–	–	–	–	–	5
JNH	1	Jump if C = 0 or Z = 1	–	–	–	–	–	–	5
JV	1	Jump if V = 0	–	–	–	–	–	–	5
JNV	1	Jump if V = 1	–	–	–	–	–	–	5
JVT	1	Jump if VT = 1; Clear VT	–	–	–	–	0	–	5
JNVT	1	Jump if VT = 0; Clear VT	–	–	–	–	0	–	5
JST	1	Jump if ST = 1	–	–	–	–	–	–	5
JNST	1	Jump if ST = 0	–	–	–	–	–	–	5
JBS	3	Jump if Specified Bit = 1	–	–	–	–	–	–	5,6
JBC	3	Jump if Specified Bit = 0	–	–	–	–	–	–	5,6
DJNZ/ DJNZW	1	D ← D − 1; If D ≠ 0 then PC ← PC + 8-bit offset	–	–	–	–	–	–	5
DEC/DECB	1	D ← D − 1	✔	✔	✔	✔	↑	–	
NEG/NEGB	1	D ← 0 − D	✔	✔	✔	✔	↑	–	
INC/INCB	1	D ← D + 1	✔	✔	✔	✔	↑	–	
EXT	1	D ← D; D + 2 ← Sign (D)	✔	✔	0	0	–	–	2
EXTB	1	D ← D; D + 1 ← Sign (D)	✔	✔	0	0	–	–	3
NOT/NOTB	1	D ← Logical Not (D)	✔	✔	0	0	–	–	
CLR/CLRB	1	D ← 0	1	0	0	0	–	–	
SHL/SHLB/SHLL	2	C ← msb - - - - - lsb ← 0	✔	✔	✔	✔	↑	–	7
SHR/SHRB/SHRL	2	0 → msb - - - - - lsb → C	✔	✔	✔	0	–	✔	7
SHRA/SHRAB/SHRAL	2	msb → msb - - - - - lsb → C	✔	✔	✔	0	–	✔	7
SETC	0	C ← 1	–	–	1	–	–	–	
CLRC	0	C ← 0	–	–	0	–	–	–	

4. *High-speed output (HSO) unit*: The HSO unit, illustrated in Fig. 6.12, can generate events at specified values of timer 1 or timer 2. Up to eight pending events can be stored in the content addressable memory (CAM) of the HSO unit at one time. Commands are placed into the HSO unit by first writing to HSO command with the event to occur, and then to HSO time with the timer match value.

TABLE 6.1 Instruction Summary *(Continued)*

Mnemonic	Operands	Operation (Note 1)	Flags						Notes
			Z	N	C	V	VT	ST	
CLRVT	0	VT ← 0	–	–	–	–	0	–	
RST	0	PC ← 2080H	0	0	0	0	0	0	8
DI	0	Disable All Interupts (I ← 0)	–	–	–	–	–	–	
EI	0	Enable All Interupts (I ← 1)	–	–	–	–	–	–	
DPTS	0	Disable all PTS Cycles (PSE = 0)	–	–	–	–	–	–	
EPTS	0	Enable all PTS Cycles (PSE = 1)	–	–	–	–	–	–	
NOP	0	PC ← PC + 1	–	–	–	–	–	–	
SKIP	0	PC ← PC + 2	–	–	–	–	–	–	
NORML	2	Left shift till msb = 1; D ← shift count	✔	✔	0	–	–	–	7
TRAP	0	SP ← SP − 2; (SP) ← PC; PC ← (2010H)	–	–	–	–	–	–	9
PUSHA	1	SP ← SP-2; (SP) ← PSW; PSW ← 0000H; SP ← SP-2; (SP) ← IMASK1/WSR; IMASK1 ← 00H	0	0	0	0	0	0	
POPA	1	IMASK1/WSR ← (SP); SP ← SP + 2 PSW ← (SP); SP ← SP + 2	✔	✔	✔	✔	✔	✔	
IDLPD	1	IDLE MODE IF KEY = 1; POWERDOWN MODE IF KEY = 2; CHIP RESET OTHERWISE	–	–	–	–	–	–	
CMPL	2	D-A	✔	✔	✔	✔	↑	–	
BMOV, BMOVi	2	[PTR_HI] + ← [PTR_LOW] + ; UNTIL COUNT = 0	–	–	–	–	–	–	

NOTES:
1. If the mnemonic ends in "B" a byte operation is performed, otherwise a word operation is done. Operands D, B, and A must conform to the alignment rules for the required operand type. D and B are locations in the Register File; A can be located anywhere in memory.
2. D,D + 2 are consecutive WORDS in memory; D is DOUBLE-WORD aligned.
3. D,D + 1 are consecutive BYTES in memory; D is WORD aligned.
4. Changes a byte to word.
5. Offset is a 2's complement number.
6. Specified bit is one of the 2048 bits in the register file.
7. The "L" (Long) suffix indicates double-word operation.
8. Initiates a Reset by pulling RESET low. Software should re-initialize all the necessary registers with code starting at 2080H.
9. The assembler will not accept this mnemonic.

Flat Settings. The modification to the flag setting is shown for each instruction. A checkmark (✔) means that the flag is set or cleared as appropriate. A hyphen means that the flag is not modified. A one or zero (1) or (0) indicates that the flag will be in that state after the instruction. An up arrow (↑) indicates that the instruction may set the flag if it is appropriate but will not clear the flag. A down arrow (↓) indicates that the flag can be cleared but not set by the instruction. A question mark (?) indicates that the flag will be left in an indeterminant state after the operation.

SOURCE: Intel Corporation.

5. *Serial port*: The serial port is functionally compatible with the serial port on the MCS-51 and MCS-96 families of microcontrollers. One synchronous and three asynchronous modes are available. The asynchronous modes are full duplex.

6. *A/D converter*: The A/D converter consists of a sample-and-hold circuit, an eight-channel multiplexer, and an 8- or 10-bit successive approxima-

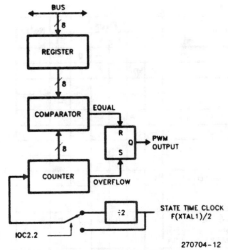

• Duty Cycle Programmable in 256 Steps

Figure 6.10   PWM block diagram.
(Courtesy Intel Corporation)

HSI Trigger Options

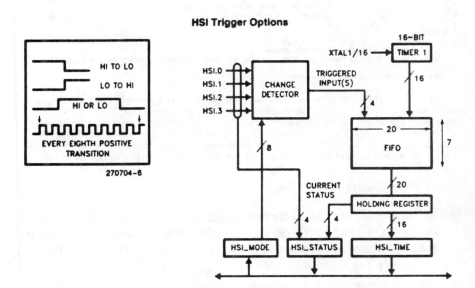

Figure 6.11   HSI block diagram. *(Courtesy Intel Corporation)*

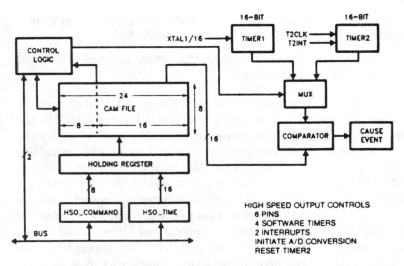

Figure 6.12   HSO block diagram. *(Courtesy Intel Corporation)*

tion A/D converter. Analog signals can be sampled by any of the eight analog input pins ACH0 through ACH7 that are shared with Port 0. An A/D conversion is performed on one input at a time.

Analog outputs can be generated by either using the PWM or the HSO output, as illustrated in Fig. 6.13.[2] Either device will generate a rectangular pulse train. If a smooth analog signal is desired, the rectangular waveform must be filtered. In most cases this filtering is best done after the signal is buffered to make it swing from 0 to 5 volts since both of the outputs are guaranteed only to low current levels.

There are five 8-bit ports on the 196KC:

1. Port 0 is an input port that is also the analog input for the A/D converter.

2. Port 1 is a quasi-bidirectional port. Its three most significant bits are multiplexed with the HOLD#/HLDA# functions. The two extra PWM outputs are multiplexed on bits 4 and 3 of Port 1.

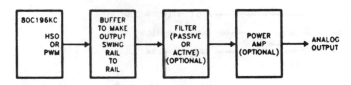

Figure 6.13   D/A Buffer block diagram. *(Courtesy Intel Corporation)*

3. Port 2 contains three types of pins: quasi-bidirectional, input and output. Its input and output lines are shared with other functions.

4. Ports 3 and 4 are bidirectional ports that share their pins with the multiplexed address/data bus.

The 196KC recognizes 28 sources of interrupt. These sources are organized as 15 vectors plus special vectors for nonmaskable interrupt (NMI), the TRAP instruction, and unimplemented opcodes. Figure 6.14 shows the routing of the interrupt sources into their vectors as well as the control bits that enable some of the sources. The NMI vectors indirectly through location 203EH. The TRAP instruction causes an indirect vector through location 2010H. It provides a single instruction interrupt useful in debugging software. The TRAP instruction prevents the acknowledgment of interrupts until after execution of the next instruction. Opcodes that are not implemented on the 196KC will cause an indirect vector through location 2012H if attempted. A block diagram of the interrupt structure is shown in Fig. 6.15.[2] The transition detector looks for 0 to 1 transitions on any of the

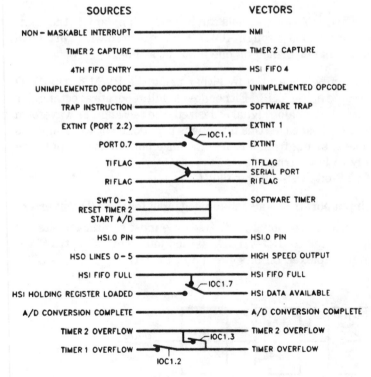

Figure 6.14 80C196KC interrupt sources. *(Courtesy Intel Corporation)*

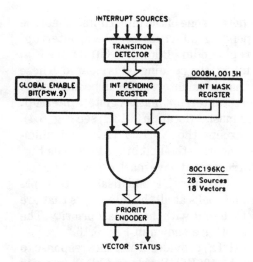

Figure 6.15 80C196KC interrupt structure block diagram. *(Courtesy Intel Corporation)*

Number	Source	Vector Location	Priority
INT15	NMI	203EH	15
INT14	HSI FIFO Full	203CH	14
INT13	EXTINT1	203AH	13
INT12	TIMER2 Overflow	2038H	12
INT11	TIMER2 Capture	2036H	11
INT10	4th Entry into HSI FIFO	2034H	10
INT09	RI	2032H	9
INT08	TI	2030H	8
SPECIAL	Unimplemented Opcode	2012H	N/A
SPECIAL	Trap	2010H	N/A
INT07	EXTINT	200EH	7
INT06	Serial Port	200CH	6
INT05	Software Timer	200AH	5
INT04	HSI.0 Pin	2008H	4
INT03	High Speed Outputs	2006H	3
INT02	HSI Data Available	2004H	2
INT01	A/D Conversion Complete	2002H	1
INT00	Timer Overflow	2000H	0

Figure 6.16 80C1968C interrupt priorities. *(Courtesy Intel Corporation)*

interrupt sources. When hardware detects one of the interrupts, it sets the corresponding bit in one of two pending interrupt registers, interrupt pending (location 0009H) and interrupt pending 1 (location 0012H; see Fig. 6.4). When the interrupt vector is taken, the pending bit is cleared. These registers can be read or modified. Individual interrupts can be enabled or disabled by setting or clearing bits in the interrupt mask registers interrupt mask (location 0008H) and interrupt mask 1 (location 0013H; see Fig. 6.4). The processing of all interrupts except the NMI, TRAP, and unimplemented opcode can be disabled by clearing the I bit in the PSW (see Fig. 6.9). Setting the I bit will enable interrupts that have mask register bits set. The I bit is controlled by the enable interrupts (EI) and disable interrupts (DI) instructions. The priority encoder looks at all the interrupts that are both pending and enabled, and selects the one with the highest priority. The priorities (15 the highest, 0 the lowest) are shown in Fig. 6.16.[2]

The peripheral transaction server (PTS) provides DMA-like response to an interrupt. It is a new feature on the 196KC. Single and block transfer modes are supported, as well as special modes to service the A/D converter and the high speed I/O (HSI/O). Any of the 15 interrupt vectors can be

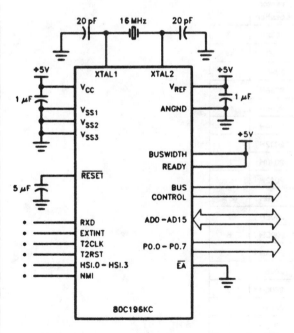

NOTE:
*Inputs must be driven high or low.

Figure 6.17 80C196KC minimum hardware connections. *(Courtesy Intel Corporation)*

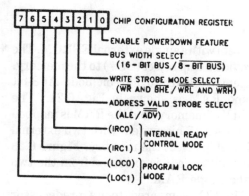

Figure 6.18 Chip configuration register.
(Courtesy Intel Corporation)

alternatively mapped to its PTS channel. The PTS generates a shorter interrupt service cycle, compared to a regular interrupt service routine. This enhances the interrupt response and renders the 196KC more attractive for real-time embedded applications.

Figure 6.17 shows the minimum connections needed to get the 196KC up and running. It is important to tie all unused inputs to power inputs V_{CC} or ground V_{SS}. If these pins are left floating, they can float to a mid voltage level and draw excessive current. Some pins, such as NMI or EXTINT, may generate spurious interrupts if left unconnected.

The chip configuration register (CCR) is the first byte fetched from memory following a chip reset. The CCR is fetched from the chip configuration byte (CCB) at location 2018H in either internal or external memory depending on the EA# pin. The CCR is only loaded once during the reset sequence. Once loaded, the CCR cannot be changed until the next reset. The CCR is illustrated in Fig. 6.18.[2] Bits 7 and 6 control ROM/EPROM protection. Bits 5 and 4 control the internal READY mode. Bits 3, 2, and 1 determine the bus control signals. Bit 0 enables or disables the powerdown mode. Ready control selects wait-state options for external memory (internal memory has zero wait-state) as listed in Fig. 6.19.[2]

IRC1	IRC0	Description
0	0	Limit to one waitstate
0	1	Limit to two waitstates
1	0	Limit to three waitstates
1	1	Wait states not limited internally

Figure 6.19 Ready control modes. *(Courtesy Intel Corporation)*

Examples of External Memory
Interconnections to 196KC

Figure 6.20 shows a simple 8-bit system with a single external EPROM. The ADV# mode can be selected to provide a chip select (CS#) to the memory. By setting bit 1 of CCR to 0, the system is locked into the 8-bit mode. An 8-bit system with external EPROM and RAM is shown in Fig. 6.21.[2] The EPROM is mapped into the lower half of memory, and the RAM is mapped into the upper half. Figure 6.22 shows a 16-bit system with two external EPROMs. Again, ADV# is used to chip select the memory. Figure 6.23 shows a system with dynamic bus width. Code is executed from the two EPROMs, and data is stored in the single RAM.

The new features on the 80C196KC, compared to its predecessor 80C196KB, can be summarized as follows:

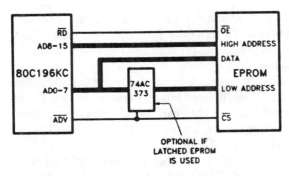

Figure 6.20   8-bit system with EPROM. *(Courtesy Intel Corporation)*

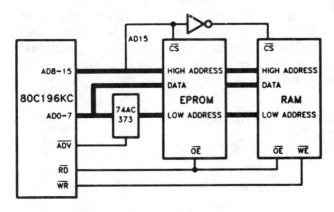

Figure 6.21   8-bit system with EPROM and RAM. *(Courtesy Intel Corporation)*

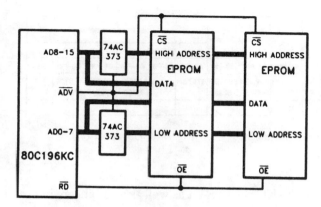

Figure 6.22 16-bit system with EPROM. *(Courtesy Intel Corporation)*

1. The KC has 488 bytes of RAM and is available with 16 kbytes of EPROM or ROM. The KB has 232 bytes of RAM and 8 kbytes of EPROM or ROM. A vertical windowing scheme has been implemented on the KC to allow the extra 256 bytes of RAM to be accessed as registers.

2. The KC has two additional PWM outputs.

3. The KC has the PTS.

4. Timer 2 on the KC can be clocked internally as well as externally (on KB externally only).

5. The HSO on KC has one new command that allows all of the pins to be addressed simultaneously.

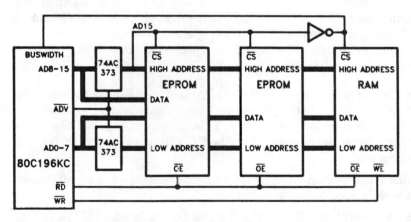

Figure 6.23 16-bit system with dynamic buswidth. *(Courtesy Intel Corporation)*

	Factory Masked ROM			CPU			User Programmable					
							EPROM			OTP		
	68-Pin	64-Pin	48-Pin	68-Pin	64-Pin	48-Pin	68-Pin	64-Pin	48-Pin	68-Pin	64-Pin	48-Pin
ANALOG	8397BH 8397JF	8397BH 8397JF	8395BH 8396	8097BH 8097JF	8097BH 8097JF	8095BH 8096	8797BH		8795BH 8796	8797BH 8797JF	8797JF 8797BH	8798
NO ANALOG	8396BH			8X9X								

Figure 6.24 HMOS MCS®96 packaging. *(Courtesy Intel Corporation)*

6. The A/D converter on the KC has an 8-bit as well as a 10-bit conversion mode (8-bit only on KB).

7. The KC has two UPROM (unerasable PROM) for additional security enhancements.

Earlier CMOS members of the MCS-96 family (8x9x) are summarized in Fig. 6.24.[3] A detailed description of the 8096 can be found in Peatman.[1] A design example featuring an 80C196KB-based system is presented in App. A.

6.2 MOTOROLA MC68332 MICROCONTROLLER

The Motorola MC68332 microcontroller is marketed as a 32-bit microcontroller.[7,8] It has an internal 32-bit data path and 32-bit CPU registers. However, its external data bus is 16 bits, and its external address bus is 24 bits (as on MC68000 and MC68010, the 16-bit members of the Motorola M68000 family). For this reason the MC68332 is classified as a 16-bit microcontroller in this text. It is the first product in the Motorola M68300 family of embedded microcontrollers.

The primary design objectives of the MC68332 design team were as follows:[6]

1. Increase the performance of the total system (CPU and peripherals) by adding intelligence to the system peripherals so they can process simple events with their own resources. This eliminates much of the event servicing normally performed by the CPU, thus enhancing the overall performance.

2. Define a high-performance CPU by implementing the already successful and widely used M68000 architecture.[4]

3. Design a microcontroller family that could be easily adapted to a variety of individual applications. This goal required a flexible design that formalized interconnections and reduced logic interdependencies. This

was achieved by adopting a modular approach; several functional modules were designed simultaneously and independently. A standard intermodule bus (IMB) interface was created, freeing the designers from having to know unnecessary details about other modules. The MC68332 was actually designed in Texas, California, and Israel, with overall responsibility lodged in Austin, Texas.[6]

Figure 6.25[8] illustrates the modules and external interconnections of the MC68332. The main components are as follows:

1. An M68000 family CPU, called CPU32.

2. Intermodule bus (IMB), containing 24 address and 16 data lines along with associated control signals for data transfer handshaking, interrupt, and bus mastership arbitration. The IMB is a synchronous, multimaster, two-clock-cycle bus.[6]

3. A system integration module (SIM), connected to the IMB, that contains programmable chip select logic, a system clock, a periodic interrupt function, and system protection features. The SIM module contains the external bus interface (EBI), interfacing between the IMB and the external bus (see Fig. 6.25).

4. A queued serial module (QSM) that contains both asynchronous and high speed synchronous serial submodules.

5. A 2-kbyte (1024 × 16) static RAM (SRAM).

6. A time processing unit (TPU) that processes time-based, high-frequency I/O functions.[6]

The MC68332 is a 1-micron, HCMOS, 132-pin, 422,000-transistor, 16.78-MHz chip. It has 16 independent programmable channels and pins. Any channel can perform any time function such as input capture, output compare, pulse width modulation, stepper motor activation, and others. There are two timer count registers with programmable prescalers and selectable channel priority levels. There are two serial I/O subsystems:

1. Enhanced serial communications interface (SCI) universal asynchronous receiver transmitter (UART) with parity and programmable baud rate module counter

2. Enhanced serial peripheral interface with I/O queue (QSPI).

The on-chip programmable chip select logic can handle up to 12 signals for memory and peripheral I/O. The MC68332 has up to 32 discrete I/O pins.[7]

The MC68332 CPU, called CPU32, is illustrated in Fig. 6.26.[7] Its architecture is essentially the same as that of the M68000 family.[4] The MC68332 programming model, illustrated in Fig. 6.27, is most similar to

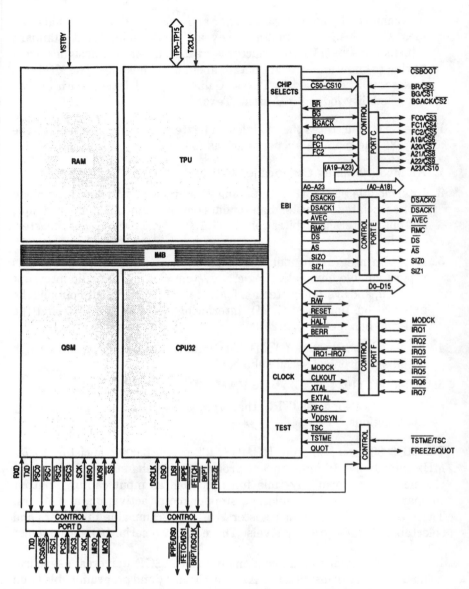

Figure 6.25 MC68332 block diagram. *(Courtesy of Motorola, Inc.)*

that of the MC68010. The details of the status register are shown in Fig. 6.28, and the MC68332 instruction set is listed in Table 6.2.[7]

A later member of the M68300 family, the MC68340,[8] features a two-channel on-chip DMA controller in addition to the MC68332 features described earlier.

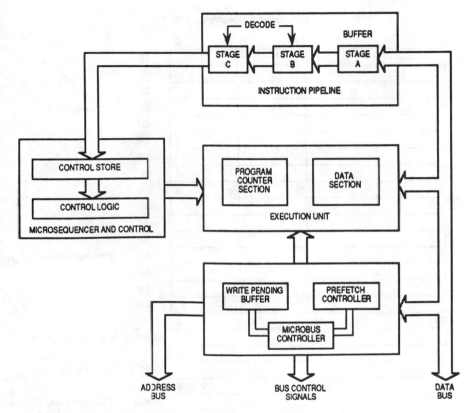

Figure 6.26   CPU32 block diagram. *(Courtesy of Motorola, Inc.)*

REFERENCES

1. J. B. Peatman, *Design with Microcontrollers*. New York: McGraw-Hill, 1988.
2. *80C196KC User's Guide*. Intel Corporation, #270704-003, October 1990.
3. *16-bit Embedded Controllers*. Intel Corporation, #270646-003, 1991.
4. D. Tabak, *Advanced Microprocessors*. New York: McGraw-Hill, 1991.
5. D. Tabak, *RISC Systems*. U.K.: Research Studies Press; and New York:Wiley, 1990.
6. J. Jelemensky, V. Goler, B. Burgess, J. Eifert, and G. Miller, "The MC68332 Microcontroller," *IEEE MICRO*, 9(4):31–50, August 1989.
7. R. Wilson, "Motorola Spotlights Bandwidth in 68340 MCU Entry," *Computer Design*, p. 124, October 1, 1990.
8. *MC68332 Technical Summary*. Motorola Document BR756/D, 1990.

PROBLEMS

Problems 1 to 15 are for the 80C196KC.

1. Add five top word values of the stack. Push the result on top of the stack.

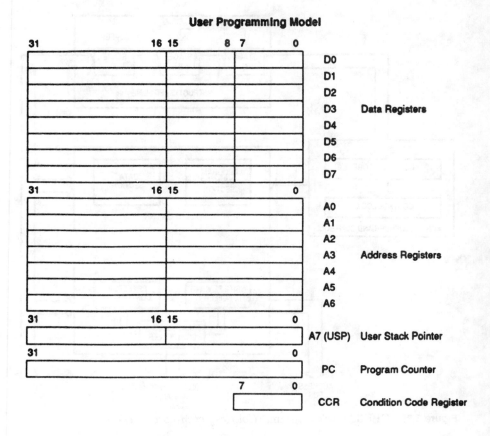

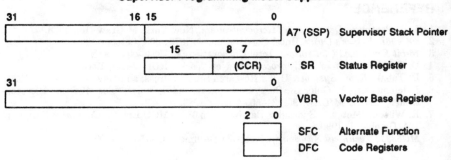

Figure 6.27   MC68332 programming model. *(Courtesy of Motorola, Inc.)*

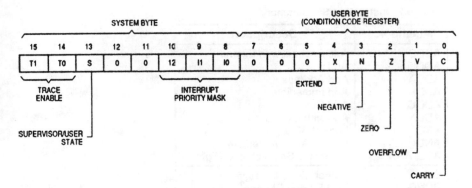

Figure 6.28 Status register. *(Courtesy of Motorola, Inc.)*

2. Receive 10 word values from an input device at address INP1 and push them on stack.

3. Scan 16 word items in array DATA1 and store all items ≥ 25 in array D25.

4. Arrays A1 and B1 contain 10 word components of two vectors. Calculate the scalar product of the two vectors and push it on stack.

5. The PSW content is 86H. Provide a complete interpretation of the setting.

6. The result of the last arithmetic operation was negative, there was a carry out, and the PTS is enabled. What will be the PSW content?

7. The value stored in the PWM register (Fig. 6.10) is 16. Provide a complete description of the PWM output, including a timing diagram.

8. What should be the content of the PWM register if the desired PWM output is a square wave whose high time interval is equal to its low time interval?

9. The HSI inputs (Fig. 6.11) HSI.0 to HSI.3 provide periodic pulses every two clock cycles consecutively: HSI.0 provides a pulse, then HSI.1 after two cycles, and so on. Specify the FIFO 20-bit sequence for a period of 20 clock cycles.

10. Repeat Problem 9 for a case when only HSI.3 provides an input pulse every five clock cycles.

11. It is necessary to cause an output event every four clock cycles, using timer 1 of the HSO (Fig. 6.12). What should be the sequence of values fed into the time part (16-bit) of the CAM file?

12. Draw a detailed wiring diagram for interfacing the 196KC, in 8-bit mode, with an 8-kbyte EPROM chip.

13. Repeat Problem 12 adding an 8-kbyte RAM chip.

14. Repeat Problem 12 in 16-bit mode, using two 8-kbyte EPROM chips.

15. Repeat Problem 14 adding two 8-kbyte RAM chips.

TABLE 6.2 Instruction Set Summary

Mnemonic	Description
ABCD	Add Decimal with Extend
ADD	Add
ADDA	Add Address
ADDI	Add Immediate
ADDQ	Add Quick
ADDX	Add with Extend
AND	Logical AND
ANDI	Logical AND Immediate
ASL, ASR	Arithmetic Shift Left and Right
Bcc	Branch Conditionally
BCHG	Test Bit and Change
BCLR	Test Bit and Clear
BGND	Background
BKPT	Breakpoint
BRA	Branch
BSET	Test Bit and Set
BSR	Branch to Subroutine
BTST	Test Bit
CHK, CHK2	Check Register Against Upper and Lower Bounds
CLR	Clear
CMP	Compare
CMPA	Compare Address
CMPI	Compare Immediate
CMPM	Compare Memory to Memory
CMP2	Compare Register Against Upper and Lower Bounds
DBcc	Test Condition, Decrement and Branch
DIVS, DIVSL	Signed Divide
DIVU, DIVUL	Unsigned Divide
EOR	Logical Exclusive OR
EORI	Logical Exclusive OR Immediate
EXG	Exchange Registers
EXT, EXTB	Sign Extend
LEA	Load Effective Address
LINK	Link and Allocate
LPSTOP	Low Power Stop
LSL, LSR	Logical Shift Left and Right
ILLEGAL	Take Illegal Instruction Trap
JMP	Jump
JSR	Jump to Subroutine

SOURCE: Courtesy of Motorola, Inc.

TABLE 6.2 Instruction Set Summary *(Continued)*

Mnemonic	Description
MOVE	Move
MOVE CCR	Move Condition Code Register
MOVE SR	Move Status Register
MOVE USP	Move User Stack Pointer
MOVEA	Move Address
MOVEC	Move Control Register
MOVEM	Move Multiple Registers
MOVEP	Move Peripheral
MOVEQ	Move Quick
MOVES	Move Alternate Address Space
MULS, MULS.L	Signed Multiply
MULU, MULU.L	Unsigned Multiply
NBCD	Negate Decimal with Extend
NEG	Negate
NEGX	Negate with Extend
NOP	No Operation
OR	Logical Inclusive OR
ORI	Logical Inclusive OR Immediate
PEA	Push Effective Address
RESET	Reset External Devices
ROL, ROR	Rotate Left and Right
ROXL, ROXR	Rotate with Extend Left and Right
RTD	Return and Deallocate
RTE	Return from Exception
RTR	Return and Restore Codes
RTS	Return from Subroutine
SBCD	Subtract Decimal with Extend
Scc	Set Conditionally
STOP	Stop
SUB	Subtract
SUBA	Subtract Address
SUBI	Subtract Immediate
SUBQ	Subtract Quick
SUBX	Subtract with Extend
SWAP	Swap Register Words
TBLS,TBLSN	Table Lookup and Interpolate (Signed)
TBLU, TBLUN	Table Lookup and Interpolate (Unsigned)
TAS	Test Operand and Set
TRAP	Trap
TRAPcc	Trap Conditionally
TRAPV	Trap on Overflow
TST	Test Operand
UNLK	Unlink

Problems 16 to 20 are for the MC68332.

16. When a byte (.B; 8-bit) or word (.W; 16-bit) operation is performed, the upper bits of any data register involved are not affected. What will be the content of the registers and the flag settings at the completion of the following operations? (The $ denotes hexadecimal in Motorola assembly.)

   a. ADD.B D2, D4; initially (D2) = $FA, (D4) = $0F

   b. ADD.W D0,D1; initially (D0) = $ABCD, (D1) = $F000

   c. CMP.B D4,D5; initially (D4) = $FF, (D5) = $FF

   d. CMP.B D4,D5; initially (D4) = $A1, (D5) = $B1

   e. CMP.B D4,D5; initially (D4) = $87, (D5) = $0A

17. Receive a sequence of 10 word items from an input device DEV1 and push them on stack, pointed to by A7 (SP).

18. Add the two top values on stack and push the result on top of stack, without destroying the operands.

19. The content of the status register is $0619. Provide a complete interpretation.

20. The MC68332 executes in supervisor mode. The currently recognized interrupt level is 7, and there were an overflow and a carryout in the preceding operation. The result was positive. There is no tracing. What will be the status register settings?

Design Example

80C196KB-BASED COMMAND AND TELEMETRY INTERFACE UNIT (CTIU)*

The CTIU was designed as a standard component of a spacecraft command and data handling system. The CTIU is used to provide a two-way communication between different users throughout the spacecraft and the central computer (CC). The CC performs the functions of command decoding and distribution and telemetry format generation. The CC communicates with the ground station via the radio frequency (RF) equipment. The CTIU provides one or more users (electronic packages) with access to the CC via RS-232C circuitry. The CTIU accepts commands from the CC via RS-232C (commands are issued either by ground or CC) and relays them to the users. The CTIU has 64 channels available for discrete command distribution and 8 channels for serial digital command distribution. The CTIU requests telemetry data from user channels based on the preprogrammed telemetry format. It accepts, converts, and conditions the data as necessary and transmits it via RS-232C to the CC for insertion into the telemetry downlink stream. Telemetry synchronization pulses are also supplied by the CTIU to the users. The CTIU block diagram is shown in Fig. A.1.

The CTIU is composed of an 80C196KB microcontroller (referred to as simply 196KB), peripherals, external memory, drivers/receivers, and additional logic hardware. The 196KB serves as the CTIU's CPU. It receives commands from the CC, decodes them, generates the appropriate signals to transmit the commands and receive the telemetry signals, checks the

---

*Worked out by GMU senior student Kamdin Shakoorzadeh.

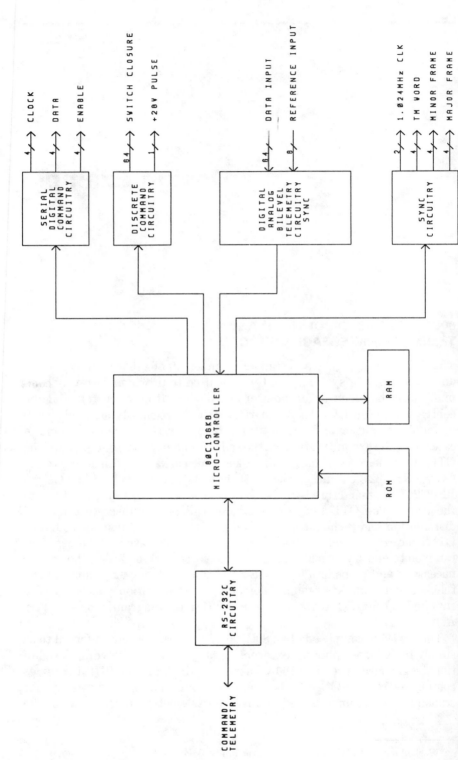

Figure A.1 CTIU top level block diagram.

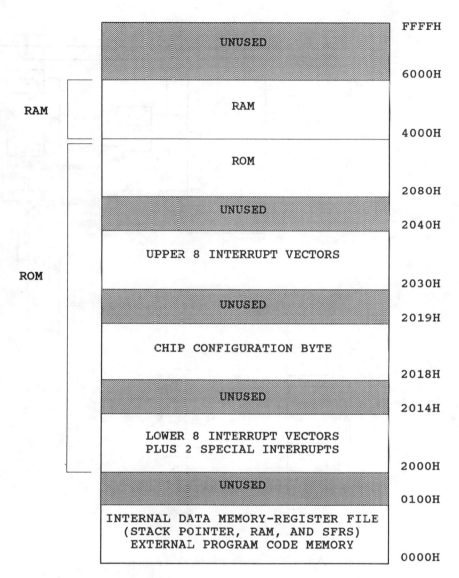

Figure A.2 CTIU memory map.

telemetry words, and transmits them to the CC via RS-232C. Commands are received from the spacecraft's CC in 25-bit messages called command words (CW), which are transmitted over the RS-232C in their ASCII hexadecimal representation. Upon arrival of CWs at the port, the 196KB will be interrupted and it will start to receive characters into a buffer. The command-receiving procedure will end when a carriage return character is

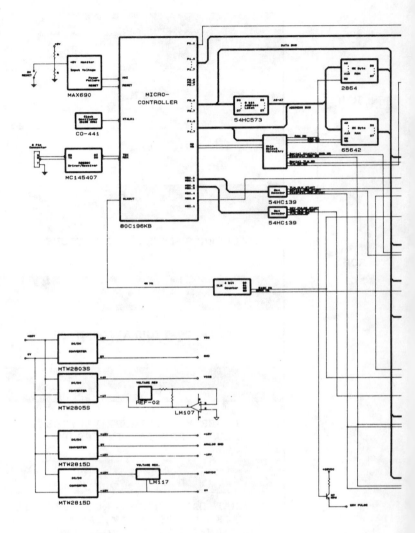

Figure A.3 CTIU schematic block diagram.

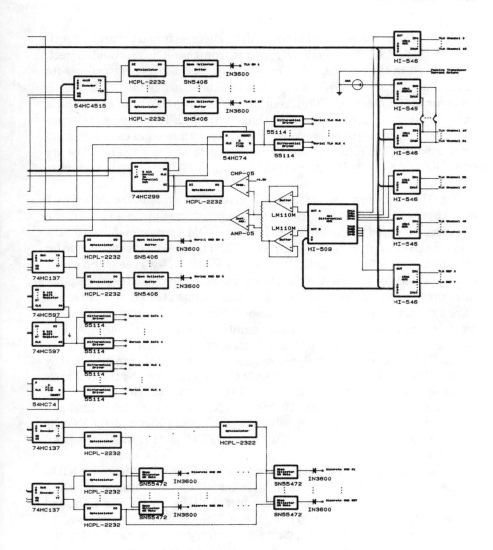

received. Then, the received string will be checked for structural correctness and validity. If the command received from the CC is valid and correct, the content of the command buffer will be passed to the next process for execution, and a received-message signal will be issued to the CC. If there is an error in the received CW, a request for retransmission of command is issued by sending an error-message signal to the CC. The collected telemetry data from the users are packetized in telemetry messages called telemetry words (TW), which are similar in format to the CWs. The TWs are generated by the 196KB and passed to the RS-232C circuitry to be sent to the CC.

The CTIU memory map is shown in Fig. A.2. The system is equipped with an 8-kbyte EEPROM (electrically erasable programmable read-only memory), implemented by a 28C64 chip. The total ROM memory required for the CTIU board is approximately 6 kbytes, which leaves a 25 percent margin for future software updates in EEPROM. An EEPROM was selected as the system ROM to provide the flexibility for software updates during the development phase. The 28C64 chip is radiation protected and proven to have a good resistance to the electron and proton radiation, which may cause single-event upsets and latch-ups at the altitude of the CTIU mission. This property certainly influenced the selection of the 28C64.

The CTIU board is also equipped with an 8-kbyte SRAM, implemented by the 65642 ($8K \times 8$) chip. The total RAM currently required by the system is approximately 4 kbytes, which leaves a 50 percent margin for any dynamic table loads that may be required to reside in RAM. The 65642 chip also has a good resistance against ionized radiation. As shown in Fig. A.2, the ROM is mapped into the address interval 2000H to 3FFFH, and the RAM is mapped into 4000H to 5FFFH.

A 4-bit counter is provided at the clock output (CLKOUT) of the 196KB to divide the output frequency of 4 MHz for slower operations. It is implemented by a 74HC393 chip and it produces 2-MHz, 1-MHz, 512-KHz, and 256-KHz clock trains. The 512-KHz and 256-KHz outputs are used to synchronize the serial digital command/telemetry data and clock operation of the interface circuitry. The overall system diagram is shown in Fig. A.3.

32-Bit Microcontrollers

One of the main reasons for developing 32-bit microcontrollers is that many applications, such as avionics, aerospace, robotics, image processing, and others, require intensive and precise computations when 32-bit data are involved. High-performance requirements, particularly in complicated real-time applications, do not allow performing 32-bit operations by splitting them into subsequent 16-bit operations because multiprecision computations are too slow. Thus, only full-scale 32-bit systems can satisfy the operational requirements in many cases. The development of 32-bit microcontrollers came to satisfy the aforementioned requirements. Four 32-bit microcontrollers, designed to serve as embedded processors, among other things, are presented in this chapter. Intel 80960CA, LSI Logic LR33000, AMD Am29050, and National Semiconductor NS32SF641. The 80960CA was selected as a primary example in this text to be presented in more detail; there is not enough space to present all of the systems on an equal basis. All of these processors are compared at the end of this chapter.

7.1 INTEL 80960CA SUPERSCALAR EMBEDDED PROCESSOR

Following its 8-bit (Chap. 5) and 16-bit microcontrollers (Chap. 6), Intel developed a family of 32-bit microcontrollers—the 80960 family.[1,2] Some additional goals Intel pursued in the development of the 80960 family were to combine the best architectural features of advanced CISC (complex instruction set computers) systems with the implementation efficiencies of newer RISC (reduced instruction set computer)[2] designs.[3] Another Intel goal was to define an architecture that could support a broad family of product designs and span a wide range of different implementation technologies, such as industrial robotics, office automation, medical instrumentation, avionics, aerospace, and others.

The 80960 family (sometimes identified simply as 960 or i960) started with the 80960KA chip, which constituted the realization of the basic RISC-type core architecture, designed for the 960 family.[1] The 80960KA was designed for integer processing, implementing an architecture with a number of distinct RISC-type properties[2] such as memory access by load and store instructions only and a relatively large CPU register file. Some other properties of the core architecture, such as close to 200 instructions and 9 addressing modes,[1] can be regarded as a deviation from RISC principles, which recommend a smaller variety of instructions and addressing modes.[2] The basic core architecture was implemented (sometimes with slight modifications) on all subsequent products of the 960 family. The details of this architecture are presented in Sec. 7.1.3. A floating-point unit was added to the basic core architecture, yielding a new version called *Numerics Architecture*, realized on the 80960KB chip.[1,2] It implemented the floating-point IEEE 754-1985 standard single-precision (32 bits) and double-precision (64 bits), as well as the extended-precision 80-bit format. Adding a memory management and protection capability, protected architecture was attained, realized on the 80960MC chip, particularly designed for military applications.[1,2]

The latest (in 1990) product of the 960 family is the Superscalar 80960CA processor.[3,7] It is also denoted in Intel's literature as 960CA or i960CA. Most of the space in this text is dedicated to the 80960CA. A table comparing it with its predecessors in the 960 family is given later in this chapter. At this point, the term *Superscalar* must be defined.

Superscalar processor: A processor capable of fetching, decoding, and executing more than one instruction at a time

In a superscalar processor, a number of instructions at a time (four on the 80960CA) is fetched from the instruction cache (or from the main memory in case of a miss) into the CPU. This is arranged by providing enough data lines between the cache and the CPU. By providing a sufficient amount of decoding logic circuitry, all of the fetched instructions are decoded simultaneously. To execute the fetched instructions concurrently, there is a need for a sufficient number of execution units capable of working in parallel. For instance, one can have separate integer and floating-point execution units, and separate branch control and address generation units. Within each of the integer and floating point units, we can have separate add/subtract and multiply/divide units. In this case, if the four subsequent instructions, fetched simultaneously are as follows:

Integer add

Floating-point add

Integer multiply

Branch.

The instructions can be executed concurrently, given the foregoing execution units, and provided that the operands can be presented to the execution units on time. The most expedient way of providing the operands is to have them preloaded into CPU registers, as is done in the RISC systems.[2] The next best thing is to have the operands in a data cache (also a common practice in modern RISC and CISC systems).

Problems arise when we have a sequence of identical instructions, such as four floating-point adds, fetched simultaneously. They can be decoded concurrently, but since we would usually have only one floating-point adder, they will be executed sequentially. In this case, the superscalar capability will not be fully utilized. This can be avoided by judicious programming and by installing appropriate measures in an optimizing compiler, trying to avoid sequences of identical instructions, if possible.

The obvious advantage of a superscalar processor is faster execution. If n instructions are fetched from cache at a time and if appropriate n executing resources are available, the processor will execute n times faster than a regular processor. As argued, the n performance ratio may not be achievable at all times because of unavailability of sufficient executing resources. However, on the average, a ratio of well over one instruction per cycle can be achieved, as it is on the 80960CA.

The details of the 80960CA superscalar structure are presented in Sec. 7.1.1, and its architecture in Sec. 7.1.2. A summarizing review of the 80960 family and its auxiliary resources is given in Sec. 7.1.3.

7.1.1 Superscalar Organization

The 80960CA is a 1-micron, CHMOS-IV, 600,000 transistor, 33 MHz (in 1990) chip. It was designed for fast, superscalar, embedded processor operation. The 80960CA block diagram is shown in Fig. 7.1. The 80960CA contains an on-chip 1-kbyte, two-way, set-associative instruction cache, an instruction scheduler with three execution pipelines, a six-port CPU register file, 1.5-kbyte on-chip data RAM, integer execution unit, multiply/divide unit, address generation unit, interrupt controller, DMA (direct memory access) controller, a 4-kbyte microcode storage ROM (part of the 80960CA instructions are microcoded), and a bus controller.

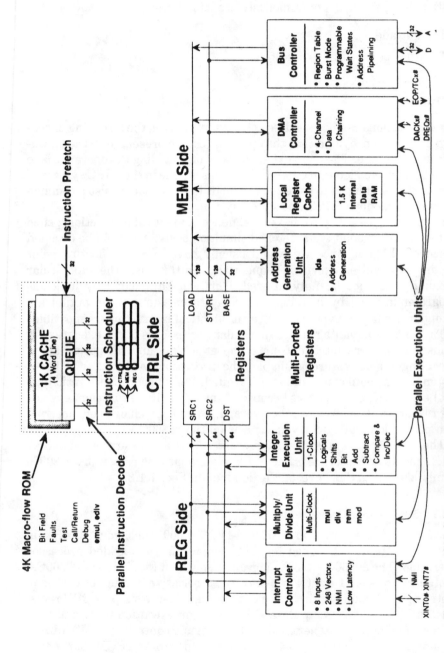

Figure 7.1 1960 CA processor-parallel units. *(Courtesy Intel Corporation)*

Instruction cache. The 80960CA has a 1-kbyte, two-way, set-associative (two lines/set),[8,9] on-chip instruction cache. The line size is 32 bytes, or 8 words (1 word = 4 bytes on the 80960CA). Recent cache performance studies[10,11] have shown that the aforementioned parameters (two lines/set, 32 bytes/line) fall within the parameter values region yielding higher cache hit ratios and better overall performance. A four lines/set design may yield an even higher hit ratio; however, it will complicate the cache control circuitry. Moreover, two lines/set is more suitable for a relatively small cache of 1 kbytes. On a miss, lines within the cache are replaced according to a least recently used (LRU) algorithm.[8,9] The 80960CA designers have introduced a further refinement in the cache refill policy: pairs of words (8 bytes) can be replaced separately to reduce refill delay following program branches or calls.[3]

During the first iteration of a program loop, instructions are fetched from external memory (main memory) using *four-word burst-mode* data transfers (only if memory region is set up for bursts). Each instruction is saved within the cache as it is executed. On successive iterations of the loop, up to four instructions will be retrieved from the cache at a time to be decoded within the instruction scheduler.[3] As can be seen in Fig. 7.1, there are four parallel 32-bit data busses between the instruction cache and the instruction scheduler.

At initialization time, software can split the cache in two. Either or both halves (512 bytes each) can be preloaded with time-critical code, and then "locked" to prevent the code from being replaced later. This ensures that when a high-priority interrupt-service routine is invoked (such as in a real-time, strict deadline application), the critical portions of the routine will begin executing instantly, without first needing to retrieve instructions from a slower main memory outside of the chip. When half of the cache is locked, the other half continues functioning as a conventional 512-byte, two-way, set-associative cache.[3]

The instruction data path from the instruction cache and the operand data path from the data RAM or from the main memory through the bus controller are completely separate, using physically different busses. Thus, instructions and operand data can be accessed concurrently, facilitating pipelined operation[8] and increasing the overall performance.

Instruction sequencer. The instruction sequencer decodes the four instructions, fetched from the instruction cache simultaneously, and schedules them to be executed on three execution pipelines (see Fig. 7.1.), each independent of the others and each dedicated to a different class of instructions. Each pipeline can begin processing a new instruction on every clock cycle, for an instantaneous processing rate (in principle) of three instructions per cycle. In practice, the rate is about two instructions per cycle, since it is difficult to produce three consecutive instructions befitting the three pipelines at all times.[3,4] The three parallel pipelines are illustrated in Fig. 7.2. The pipeline tasks are as follows:

1. *CTRL*: Two-stage, implements program control instructions, such as conditional and unconditional branches and subroutine calls and returns.

2. *MEM*: Three stage, coordinates all memory transfer operations, including address calculations and internal and external data fetches (loads) and stores.

3. *REG*: Three stage, performs most register-to-register arithmetic operations, such as addition, multiplication, logical operations, and bit-field control.

Generally speaking, since four instructions are fetched and decoded simultaneously, it would be reasonable to place four pipelines in the system. This would certainly call for a higher investment in the silicon area taken up by the execution units and would complicate the on-chip logic. Since it is even difficult to feed three pipelines at a time with appropriately matched instructions and data, some of the execution hardware would be idle, and adding a fourth pipeline does not justify the cost to be incurred.

In addition to the decoding and scheduling, the instruction scheduler dynamically detects dependencies between adjacent instructions that prevent those instructions from being executed incorrectly in parallel.[4] The instruction scheduler proceeds to resolve any dependency hazards that were detected.

Example   Assume that the following sequence of four instructions is fetched from the instruction cache into the instruction sequencer.[4]

```
addi    4,r1,r2;      4+(r1) --> r2      add integer
ld      (r3),r4;      M[(r3)] --> r4     load from memory
addi    r5,r6,r7;     (r5)+(r6) -> r7    add integer
muli    r6,r8,r9;     (r6)×(r8) -> r9    multiply integer signed
```

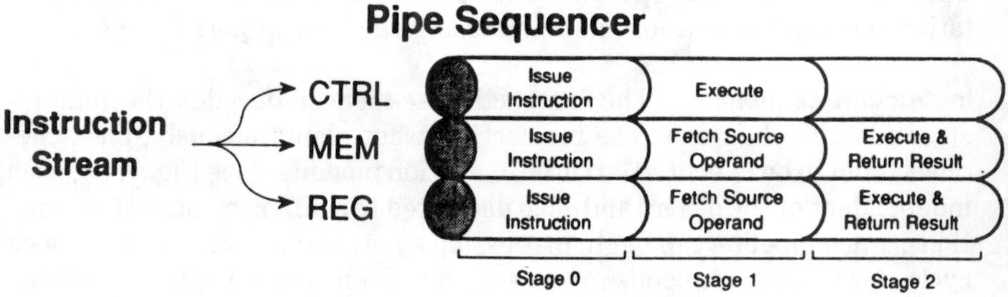

Figure 7.2   1960 CA processor-multiple parallel pipelines. *(Courtesy Intel Corporation)*

```
INSTRUCTION        Clock cycle: 1  2  3  4  5     PIPELINE

addi   4,r1,r2                  II FO ER               REG
ld     (r3),r4                  II FO ER               MEM
addi   r5,r6,r7                    II FO ER            REG
muli   r6,r8,r9                       II FO ER         REG
                            (a)

INSTRUCTION        Clock cycle: 1  2  3  4  5  6  7  8  9

addi   4,r1,r1                  II FO ER
ld     (r1),r4                     II FO ER
addi   r4,r6,r7                       II FO ER
muli   r6,r7,r8                          II FO ER
                            (b)
```

Figure 7.3 Example of an instruction sequence: (a) sequence of independent instructions; (b) sequence of dependent instructions.

where (x) denotes the contents of x, and M[(ri)] denotes the contents of a memory location, whose address is in register ri.

There are no dependencies between the four instructions. Had there been enough resources (pipelines and execution units), the sequencer could schedule all four instructions to be handled simultaneously. The ld instruction should be issued to the MEM pipeline and the other three to the REG pipeline. Since there is only one REG pipeline, the scheduling will be as shown in Fig. 7.3a. The pipeline stages are denoted as follows:

II = issue instruction (Fig. 7.2)

FO = fetch source operand

ER = execute and return result.

Now assume a different set of operands in the same instruction sequence, as shown in Fig. 7.3b. Each of the foregoing instructions uses as a source operand the destination of the preceding instruction. For this reason, the instruction sequencer must delay the issue of the subsequent instruction in such a way that in the beginning of the FO cycle the operand updated by the previous instruction should be available. A hardware resource facilitating the resolution of dependencies is a *scoreboarding* register, a feature practiced in a number of RISC systems.[2] Each bit of the scoreboarding register represents a CPU register. For instance, bit 4 represents register r4. During the execution of the ld (load) instruction, bit 4 is set, preventing the use of r4 by any other instruction. After the load is completed, bit 4 is reset, and register r4 can be used by the next instruction.[2,4]

CPU register file.   The 80960 register file is composed of thirty-two 32-bit registers: 16 global and 16 local. The global registers are used by all procedures running on the 80960. The 16 local registers are allocated for use by the currently running procedure. When another procedure is called, the content of the local registers is stored in the on-chip RAM, and the called procedure uses the local registers. Upon return, the calling procedure resumes using the local registers. The register file has six ports, shown in Fig. 7.1, allowing parallel access of the register set by several 80960 subunits. This parallel access results in an ability to execute one simple logic or arithmetic instruction, one memory access operation (load or store), and one address calculation per clock cycle.[5]

The MEM pipeline subunit (Fig. 7.1) interfaces to the register file with a 128-bit-wide (4 word) LOAD bus and a 128-bit wide STORE bus. This enables the movement of up to four words per clock cycle to and from the register file. It also allows loading data from a previous read access and storing data from a current write access to be processed in the register file simultaneously. An additional 32-bit port (BASE, Fig.7.1) allows an address to be simultaneously fetched by the address generation unit. The REG pipeline subunit interfaces to the register file with two 64-bit source busses (SRC1 and SRC2) and a single 64-bit destination bus (DST). With this bus structure, two source operands are simultaneously issued to the REG pipeline when an instruction is issued. A 64-bit destination bus allows the result from the previous operation to be written to the register file at the same time that the current operation's source operands are issued.[7] The 64-bit bus width allows fast double-precision operation and facilitates future extension to single- and double-precision floating-point operation.

Integer execution unit.   The integer execution unit is the 32-bit ALU of the 80960CA. It can be viewed as a self-contained REG pipeline processor with its own instruction set. As such, the execution unit is responsible for executing or supporting the execution of all the integer and ordinal arithmetic instructions, the logic and shift instructions, the move instructions, the bit and bit-field instructions, and the compare operations. The integer execution unit performs any arithmetic or logical instructions in a single clock cycle.[7]

Multiply/divide unit.   The multiply/divide unit is a REG pipeline subunit that performs integer and ordinal multiply, divide, remainder, and modulo operations. The unit detects integer overflow and divide-by-zero errors. It is optimized for multiplication, performing 32-bit multiply operations in four clock cycles. The multiply/divide unit operates in parallel (concurrently) with the integer execution unit.[7]

Address generation unit.   The address generation unit is a subunit of the

MEM pipeline (Fig. 7.1) that computes the effective address for memory access operations. It directly executes the load address instruction and calculates addresses for load and store instructions based on the addressing mode specified in these instructions. The address calculations are performed in parallel with the integer execution unit.[7]

Internal data ram.   The 80960CA contains an on-chip 1.5-kbyte (1536 bytes) data Static RAM (SRAM). One kbyte of this SRAM is mapped into the 80960CA address space from location 0000 0000H to 0000 03FFH. In other words, the first kilobyte of the 80960CA main memory is on chip. The remaining 512 bytes of the on-chip SRAM are dedicated to internal system functions and are not visible to user software. Internally the SRAM is 128 bits wide and is directly connected to the 128-bit LOAD and STORE busses, which access the register file (Fig. 7.1). Thus, up to four 32-bit words can be transferred between the register file and the SRAM, in either direction, in a single clock cycle.

Part of the SRAM area operates as a *local register cache* (LRC). When a program executes a subroutine call, 16 local registers of the register file are automatically copied to the LRC. On subroutine returns, the local register values are automatically restored into the local registers of the register file. Using quad-word (four word) transfers, all 16 registers can be saved or restored in just four clock cycles. By default, the LRC is configured to hold 5 register sets (5 × 16 = 80 registers = 320 bytes), which should satisfy the majority of all save and restore sequences. For specialized applications, with deeper subroutine nesting, the LRC can be extended into the high end of the program-visible 1-kbyte data RAM, to hold up to a total of 15 register sets (15 × 16 = 240 registers = 960 bytes; 960 − 320 = 640 bytes taken up from the 1024-byte RAM). Only five sets can reside in LRC for a total of 320 bytes; any extension of over five sets must go into the 1-kbyte RAM starting from the upper address down. If the LRC is full when a call is made, the oldest register set is copied to external main memory to make room for the new set; the values will be restored later from main memory automatically, when space in the LRC becomes available upon subroutine returns.

The first 256 bytes of the internal SRAM are user write-protected to prevent inadvertent corruption from user applications software, and to ensure operating system data integrity. Remaining locations can be write-protected under software control according to system requirements.[3,4]

Interrupt controller.   The on-chip interrupt controller operates in several modes. The controller coordinates, prioritizes, and latches up to 248 separate requests from internal and external interrupt sources. Interrupt-service vectors (starting addresses of interrupt-service routines) may be contained in internal data RAM, external main memory, or both. Eight input pins (XINT0#-XINT7#; Fig. 7.1; # denotes low asserted) can be configured as separate dedicated interrupt requests, as an encoded 8-bit service request index, or as three dedicated requests and 31 encoded indices. A ninth input pin requests nonmaskable interrupts (NMI). When an interrupt request is received, the prior state of the executing program may be saved in the on-chip SRAM and LRC. The service routine vector can be retrieved from internal SRAM, and the service routine instruction fetched from a previously loaded section of the locked instruction cache. Thus, an entire time-critical service routine can complete in a worst-case interval calculated in advance, completely independent of the main memory performance or memory bus loading at the time the service routine is invoked.[3] This feature is particularly important in real-time applications, where better predictability of performance can assure the meeting of restrictive deadlines.

Direct memory access (DMA) controller. The 80960CA contains a four-channel, on-chip, I/O processor (IOP) and DMA controller. The IOP can copy large blocks of data between memory regions while leaving the CPU free to address more general programming problems. Data can be copied between memory and peripheral devices without disrupting background program execution by the CPU, and with a service delay significantly lower than that of an interrupt service routine. DMA transfers operate in various modes. Block-mode transfers are free-running, that is, the associated channel is initialized with the number of bytes to transfer and continues transferring the data until the bytes to be transferred counter is reduced to zero. Demand-mode transfers are synchronized to external events. A source or destination I/O port indicates when it is ready to provide or absorb more data, and the DMA channel performs a single transfer to satisfy each request.[3]

Microcode ROM. The 80960CA uses a 4-kbyte microcode ROM to implement complex instructions and functions. This includes calls, returns, DMA transfers, and initialization sequences. This feature contradicts RISC principles,[2] but then the 80960CA is not really a RISC-type system; it only uses a limited number of RISC principles as argued in the beginning of this chapter. When the instruction sequencer encounters a microcoded instruction, it automatically branches to the microcode routine. The 80960CA performs this microcode branch in less than a clock cycle.[7]

Bus controller. The 80960CA bus interface hardware provides separate, unmultiplexed 32-bit busses for address and datasic single-word transfers can complete in two clock cycles. *Burst-mode* transfers can be enabled to significantly increase data-bus bandwidth. In burst mode, a single address is sent by the CPU to retrieve up to four successive data words in five clock cycles. Both individual-word and burst-mode transfers may also be pipelined, in which case the CPU sends out the next data address during a previous transfer data cycle, to keep the data bus in continuous use. Memory devices interfaced to the bus controller may be 8, 16, or 32 bits wide, in any combination. The bus controller automatically converts each 16- or 32-bit transfer into the proper sequence of partial-word read or write transfers when referencing less wide memories.[3] The bus controller can only be configured for the different size accesses on region-wide (256 Mbyte) boundaries. The 80960CA is a 168-pin ceramic pin grid array (PGA) package. A detailed description of the 80960CA pins is given in Appendix A at the end of this chapter.

7.1.2 80960CA Architecture

CPU registers.   The 80960CA has thirty-two 32-bit general-purpose registers, illustrated in Fig. 7.4. The registers are subdivided into two equal groups: 16 global and 16 local registers. The global registers are denoted: g0, g1, . . . , g15. Registers g0 through g14 are general-purpose global registers; no specific task is assigned to them. One exception is register g14, which stores the address of the next instruction following a branch and link (bal) instruction. Register g15 is reserved as the current *frame pointer* (FP), discussed later in this section. The FP or g15 register contains the address of the first byte in the current stack frame.

The local registers are denoted r0, r1, . . . , r15. Registers r3 through r15 are general purpose local registers. Registers r0, r1, and r2 are reserved for special functions as follows:

r0 contains the *previous frame pointer* (PFP)

r1 contains the stack pointer (SP)

r2 contains the return instruction pointer (RIP).

The PFP, SP, and RIP registers manage stack frame linkage for the 80960CA's procedure call and return mechanism, discussed later in this section. When a procedure call or return is executed, the content of global registers is preserved across procedure boundaries. In other words, the same set of global registers is used for each procedure. The content of the local registers is stored in the local register cache (LRC) on each procedure call. The 80960CA also has three *special function registers* (SFR) for communicating with on-chip peripherals. The SFRs are designated as sf0, sf1, and sf2. The SFRs can be accessed as source operands by most of the 80960CA instructions. The registers serve as part of the user's interface to the DMA and interrupt controllers. There are also some control registers that, like SFRs, are used to communicate with the on-chip peripherals.

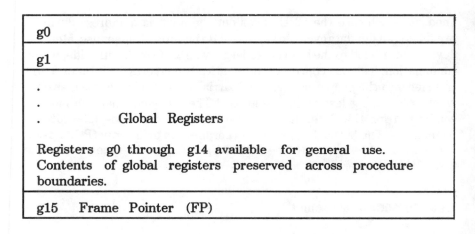

Figure 7.4   80960CA CPU general purpose register file.

Configuration information for the peripherals is generally stored in the control registers. Control registers can only be accessed by using the system control (sysctl) instruction, which can only be accessed from supervisor mode. The sysctl instruction is used to load an internal control register from a table in external memory called the *control table*. To simplify the process of peripheral configuration, the control registers are automatically loaded from the control table at initialization.[7]

Address space.   The 80960 family uses a 32-bit address. The virtual address space is therefore 2^{32} bytes, or 4 Gbytes in size. This address space is linear (unsegmented); therefore, code, data, and memory-mapped peripherals may be placed anywhere in the usable space. Some memory locations on the 80960CA are reserved or are assigned special functions, as shown in Fig. 7.5. The first 1024 bytes (addresses 0 through 1023), shown in Fig. 7.5, belong to the on-chip SRAM, discussed in the previous section. When the DMA controller is in use, 32 bytes of this SRAM are reserved for each DMA channel, for a total of $32 \times 4 = 128$ bytes (locations 64 through 255; Fig. 7.5). Additionally, 64 bytes are reserved for 16 interrupt vectors (locations 0 through 63). The availability of these interrupt vectors in the on-chip SRAM reduces interrupt delay, since they can be accessed much faster than in the external memory.[7] The 80960CA can execute in the user mode or in the supervisor mode. The first 256 bytes of the address space (locations 0 to 255; Fig. 7.5) are always write protected when a program is executing in user mode but may always be written to when executing in supervisor mode. The remainder of the on-chip SRAM can be programmed for this protection feature. The user and supervisor modes are discussed later in this section.

The upper 16 Mbytes of memory (locations FF00 0000H through FFFF

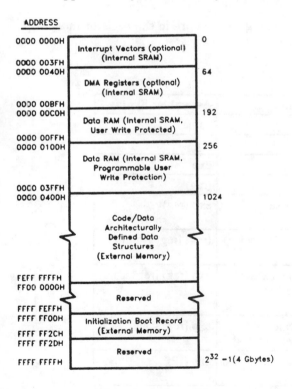

Figure 7.5 Address space. (*Courtesy Intel Corporation*)

FFFFH; Fig. 7.5) are reserved for specific functions and extensions to the 80960 architecture. The 12 words in reserved space (locations FFFF FF00H through FFFF FF2CH) are used to start up the processor when it comes out of reset. These 12 words are called the *initialization boot record*.

Addressing modes. The addressing modes of the 80960CA are summarized in Table 7.1. and are as follows:

1. *Literal*: Essentially the same as immediate in the range of 0 to 31 (5 bits). When a literal is used as an operand, the processor expands it to 32 bits by adding leading zeros. If the instruction defines an operand larger than 32 bits, the processor zero extends the literal to the operand size.

2. *Register*: The operand is the content of a CPU register. The register is referenced by specifying the register number: g0, g6, r1, and so on. For example,

```
addi 2,g7,r4; 2 + (g7) --> r4
```

2 is in the literal mode, while g7 and r4 are in the register mode. (g7) denotes the content of register g7.

TABLE 7.1 Addressing Modes

Mode	Description	Assembler Syntax
Literal	value	value
Register	register	reg
Absolute offset	offset	exp
Register Indirect	abase	(reg)
Register Indirect with offset	abase + offset	exp (reg)
Register Indirect with index	abase + (index*scale)	(reg) [reg*scale]
Register Indirect with index and displacement	abase + (index*scale) + displacement	exp (reg) [reg*scale]
Index with displacement	(index*scale) + displacement	exp [reg*scale]
IP with displacement	IP + displacement + 8	exp (IP)

Where reg is register and exp is expression.
SOURCE: Courtesy Intel Corporation.

3. *Absolute*: Absolute addressing is used to reference a memory location directly as an offset from address 0 of the address space. Typically, an assembler will allow absolute addresses to be specified through arithmetic expressions (xx + 10), symbolic labels (ALPHA), and absolute values (FF00H).

4. *Register indirect*: The content of the register specified by the mode is the address of the operand. The value in the register is referred to as the *address base* or *abase*. The register is specified in round brackets, such as (r8).

5. *Register indirect with offset*: The operand address is the sum of the register content and an offset. An offset can be specified explicitly, or as an expression, or as a symbolic label. The offset is specified in front of the register brackets. For example,

    ld 8(r6),g3; M[8 + (r6)]--> g3

An operand is fetched from a memory location at an address calculated as the sum of 8 and the content of r6, and loaded into register g3.

6. *Register indirect with index*: The content of another register, denoted as the index, is multiplied by a scale factor and added to the content of the register indirect (abase) to form the address of the operand. The scale factor can be 1, 2, 4, 8, or 16. The scaled index is specified within square brackets, following the register indirect. For example,

    ld (r5)[r7*4],r3; M[(r5) + (r7)*4]--> r3

The content of index register r7 is multiplied by 4 and added to the content of r5 to form an operand address in memory. The operand is loaded into register r3. A useful aspect of the index mode is that it *automatically* scales to the size of the memory instruction being executed. For example,

    ldq (g5)[g1],r6; load quadword (16 bytes)

When this instruction is assembled, the [g1] part will expand to [g1*16]. This feature makes it easy to write loop counters for unaligned word accesses.

7. *Register indirect with index and displacement*: A displacement value is added to the operand address calculated in the register indirect with index mode.[6] The displacement is specified explicitly or symbolically in front of the register indirect (abase), as shown in Table 7.1.

8. *Index with displacement:* A displacement is added in front of the index specification, as shown in Table 7.1. The displacement is specified as in item 7.

9. *Instruction pointer (IP) with displacement:* This mode is used with load and store instructions to make them IP relative. The IP contains the address of the instruction to be executed (same as the program counter, or PC, in other systems). The effective address (EA) is

$$EA = (IP) + \text{displacement} + 8$$

The 80960 architecture features a generous number of addressing modes, compared to RISC-type systems, which usually have no more than four.[2] On the other hand, the 80960 does not feature some addressing modes, such as memory deferred, autoincrement, and autodecrement, that can be found on other systems, such as the VAX.[9] The reason why the memory deferred mode is not featured is that it is used infrequently, it would add considerable complexity to the instruction decoder, and it would slow down the operation, since it requires an extra memory access to fetch an operand. This is unacceptable for a processor intended to be used as an embedded controller. The omission of autoincrement and autodecrement modes was also motivated by a goal to reduce complexity. The inclusion of such modes in a superscalar processor such as the 80960CA would have made it much more difficult to detect dependencies among multiple instructions. Having load and store instructions with varying values in index registers would complicate dependence detection significantly.

Instruction formats.   The 80960CA has five instruction formats, illustrated in Fig. 7.6: REG (register), COBR (control and branch), CTRL (control), MEMA (memory A), and MEMB (memory B). In all instruction formats the upper 8 bits, 24 through 31, are dedicated to the opcode. The REG format, used by most of the instructions, is intended for operations on data contained in the global and local CPU registers. In addition to bits 24–31, bits 7–10 are also dedicated to the opcode. Thus, the REG format opcode takes up 12 bits. The 5-bit src1 and src2 fields specify source operands for the instruction. The operands can be either registers or literals. The distinction between registers and literals is done with the help of mode bits M1 (for src1) and M2 (for src2). If the mode bit is zero, a register is specified. If it is 1, a literal is specified. The third mode bit, M3, distinguishes between integer (M3 = 0) and floating-point (M3 = 1) operations in 80960 products that have the floating-point capability (KB and MC). It is not in use on the 80960CA. The src/dst field specifies either a source or a destination register, depending on the instruction.[12] The local and global registers are distinguished by the leading bit: the 0xxxx encoding represents local registers, and 1xxxx

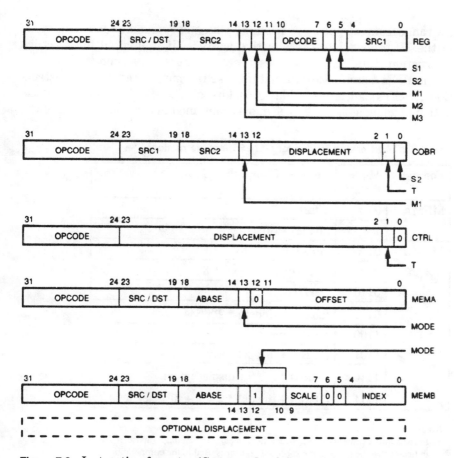

Figure 7.6 Instruction formats. *(Courtesy Intel Corporation)*

represents the global registers. Bits 5 and 6 are reserved for future extensions.

The COBR format is used for control, branch, and test instructions. The src1 and src2 fields specify source operands for the instruction. The src1 field can specify either a register or a literal, using mode bit M1, as in the REG format. The src2 field can only specify a register. Bit 0 serves as an S2 bit, allowing src2 to be an SFR (for instructions such as bit comparisons where src2 is an SFR, such as BBC or BBS). The 11-bit displacement field contains a two's complement value in the range from −1024 to 1023, and it is used for effective address calculation by several addressing modes. Bits 0 and 1 are reserved for future extension.[12]

The CTRL format is used for branch to a new IP, call, and return instructions. The 22-bit displacement field is a two's complement value in the range from -2^{21} to $2^{21} - 1$. The displacement value is added to the value in the IP to calculate the effective address. It is ignored for a return instruction. Bits 0 and 1 are reserved for future expansion.[12]

The MEMA and MEMB formats are used for memory addressing instructions, such as load, store, or lda (load address). The MEMB format offers the option of using a 32-bit displacement, contained in the second word of the

TABLE 7.2   Addressing Modes for MEM Format Instructions

Format	Mode Bit(s)	Address Computation
MEMA	0	offset
	1	(abase) + offset
MEMB	0100	(abase)
	0101	(IP) + displacement + 8
	0110	reserved
	0111	(abase) + (index) * 2^{scale}
	1100	displacement
	1101	(abase) + displacement
	1110	(index) * 2^{scale} + displacement
	1111	(abase) + (index) * 2^{scale} + displacement

Notes:
1. In the address computations above, a field in parentheses (e.g., (abase)) indicates that the value in the specified register is used in the computation.
2. The use of a reserved encoding causes an invalid opcode fault to be signaled.

Encoding of Scale Field

Scale	Scale Factor (Multiplier)
000	1
001	2
010	4
011	8
100	16
101 to 111	reserved

Note:
The use of a reserved encoding causes an invalid opcode fault to be signaled.

source: Courtesy Intel Corporation.

instruction. Using this option yields a 64-bit instruction, double the standard 32-bit length. This contradicts one of the RISC principles of uniform instruction length,[2] however, the user always has the option of not using this extension in the program. The distinction between the MEMA and MEMB formats is done by bit 12: clear (0) for MEMA, set (1) for MEMB. The src/dst field specifies the source register for store and the destination register for load instructions. The abase field specifies the register used for the register indirect addressing mode. The mode bit in MEMA and the mode field (bits 10–13) in MEMB specify the addressing mode according to the encoding in Table 7.2. The scale field in MEMB specifies the index scale factor, according to the encoding in Table 7.2. The offset field in MEMA (bits 0–11) specifies an unsigned byte offset from 0 to 4095. Bits 5 and 6 in MEMB are reserved for future extension.[12]

Data types. The 80960 architecture recognizes the following data types, illustrated in Fig. 7.7:

1. Integer of 8, 16, 32, and 64 bits

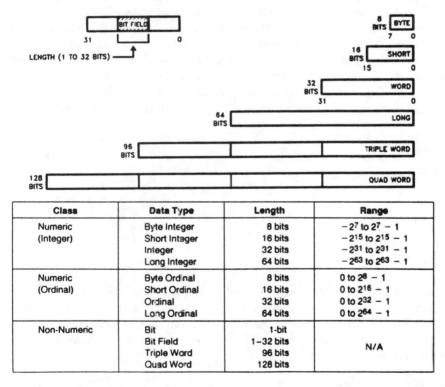

Class	Data Type	Length	Range
Numeric (Integer)	Byte Integer	8 bits	-2^7 to $2^7 - 1$
	Short Integer	16 bits	-2^{15} to $2^{15} - 1$
	Integer	32 bits	-2^{31} to $2^{31} - 1$
	Long Integer	64 bits	-2^{63} to $2^{63} - 1$
Numeric (Ordinal)	Byte Ordinal	8 bits	0 to $2^8 - 1$
	Short Ordinal	16 bits	0 to $2^{16} - 1$
	Ordinal	32 bits	0 to $2^{32} - 1$
	Long Ordinal	64 bits	0 to $2^{64} - 1$
Non-Numeric	Bit	1-bit	
	Bit Field	1–32 bits	N/A
	Triple Word	96 bits	
	Quad Word	128 bits	

Figure 7.7 Data types. *(Courtesy Intel Corporation)*

2. Ordinal of 8, 16, 32, and 64 bits

3. Bit

4. Bit field, up to 32 bits

5. Triple word, 96 bits

6. Quad word, 128 bits.

The basic word is 32 bits. A 16-bit data item is called a *short integer*, and a 64-bit data item is called a *long integer*. Integers and ordinals are considered numeric data types since the processor performs arithmetic operations with this data. The integer data type is a signed binary value in two's complement representation. The ordinal data type is an unsigned binary value. The remaining data types (bit field, triple word, and quad word) represent groupings of bits or bytes that the processor can operate on as a whole, regardless of the nature of the data contained in the group. These data types facilitate the moving of blocks of bits or bytes from one location to another.[7]

Instruction set. The instruction set of the 80960CA, subdivided into groups of different types, is shown in Table 7.3. The data movement instructions include load (ld), store (st), and move (mov) operations. The load instructions transfer (copy) bytes, short data, words, or multiple words from memory to a selected register or group of registers, depending on the size of the moved operand. The store instruction transfers the same types of operands in the opposite direction, from a register (or registers) into memory. The ld and st instructions operate on a word by default. For other operand sizes, the following should be added to the ld or st mnemonics:

1. ob for ordinal byte

2. os for ordinal short

3. ib for integer byte

4. is for integer short

5. l for long

6. t for triple

7. q for quad.

The move instructions copy data between registers. The mov mnemonic refers to words. One has to append to it l, t, q to copy long, triple, quad words respectively.[7] Two-word operands must be contained in even, odd register pairs and are denoted by specifying the number of the lowest (even) register. For example,

```
movl  g2,r6;    g2,g3 --> r6,r7
```

TABLE 7.3 Summary of the 80960CA Instruction Set

Data Movement	Arithmetic	Logical	Bit, Bit Field, and Byte
Load	Add	And	Set Bit
Store	Subtract	Not And	Clear Bit
Move	Multiply	And Not	Not Bit
Load Address	Divide	Or	Alter Bit
	Extended Multiply	Exclusive Or	Scan For Bit
	Extended Divide	Not Or	Scan Over Bit
	Remainder	Or Not	Extract
	Modulo	Nor	Modify
	Shift	Exclusive Nor	Scan Byte For Equal
	* Extended Shift	Not	
	Rotate		
Comparison	**Branch**	**Call/Return**	**Fault**
Compare	Unconditional	Call	Conditional Fault
Conditional	Branch	Call Extended	Synchronize Faults
Compare	Conditional	Call System	
Check Bit	Branch	Return	
Compare and	Compare and	Branch and Link	
Increment	Branch		
Compare and			
Decrement			
Test Condition			
Code			
Debug	**Atomic**	**Processor**	
Modify Trace	Atomic Add	Flush Local	
Controls	Atomic Modify	Registers	
Mark		Modify Arithmetic	
Force Mark		Controls	
		Modify Process	
		Controls	
		*System Control	
		*DMA Control	

NOTE: Instructions marked by (*) are 80960CA extensions to the 80960 instruction set.

SOURCE: Courtesy Intel Corporation.

Three- and four-word operands must be contained in consecutive registers, the first of which must have a number that is a multiple of four, such as g0, g4, g8, g12, r0, r4, r8, or r12.[12] For example,

```
movq r12,r4; r12,r13,r14,r15 --> r4,r5,r6,r7
```

The load address (lda) instruction causes a 32-bit address to be computed (according to the specified addressing mode) and placed in a destination register. The lda instruction is useful for loading a 32-bit constant into a register. Logical instructions perform bit-wise boolean operations on operands in registers. Since this group of instructions performs only bit-wise manipulation of data, separate logical instructions for integer and ordinal data types are not necessary and therefore are not featured. The logical instructions implemented on the 80960CA are shown in Table 7.3. Considering that most other computing systems offer only AND, OR, Exclusive OR (XOR), and NOT logical instructions, the logical instruction subset of the 80960CA can be considered excessive. After all, a variety of logical functions can be constructed from just a few.[13]

Arithmetic instructions (Table 7.3) perform add, subtract, multiply, divide, and shift operations on integer or ordinal operands in registers. An i is added to the instruction mnemonic on the right for integer data, and an o is similarly added for ordinal data. Thus, *addi* means add integer data, and *addo* means add ordinal data. Other arithmetic instructions (sub, mul, div) are dealt with in a similar manner. The 80960CA instruction set also features extended arithmetic instructions that facilitate computation on ordinals and integers longer than 32 bits. In add-with-carry and subtract-with-carry instructions, the carryout from the previous arithmetic instruction is used in the computation. The extended multiply instruction multiplies two ordinal source operands producing a long ordinal result (64 bits). The extended divide instruction divides a long ordinal dividend by an ordinal divisor and produces a 64-bit result.[7]

The atomic instructions (atomic add—atadd, atomic modify—atmod; Table 7.3) perform read-modify-write operations on operands in memory. They allow a system to ensure that when an atomic operation is performed on a specified memory location, the operation will be completed before another potential bus master is allowed to perform an operation to the same memory. These instructions are required to enable synchronization between different processors and processes, sharing the same memory in a multiprocessing system.[7,14] The atadd instruction performs the following operation:

atadd a,b,c;   (a) + b --> c

The atmod performs:

atmod a,b,c;   ((c) AND (b)) OR ((a) AND (b)) --> c

The bit instructions (Table 7.3) operate (set, clear, not, and so on) on a specified bit in a register. Bit field instructions operate (extract, modify, scan for byte) on a specified contiguous group of bits in a register. This group of bits can be from 0 to 32 bits in length.[7]

In the regular branch instructions of the 80960CA (Table 7.3), the target address is specified as a displacement to be added to the current value of IP (instruction pointer). The extended branch instructions allow IP value calculation using any addressing mode. The extended branch option is selected by adding x to the instruction mnemonic on the right. Thus, b is a regular unconditional branch, while bx is an extended unconditional branch. The branch-and-link instruction automatically saves a return instruction pointer (RIP) in r2 (Fig. 7.4) before the branch is taken. The RIP contains the address of the instruction following the branch-and-link. The branch-and-link instruction can be regular (bal) or extended (balx). Conditional branch instructions alter program flow only if the condition code flags in the arithmetic control register (specified later in this section) match a value specified by the instruction. The condition code flags indicate conditions of equality or inequality between two operands in a previously executed instruction.[7]

The 80960CA architecture has a special feature, called *branch prediction*, that allows the user to insert into the program a prediction of whether a given conditional branch will be taken. Based on a *branch prediction flag* located in bit 1 of the instruction formats COBR and CTRL,[7] the 80960CA will assume that an instruction takes or does not take a conditional branch, depending on the bit value. By executing along the predicted path of program flow, delays due to breaks in the instruction stream are often avoided. The branch prediction flag is specified at the assembly level by appending a .t for assuming a branch is taken, or an .f for assuming a branch is not taken.[7]

> *Example*   The assembler mnemonic *be.t* means that the processor will assume that this branch-if-equal instruction usually branches when encountered. On the other hand, the mnemonic *bl.f* means that the processor will assume that the branch-if-less will not be taken.[7]

Compare and conditional branch instructions compare two operands, then branch according to the immediate results. These instructions also feature the branch prediction capability. The 80960CA provides several types of instructions that are used to compare two operands (compare integer, compare ordinal, check bit). The condition code flags in the arithmetic control register are set to indicate whether one operand is less than, equal to, or greater than the other operand.[7] Conditional compare instructions (for integers and ordinals separately) test the existing status of the condition code flags before a compare is performed. These conditional compare instructions are provided to optimize two-sided range comparisons (that is, to test if a value is less than one number but greater than another).[7] The compare and increment and compare and decrement instructions (for integers and ordinals separately) set the condition code flags based on a comparison of two register sources, decrement or increment one of the sources, and store this result in a destination register.[7] The condition test instructions allow the state of the condition code flags to be tested. Based on the outcome of the comparison, a *true* or *false* code is stored in a destination register. The branch prediction flag is used in these instructions to reduce the execution time when the test outcome is predicted correctly.[7]

> *Example*[7]   The *teste.t* (test-if-equal) instruction will execute in a shorter time if the condition code flags test TRUE for the equal condition. This will happen because the user predicted ahead of time that the test will be successful by placing *.t* at the end of the mnemonic.

The 80960CA architecture features call, call extended, call system, and return instructions (Table 7.3). The call instructions and the procedure call and return mechanism are discussed later in this section. The 80960CA will fault automatically as a result of certain erroneous operations that may occur when code is executed. Fault procedures are then invoked automatically to handle the various types of faults. In addition, the fault instructions permit a fault to be generated explicitly based on the value of the condition code flags. These instructions also feature the branch prediction option. The synchronize faults (syncf) instruction causes the processor to wait for all faults to be generated that are associated with any prior uncompleted instructions. The 80960CA supports debugging and monitoring of program activity through the use of trace events. The debug instructions (Table 7.3) support debugging and monitoring software. The 80960CA provides several instructions for direct control of processor functions and for configuring the system peripherals. These instructions constitute the processor management group, listed in Table 7.3.[7]

Arithmetic control register.   The 32-bit arithmetic control (AC) register, shown in Fig. 7.8, is used primarily to monitor and control the execution of 80960CA arithmetic instructions. The processor reads and modifies bits in the AC register when performing different arithmetic operations. The AC register is also used to control fault conditions for some instructions. The *modac* instruction allows the user to directly read or modify the AC register. The processor sets the condition code flags (bits 0–2) to indicate equality or inequality as the result of certain instructions (such as the compare instructions). Other instructions, such as the conditional branch instructions, take action (branch taken or not taken) based on the value of the condition code flags (bits 0–2), whose functional assignment is listed in Fig. 7.8.[7] The integer overflow flag (bit 8) and the integer overflow mask (bit 12) are used in conjunction with the arithmetic integer overflow fault. The mask bit masks the integer overflow fault. When the fault is masked, and an integer overflow occurs, the integer overflow flag is set but no fault handling action is taken. If the fault is not masked, and an integer overflow occurs, the integer overflow fault is taken and the integer overflow flag is not set. The no

Arithmetic Condition Codes

Condition Code	Condition
001	Greater Than
010	Equal
100	Less Than

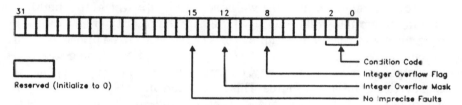

Figure 7.8 Arithmetic control register. *(Courtesy of Intel Corporation)*

imprecise fault flag (bit 15) determines if imprecise faults are allowed to occur. Fault handling and the concept of precise and imprecise faults are discussed later in this section.[7]

Process management.   Process management on the 80960CA is facilitated by the *process control* register, illustrated in Fig. 7.9 . The process control register provides access to process state information, allowing monitoring and control of certain properties of an executing process. The execution mode flag (bit 1) indicates that the processor is executing in user mode when reset (0) or supervisor mode when set (1). The trace enable bit (bit 0) and the trace fault pending (bit 10) flags control and monitor trace activity in the processor. The trace enable bit enables fault generation for trace events. The trace fault pending flag indicates that a trace event has been detected. The state flag (bit 13) determines the executing state of the processor. The processor state can be either executing when bit 13 is reset (0) or interrupted when bit 13 is set (1). The priority field (bits 16–20) indicates the current executing priority of the processor. Priority values range from 0 (lowest priority) to 31 (highest priority). The process control register can be

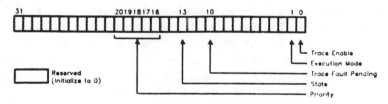

Figure 7.9 Process control register *(Courtesy Intel Corporation)*

modified by software with the modify process controls (modpc) instruction. It is a privileged instruction that may only write into the process control register when the processor is in supervisor mode. The 80960CA architecture defines a means to assign priorities to executing programs and interrupts. The current priority of the processor is stored in the priority field of the process control register, as mentioned earlier. This priority is used to determine if an interrupt will be serviced and in which order multiple pending interrupts will be serviced. Setting the priority of an executing program above that of interrupts allows critical code to be prioritized and executed without interruption. On the other hand, by assigning a certain interrupt source a higher priority, one can assure fast interrupt response to that particular interrupt source. This is important in real-time applications, where fast, predictable responses are often required. The priority field is also modified to reflect the priority of serviced interrupts. On a return from an interrupt-service routine, the priority of the processor is restored to its priority before the interrupt occurred.

As mentioned earlier, the 80960CA may execute programs in user mode or supervisor mode. The user-supervisor protection mechanism allows a system to be designed in which operating system (OS) code and data reside in the same address space as user code and data, but access to the OS procedures and data is only allowed through a tightly controlled interface. This interface is the system call table and the interrupt mechanism, to be discussed later. The processor has two operating states: executing and interrupted. In executing state, the processor can execute in user or supervisor mode. In the interrupted state, the processor always executes in supervisor mode.[7]

Call and return mechanism. The 80960CA call and return mechanism is designed to make procedure calls simple and fast and to provide a flexible method for storing and handling variables that are local to a procedure. A call instruction automatically allocates a new set of local registers and a new stack frame. All linkage information is maintained by the processor, making procedure calls and returns virtually transparent to the user. The call and return model supports efficient translation of structured high level code (such as C, or Ada) to 80960CA machine language.

At any point in a program, the processor has access to a local register set and a section of the procedure stack referred to as a *stack frame*. When a call is executed, a new stack frame is allocated in memory for the called procedure. Additionally, the current local register set is saved by the processor, freeing these registers for use by the newly called procedure. In this way, every procedure has a unique stack frame in memory and a unique set of local registers. When a return is executed, the current local register set and current stack frame are deallocated. The previous local register set and previous stack frame are restored. This call and return mechanism is

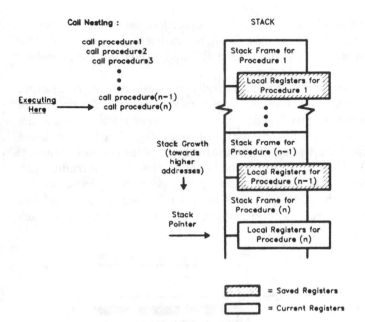

Figure 7.10 Call and return mechanism. *(Courtesy Intel Corporation)*

illustrated in Fig. 7.10.[7] The procedure stack structure is defined and supported by the 80960CA architecture. The procedure stack always grows toward higher memory addresses, and the stack pointer (SP) always points to the next available byte of the stack frame. Each stack frame must begin on a 16-byte boundary. Due to this alignment requirement, a padding space of 0 to 15 bytes may exist between adjacent stack frames in memory. When a stack frame is allocated, the first 16 words (64 bytes) are always assigned as storage for the local registers (for overflow from the local register cache; see Sec. 7.1.1). Therefore, the SP initially points to the seventeenth word in the stack frame. Although each stack frame is assigned storage space for the local registers, these locations in the stack are not guaranteed to contain the values of the saved local registers. This is because several sets (5 to 15) of local registers are stored in the local register cache (LRC; see Sec. 7.1.1). Movement of data (16 words) between the local registers and the LRC is typically accomplished in four clock cycles, with no external bus traffic. When the LRC is filled and another call occurs, we have an LRC overflow. The registers associated with the oldest stack frame are moved to the area reserved for those local registers on the stack frame in memory. The LRC is a physical extension of the on-chip data RAM, discussed in Sec. 7.1.1. The part of the data RAM used for the LRC is not visible to the user and is large enough to hold up to five sets of local registers. The LRC may be extended to hold up to 15 sets of local registers. When extended, each new register set takes up 16 words of the user's data RAM, beginning at the highest address and growing downward. The size of the LRC is selected when the processor is initialized.[7]

The 80960CA architecture automatically manages procedure linkage. One global register and three local registers are reserved for procedure linkage information. Figure 7.11 describes the pointer structure used to link frames and to provide a unique SP for each frame. Register g15 is the frame pointer (FP). The FP is the address of the first byte of the current (topmost) stack frame. The FP is always updated to point to the current frame when calls and returns are executed. Register r0 is the previous frame pointer (PFP). The PFP is the address of the first byte of the stack frame that was created prior to the frame containing this PFP. Register r1 is the stack

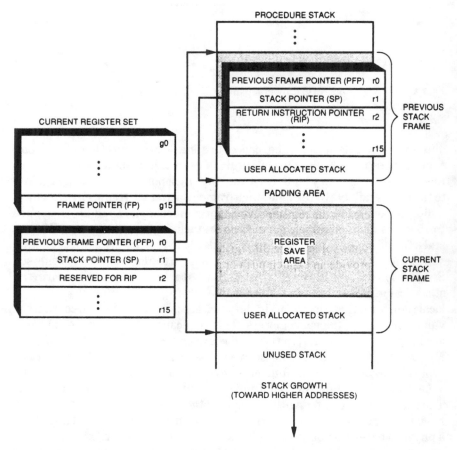

Figure 7.11 Procedure stack structure and local registers. *(Courtesy Intel Corporation)*

pointer (SP). The SP points to the next available byte of the stack frame. Register r2 is reserved for the return instruction pointer (RIP). The RIP is the address of the instruction that follows a call instruction. This is also the target address for the return from that procedure. The RIP is automatically stored in register r2 of the calling procedure when a call instruction is executed. Parameters between calling and called procedures may be passed by reference. The global registers, the stack, or predefined data structures in memory may be used to pass these parameters. The global registers provide the fastest method for passing parameters, since they can be equally used by all procedures. When a procedure is called, the values in the global registers, stored by the calling procedure, are preserved and can be used by the called procedure. If more parameters are to be passed than will fit in the global registers, additional parameters may be passed in the stack of the calling procedure, or in a data structure that is referenced by a pointer passed in the global registers. A total of 15 global registers (g0–g14) is available for parameter passing. This should accommodate a large class of programs.[7]

The 80960CA provides two methods for making procedure calls: *local calls* and *system calls*. The local call instructions initiate a procedure call using the call and return mechanism described earlier. The stack frames for these procedure calls are allocated on the *local procedure stack*. A local call is made using either of two local call instructions: *call* or *callx*. The call instruction specifies the address of the called procedure using an IP plus displacement addressing mode (explained earlier). The callx (call extended) instruction specifies the address of the called procedure using any of the 80960CA addressing modes.[7]

A system call is made using the *calls* instruction. This call is similar to a local call except that the processor fetches the IP value for the called procedure from a data structure called the *system procedure table*. The calls instruction requires a procedure number operand. This procedure number serves as an index into the system procedure table, which contains IP values for specific procedures. The system procedure table is shown in Fig. 7.12. The system call mechanism supports two types of procedure calls: *system-local calls* and *system-supervisor calls* (also referred to as supervisor calls). The system-local call performs the same action as the local call instructions with one exception: The IP target value for a system-local call is fetched from the system procedure table. The supervisor call differs from the local call as follows:

1. A supervisor call causes the processor to switch to another stack, called the supervisor stack.

2. A supervisor call causes the processor to switch to the supervisor execution mode and asserts the supervisor pin (SUP#; see App. 7.A) for all bus accesses.

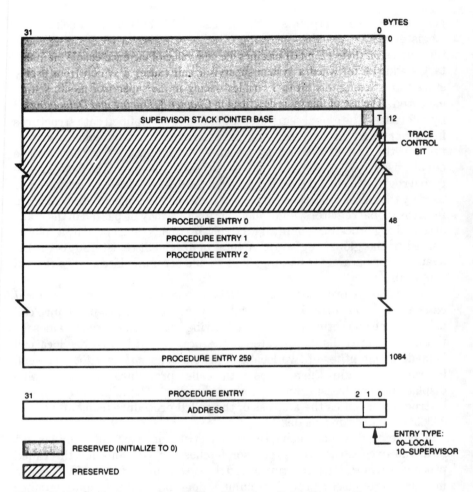

Figure 7.12 System procedure table. *(Courtesy Intel Corporation)*

The system call mechanism offers several benefits. The system call promotes the portability of application software. System calls are commonly used for OS services. By calling these services with a procedure number rather than a specific IP value, application software does not have to be changed each time the implementation of the OS service is modified. Additionally, the ability to switch to a different execution mode and stack allows OS procedures and data to be insulated from application code.[7]

Interrupts.   The 80960CA architecture defines 248 interrupt vectors, which are pointers (addresses) to interrupt handling routines. This means that 248 unique interrupt handling procedures may be used on a 960CA-based system. An 8-bit interrupt vector number is associated with each interrupt vector. This number ranges from 8 to 255. Each interrupt vector has a priority from 1 to 31, which is determined by the five most significant bits of the interrupt vector number. Priority 1 is the lowest priority and 31 is the highest. Priority 0 interrupts are not defined. The 80960CA executes with a unique priority ranging from 0 to 31. When an interrupt is serviced, the processor's priority switches to the priority corresponding to that of the interrupt request. When a return from an interrupt procedure is executed, the process priority is restored to its value prior to servicing the interrupt. The processor compares its current priority with the priority of the interrupt request. If a requested interrupt priority is greater than the processor's current priority or equal to 31, the processor services the interrupt immediately; otherwise, the processor saves (posts) the interrupt request as a pending interrupt so it can be serviced later. When the processor's priority falls below the priority of a pending interrupt, the pending interrupt is serviced. Interrupts with a priority of 0 would never be serviced with this mechanism. For this reason, vectors numbered 0 to 7 are not defined.[7]

The 80960CA interrupt vectors and other interrupt-related data are stored in an interrupt table, illustrated in Fig. 7.13. The first 36 bytes of the table are used to post pending interrupts. The 31 most significant bits in the 32-bit pending priorities field represent a possible priority (1 to 31) of a pending interrupt. When the processor posts an interrupt in the interrupt table, the bit corresponding to the interrupt's priority is set.

Example An interrupt with a priority of 12 is posted in the interrupt table. Bit 12 is set in the pending priorities field.

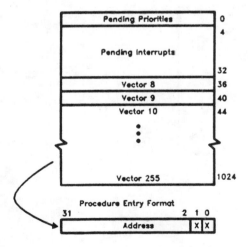

Figure 7.13 Interrupt table. *(Courtesy Intel Corporation)*

The pending interrupts field contains a 256-bit string in which each bit represents an interrupt vector. When the processor posts an interrupt in the interrupt table, the bit corresponding to the vector number of that interrupt is set. All interrupt vectors with an index whose least significant four bits are 0010 can be cached in the on-chip data RAM. The processor will automatically read these vectors from data RAM when the interrupt is serviced. This feature reduces the added delay due to an external memory access of the interrupt table for the above vectors. The nonmasked interrupt (NMI) vector is always cached on-chip, since it is usually invoked in emergency situations, such as a power failure.

Stack frames for interrupt-handling procedures are allocated on a separate *interrupt stack*. The interrupt stack can be located anywhere in the address space. The beginning address of the interrupt stack is specified when the processor is initialized. When an interrupt is serviced, the current local registers are saved. A new local register set and stack frame are allocated on the interrupt stack for the interrupt-handling procedure, and the processor switches to supervisor execution mode. The call and return mechanism described earlier also applies to context switching initiated by interrupts and faults.[7]

The 80960CA has eight interrupt pins and one NMI pin. The eight external interrupt pins can be configured in one of three modes: dedicated, expanded, or mixed, as illustrated in Fig. 7.14. In *dedicated mode*, each of the eight interrupt pins XINT0# to XINT7# acts as a separate, dedicated interrupt input, representing a separate interrupt source. When an interrupt signal is detected on an interrupt pin, a unique interrupt is requested for that pin. It is possible to map each dedicated pin to one of a number of

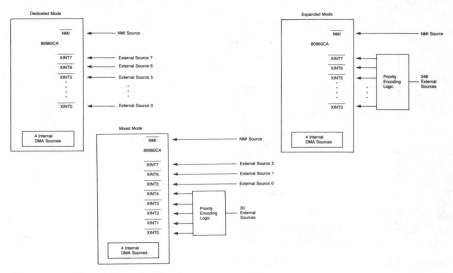

Figure 7.14   Interrupt modes. *(Courtesy Intel Corporation)*

possible interrupt vectors. This is accomplished by programming the interrupt map control registers with an 8-bit interrupt vector (value of 8 to 255) number for each pin. Only the upper 4 bits of the vector number can be programmed for a dedicated mode interrupt. The lower 4 bits are fixed at the value 0010 (binary). With four programmable bits, 1 of 15 (combination 0000 is not defined for interrupts) interrupt vectors is available for each dedicated pin. These interrupt vectors span the even priority (since bit 3 of the vector is always zero) 2 to 30. These 15 interrupt vectors can be stored in the first 64 bytes of the on-chip data RAM (see Fig. 7.5). If this interrupt vector storage is selected, the processor will automatically fetch the vector from data RAM. The DMA controller can request four interrupts to signal the end of a DMA access for each of the featured four channels. These four interrupt signals are handled by the interrupt controller in the same way as an interrupt pin configured as a dedicated input. Each of the four DMA sources may request one of 15 interrupts by programming the interrupt map control registers for that source.

In *expanded* mode, external hardware considers the interrupt pins XINT0# to XINT7# as an 8-bit binary number. This number is used directly as the interrupt vector number 8 to 255. Each of the 248 possible interrupt vectors can be referenced in this way, allowing a separate external source for each vector. External hardware is responsible for recognizing individual hardware sources (up to 248; see Fig. 7.14) and then driving the encoded, 8-bit interrupt vector number corresponding to that source onto the interrupt pins.

In *mixed* mode, the eight interrupt pins are divided into two functional sets, as shown in Fig. 7.14. One three-pin set, XINT5# to XINT7#, functions in dedicated mode, the other five-pin set, XINT0# to XINT4#, functions in expanded mode. A programmable vector number is associated with each of the XINT5# to XINT7# pins. The remaining five interrupt pins (XINT0# to XINT4#) are treated as the most significant 5 bits of the expanded mode vector number. The lower-order bits are internally preset to 010 (binary) to form the full 8-bit value for the vector number.

The interrupt controller uses two special function registers to manage interrupt requests by hardware sources. The hardware interrupt pending register (IPND) and the hardware interrupt mask register (IMSK) are addressed as sf0 and sf1, respectively. A single bit in each register corresponds to each of the eight possible external sources and four DMA sources for hardware interrupts. The IMSK register performs the function of masking hardware interrupts, and the IPND register implements posting of interrupts requested by hardware.

In addition to the maskable hardware interrupts, a single nonmaskable interrupt (NMI; see Fig. 7.14) is provided. A dedicated interrupt pin NMI is used to request this interrupt. The NMI has the highest priority level (31). The NMI handling procedure can never be interrupted and must execute the return instruction before other procedures can execute. The NMI handling procedure is entered through vector 248, stored in the on-chip data RAM at initialization. This reduces the delay in handling the NMI interrupt.

Faults. The 80960CA is able to detect various conditions in code or in its internal state that could cause the processor to deliver incorrect results. These conditions are referred to as faults. The 80960CA architecture provides fault handling mechanisms to detect and, in most cases, fully recover from a fault. Recovery from certain predefined faults is handled by special fault handling procedures. Pointers to the fault handling procedures are stored in a fault table, which may be located anywhere in the address space. The location of the fault table is specified at initialization.

7.1.3 The 80960 Family and Applications

As mentioned in the beginning of this chapter, the 80960 family has three predecessors to the 80960CA, namely 80960KA, 80960KB, and 80960MC.[1,2,6] Some of the properties of the 80960 KA, KB, MC, and CA, are summarized in Table 7.4.

A block diagram of the 80960KA is shown in Fig. 7.15. This is the basic processor of the 80960 family, implementing its core architecture. The 80960KB, whose block diagram is shown in Fig. 7.16, is an extension of the 80960KA, featuring an added on-chip floating-point unit (FPU) and four 80-bit floating-point registers. The FPU implements the IEEE 754-1985 floating-point standard, featuring the 32-bit single precision and 64-bit double precision. In addition, the FPU features the 80-bit extended precision.[1] The 80960MC, illustrated in Fig. 7.17, extends the 80960KB by including an on-chip memory management unit (MMU). The 80960MC implements the so-called *protected architecture* (PA).[1] The PA recognizes the basic concept of a *process* (or, synonymously, *task*). The state of a process is described in a data structure called a *process control block* (PCB). A process specifies a 4-Gbyte virtual address space. The overall virtual address space is subdivided into four 1-Gbyte regions. The top address region 3 (C000 0000H to FFFF FFFFH; H = hexadecimal) is shared among all processes. Region 3 contains information not associated with specific processes, such as the interrupt stack and interrupt handling routines. The lower regions 0–2 (0000 0000H to BFFF FFFFH) can be subdivided between processes running on the system.

The PA standard page size is 4 kbytes. Virtual addresses are translated to physical addresses by converting the upper 20 bits of the virtual address and using the lower 12 bits as an offset within the page frame in the physical memory. The upper two bits (30, 31) of the virtual address denote one of the four regions, the next 18 bits denote a page within that region, and the 12 least significant bits denote a byte within that page. The 80960MC MMU

TABLE 7.4   The 80960 Family

Processor	80960KA	80960KB	80960MC	80960CA
Global registers	16	16	16	16
Local registers	16	16	16	16
Floating=point registers	0	4	4	0
Special function registers	0	0	0	3
LRC sets	4	4	4	5 ÷ 15
Instruction cache (bytes)	512	512	512	1024
Cache mapping	Direct	Direct	Direct	Two=way
On=chip data RAM (Kbytes)	0	0	0	1
On=chip FPU	No	Yes	Yes	No
On=chip MMU	No	No	Yes	No
On=chip DMA	No	No	No	Yes
On=chip interrupt controller	Yes	Yes	Yes	Yes
Hardwired interrupts	4	4	4	8
PGA pins	132	132	132	168

FPU = floating = point unit; MMU = memory management unit; DMA = direct memory access; PGA = pin grid array; LRC = local register cache

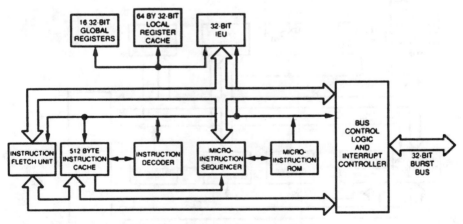

Figure 7.15   The 80960KA. *(Courtesy Intel Corporation)*

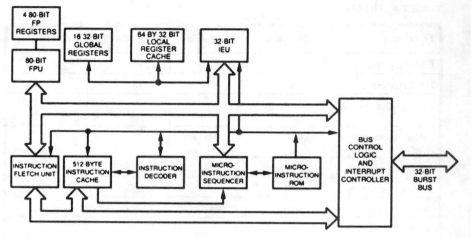

Figure 7.16   The 80960KB. *(Courtesy Intel Corporation)*

contains a 64-entry translation lookaside buffer (TLB) for fast virtual to physical address translation. For a 4-kbyte page size, the TLB maps 64 × 4096 = 256 kbytes of the address space. The 80960MC was particularly designed to function within the military temperature range from −55C to +125C.[1,2]

Intel recently announced two new versions of 80960 family processors: the 80960SA and the 80960SB, illustrated in Fig. 7.18. The internal structure and architecture of the 80960SA is essentially the same as that of

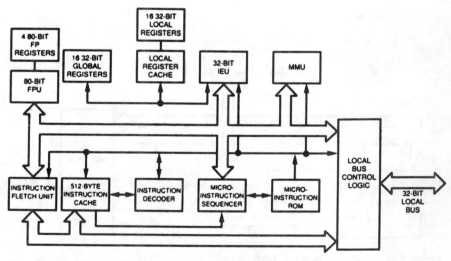

Figure 7.17   The 80960MC. *(Courtesy Intel Corporation)*

80960KA (Fig. 7.15), and the 80960SB is essentially identical to the 80960KB (Fig. 7.16). The difference is that both 80960 SA and SB feature an external 16-bit data bus, while featuring an internal 80960 32-bit architecture. The main motivation for these products is cost reduction due to a reduced number of interface devices needed to form a whole 80960-based microcontroller system (the same motivation as it was for the creation of 8088, 80188, MC68008, and other similar microprocessors).

The 80960 development tools. Intel provides a variety of software and hardware tools intended to facilitate design and development implementing processors of the 80960 family. These development tools are briefly reviewed in the following paragraphs.

ASM-960 macro assembler: Used to fine-tune sections of code for peak program execution speed on all 80960 family processors. It includes an assembler and the following features:

Macro preprocessor: Provides code generation flexibility.

Linker: Provides incremental program linking/locating and link-time optimization.

Archiver: Allows to build reusable function libraries for applications.

Disassembler: Produces assembly language from object files.

Symbol dumper: Provides symbolic information from a program file to facilitate debugging.

ROM image builder: Produces a hexadecimal file suitable for PROM programmers.

Floating-point arithmetic library (FPAL): Included for 80960 CA, KA, and SA processors that do not have an on-chip FPU. It eliminates the need to develop floating-point code by the user.

GEN-960 system generator: Used to set up data structures for stand-alone, embedded applications that use the on-chip features of the 80960 architecture. It is used with other 80960 tools to generate and refine ROM or RAM code. GEN-960 supplies a set of command and template files containing assembly code and linker control commands to set up processor control blocks, interagent communication mechanisms, system procedure tables, and other requirements for initialization. The result is a batch file containing all the commands needed to compile, assemble, and link the final target system.

C-960 Compiler (also denoted: iC-960): A fully optimizing C language compiler developed for the 80960 family. It supports the full C language as described in Kernighan and Ritchie,[15] includes standard ANSI extensions to the C language, and is used in conjunction with ASM-960 for

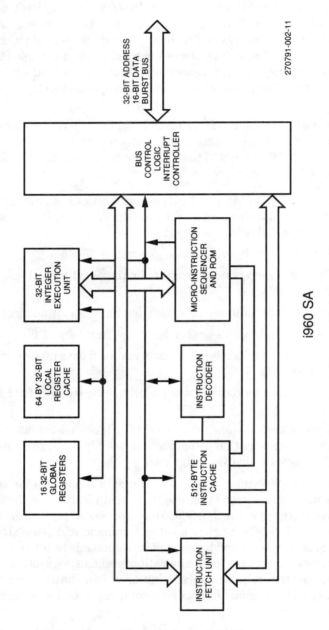

Figure 7.18 The 80960SA. *(Courtesy Intel Corporation)*

270791-002-11

i960 SA

32-BIT ADDRESS
16-BIT DATA
BURST BUS

BUS
CONTROL
LOGIC
INTERRUPT
CONTROLLER

32-BIT
INTEGER
EXECUTION
UNIT

64 BY 32-BIT
LOCAL
REGISTER
CACHE

16 32-BIT
GLOBAL
REGISTERS

MICRO-INSTRUCTION
SEQUENCER
AND ROM

INSTRUCTION
DECODER

512-BYTE
INSTRUCTION
CACHE

INSTRUCTION
FETCH UNIT

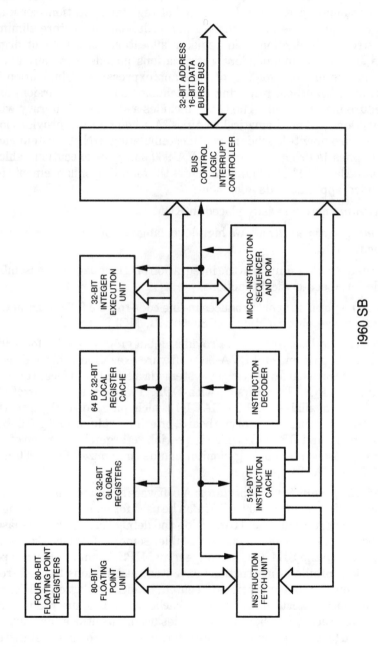

Figure 7.18 The 80960SB. *(Continued)*

i960 SB

creating object files. The C-960 compiler supports a number of processor dependent optimizations including global register allocation, constant propagation, arithmetic identity folding, redundant load/store elimination, strength reduction, and register allocation/scheduling of arguments. Processor independent optimizations include common subexpression elimination, folding of constant expressions, elimination of superfluous branches, removing unreachable code, tail recursion, and procedure incorporation. The C-960 includes a standard C library with I/O functions and mathematical routines. A second library provides low level, environment-dependent routines emulating UNIX system calls and supplies I/O routines for the EVA-960 software execution vehicle (discussed later). The C-960 also includes the following enhancements for embedded application development:

Programs may be easily placed in ROM.

Memory-mapped I/O allows high-level language access to application specific I/O.

Inline assembly simplifies the integration of C language and assembly code for speed-critical functions.

Floating point support produces in-line code to take full advantage of the floating-point capability of 80960 KB, MC, and SB.

DB-960 source-level debugger: Provides debugging capability for 80960 code executing on an Intel EVA-960 software execution vehicle (discussed later) or on a hardware target system via a serial interface. It is also available for PC ATs equipped with DOS. Two serial versions of the DB-960 are available: DB-960CADIC which plugs directly into the 80960CA socket on the prototype board, offering a "plugin and go" debug environment, and DB-960D which is a serial, retargetable version of the DB-960 whose system debug monitor can be customized for 80960 KA/KB, CA, or SA/SB operation.

SIM960CA performance Simulator: A software simulator that emulates the instruction set of the 80960CA. Can be used for benchmark testing of application performance and to develop and debug early 80960CA-based designs before target systems are available. Some performance statistics gathered with the SIM960CA are speed in MIPS, number of cycles per instruction, profile of instructions executed, instructions fetched from memory, number of instructions executed, branch prediction efficiency, total simulated execution time, stack cache overflows, instructions received from cache, number of clock cycles used, bus utilization, current clock speed (cycles/s), and wait-state information. Performance statistics being gathered can be reset at any point, allowing performance analysis between any two locations. For example, execution time for a particular procedure can be measured. SIM960CA can be hosted on IBM PC, XT, or

AT (or compatibles), under DOS version 3.3 or above. Other versions of SIM960CA can be hosted on VAX/VMS and Sun 3 systems.

EVA-960KB software execution vehicle: A single PC AT plug-in board that provides architecture evaluation, benchmarking, and software development for 80960 KB, SA, and SB-based systems. It contains a 20-MHz 80960KB CPU, 4- or 16-Mbyte DRAM, 64 kbytes SRAM, three-channel programmable interval timer, hosted debug monitor, and DOS access libraries.

DB960CADIC in-circuit debug monitor: Provides real-time hardware and software debugging capabilities for 80960CA-based embedded designs. The user can run at the full speed of the target processor. The DB960CADIC can be used by both hardware and software developers at any stage of design. Early in the development process, it allows software debugging when inserted into an existing 80960CA board. Later in the design cycle, it can be inserted into the user's target system, thus facilitating the debugging of hardware/software integration. The DB960CADIC includes 128 kbytes of memory and RS322 and RS422 communication links. It can be hosted on any DOS run system. In fact, all of the 80960 development tools are DOS compatible. ASM-960, C-960, and GEN-960 are also UNIX and VMS compatible. SIM-960CA is also VMS compatible.

A real-time OS called Vx960 was developed for the i960 family. It is Intel's offering of Wind River Systems' (Emeryville, Calif.) VxWorks OS, ported to the i960 environment. Vx960 is a real-time OS that includes extensive networking facilities and high-level development tools, all designed to work together with a developer's host computing system. The Vx960 kernel provides fundamental OS primitives and real-time features such as multitasking, preemptive scheduling, and intertask communication. The Vx960 is compatible with Unix. Vx960 and Unix can exchange files, communicate, log in to each other, and use each other as servers. Hosts supported by Vx960 include Intel i386 system V, Sun 3 and 4, IBM RS/6000, and HP9000 series 300.

Design with 80960 family and applications. Although 80960CA has a total of 2 kbytes of on-chip memory (1-kbyte instruction cache and 1-kbyte main memory RAM), it may not be sufficient for many microcontroller applications. When controlling more complicated processes more code and data may be required and external memory must be used. Thus, one of the first steps in hardware design of an 80960CA-based microcontroller is to provide adequate external memory and the appropriate interface to it. Since the 80960CA has a four-channel DMA controller and an interrupt controller on chip, additional interface chips may not be required in many applications. In many microcontroller applications dedicated to particular systems, a large part of code and data may be fixed for a significant period. For this reason, the use of an EPROM (erasable programmable ROM) is recommended. It has higher density and a lower cost per bit compared to RAM, and it is nonvolatile (no information is erased when power is down). On the other hand, the information on it can be changed off line periodically using a PROM programmer device.

Intel has developed a special EPROM chip 27960 for microcontroller implementation with the processors of the 80960 family. It is a 128-kbyte (1-Mbit) CHMOS-IIIE EPROM chip. It is organized internally in blocks of 4 bytes that are accessed sequentially. The address of the 4-byte block is latched and incremented internally. After a set number of wait states (one or two), data is sent out one word (4 bytes) at a time each subsequent clock cycle. THe 27960 was specially designed to utilize the 80960 burst busses.

An example of an 80960CA external EPROM interface design is illustrated in Fig. 7.19.[6] The total external EPROM size is 128K × 32 or 512 kbytes. The lower 17 address bits are used for internal decoding in the four 128-kbyte 27960 chips. The first 1024 (0 to 1023) addresses of the 80960CA memory are taken up by the internal 1-kbyte RAM. Thus, the external EPROM can not be assigned addresses from zero, but from any address exceeding 1023. Therefore, a number of upper address bits (depending on the address interval assigned to the EPROM) is used to decode the EPROM address, using the 85C508 decoder, and activate the chip select (CS) pins of the four 27960 chips. Each of the four data quadrants is taken from a different 27960 chip and interconnected to the 32-bit data bus. The ADS# signal (address strobe; # denotes low asserted; see Appendix A of this chapter) indicates to the EPROM that a valid address has been issued by the 80960CA CPU. The BLAST# signal indicates to the EPROM the last transfer in a bus access, for both burst and nonburst accesses.

The 80960 family was designed to interface with Local Area Networks (LAN) among other types of systems. A special LAN coprocessor, denoted 82596CA, was developed to interface with the 80960 family, particularly with 80960 CA, KA, and KB. A block diagram of the 82596CA and its interaction with an 80960 CPU is shown in Fig. 7.20.[6] The 82596CA is a CHMOS IV, 1-micron, 132-pin, 32-bit, 33-MHz chip. It is a high-performance LAN coprocessor that performs high-level commands, command chaining, and interprocessor communication via shared memory. This relieves the CPU of many tasks associated with network control; all time-critical functions are performed independently of the CPU, which greatly improves the overall performance. The 82596CA supports the IEEE 802.3 network standard. It can interface with the 82C501AD Ethernet serial interface chip. It has an 82586 LAN coprocessor software compatible mode, which offers a smooth migration path for the existing 82586 software base. The 82596CA has internal FIFO (first in, first out) queues;

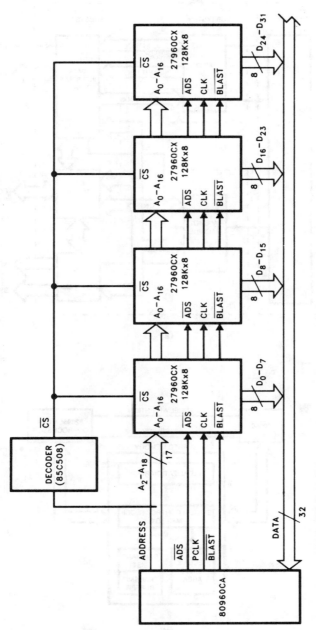

Figure 7.19 27960CX-i960 CA processor interface. *(Courtesy Intel Corporation)*

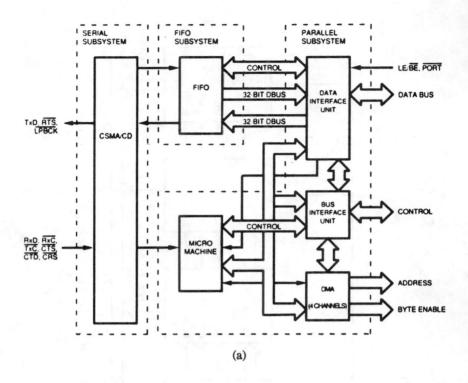

(a)

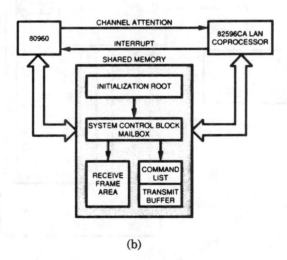

(b)

Figure 7.20 The 82596 LAN coprocessor: (a) 82596 functional block diagram; (b) 80960 and 82596 interaction. *(Courtesy Intel Corporation)*

128 bytes for receive, and 64 bytes for transmit (Fig.7.20). It also has an on-chip four-channel DMA controller.

Microcontrollers of the 80960 family are featured on a number of commercial computing boards manufactured by companies other than Intel. Heuricon Corporation (Madison,Wis.) features the HK80/V960E single-board 80960CA-based computer. It was designed for real-time performance to be used in a variety of applications ranging from intelligent communications processing to embedded process control. A block diagram of the HK80/V960E is shown in Fig. 7.21. It contains an 80960CA CPU, an 82596CA LAN coprocessor, four serial ports, a small computer serial interface (SCSI) port, VME and VSB bus interfaces, 2 or 8 Mbytes DRAM, 128 bytes nonvolatile RAM, and up to 2 Mbytes EPROM. Heuricon supports a variety of UNIX development hosts, from Sun workstations to its own SCALOS development system. The link between UNIX hosts and the HK80/V960E is implemented via the VxWorks cross development environment, which includes unique driver support for SCSI peripherals. With VxWorks, designers can develop their applications on a UNIX host such as a Sun workstation. The resulting code can then be downloaded to a target HK80/V960E board, which runs the VxWorks real-time OS. Designers can debug their code using either a source-level debugger, which resides on the host, or a symbolic debugger, which resides on the target. The link between the UNIX host and target board may be implemented via either Ethernet or a common VME backplane.

Phoenix Technologies Ltd. (Norwood,Mass.) manufactures the P960 (also called Stampede) laser printer controller, whose block diagram is shown in Fig. 7.22.[6] It uses the 80960 KA or KB as its CPU and it is adaptable to Ricoh, Canon, TEC, and other printers printing up to 30 ppm. The P960 supports multiple communication channels, including serial RS-232C/RS-422A, parallel (Centronics), Apple Talk, or SCSI. It can support 0.5- to 32-Mbyte DRAM, up to 4 Mbytes of EPROM on base controller, and up to 16 Mbytes of ROM on cartridges.

Tadpole Technology Inc. (Waltham,Mass.) manufactures an 80960CA-based single board computer whose block diagram is shown in Fig. 7.23.[6] It is called TP960V and it includes two serial I/O ports, 32 to 512 kbytes SRAM, 0.5 to 1.0 Mbytes EPROM, 2 to 8 Mbytes of dual ported DRAM, and VME and SCSI interfaces. It supports the IEEE 802.3 Ethernet interface standard. The TP960V can be used as a versatile system controller in a wide variety of applications, including real time. It implements the VxWorks real-time software, comprising a real-time kernel, compiler and debugging tools, and high-performance networking software.

CYCLONE Microsystems (New Haven,Conn.) features two 80960CA-based single-board computers. These computers are intended for diversified applications. The CVME960 contains 1 or 2 Mbytes of pipelined SRAM.

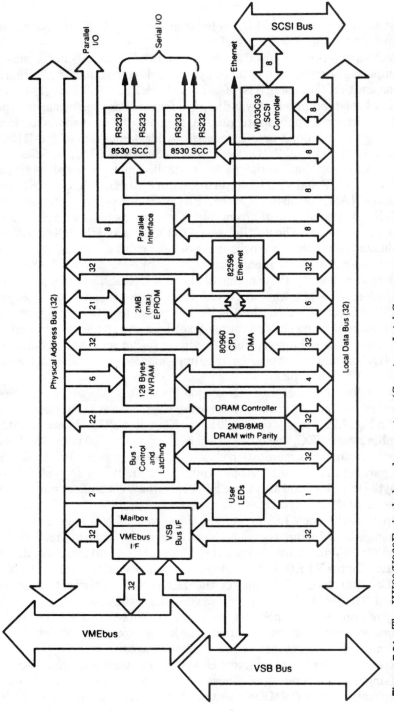

Figure 7.21 The HK80/V960E single-board computer. (*Courtesy Intel Corporation*)

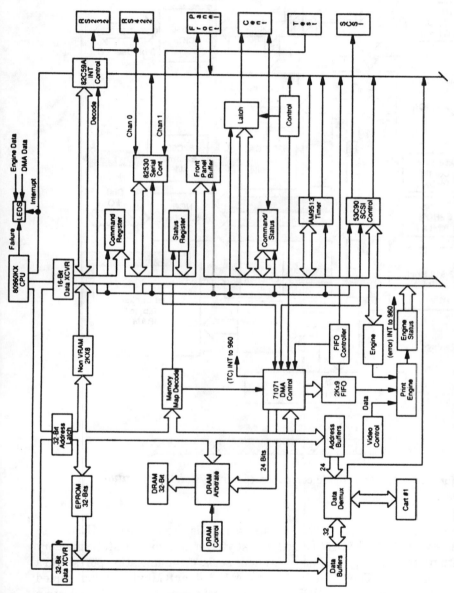

Figure 7.22 Block diagram of phoenix laser printer controller. *(Courtesy Intel Corporation)*

349

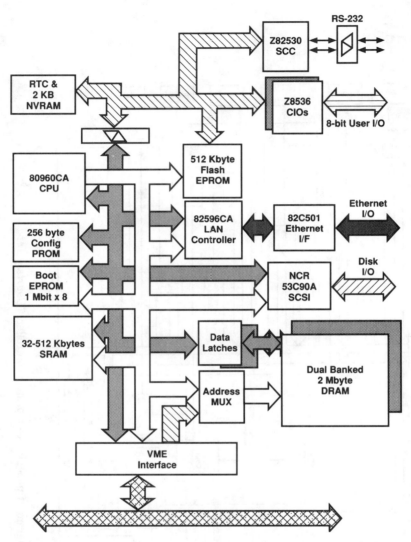

Figure 7.23 TP 960V single computer. *(Courtesy Intel Corporation)*

The SRAM is dual ported and its access time is 35 ns. The CVME960 also contains up to 1 Mbyte of EPROM, 2 kbytes of non-volatile static memory with battery backup, two asynchronous serial communication ports, and a real-time clock. The CVME961 contains 1, 2, 4, or 8 Mbytes of dual ported DRAM, up to 1 Mbyte of EPROM, two asynchronous serial communication ports, 2 kbytes of nonvolatile static memory with battery backup, and a real-time clock. Both computers feature a VME bus interconnection, and their top operating frequency is 33 MHz.

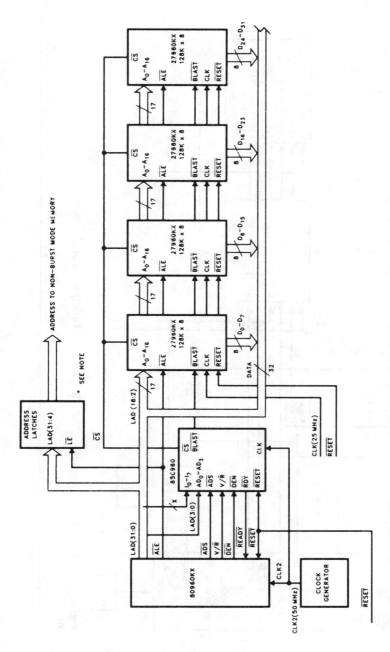

Figure 7.24 i960 KA/KB processor, 27960 KX and 86C960 system. *(Courtesy Intel Corporation)*

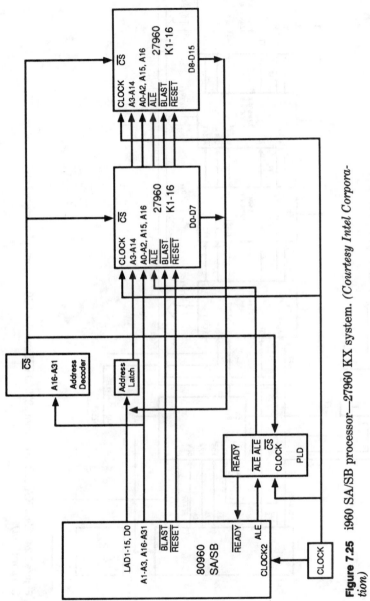

Figure 7.25 i960 SA/SB processor—27960 KX system. (*Courtesy Intel Corporation*)

Examples illustrating the interface of 80960KX and 80960SA/SB to EPROM (27960) chips are shown in Figs. 7.24 and 7.25. Additional design examples featuring 80960CA-based systems are presented in App. B.

It seems that the 80960 family, and particularly the superscalar 80960CA, has gained a sizable niche in the microcontroller applications market. New applications featuring 80960 family processors are announced on a monthly basis. This is understandable considering the high performance, versatility, and compactness of the 80960CA.

7.2 LSI LOGIC LR33000 EMBEDDED PROCESSOR

The LR33000 is an embedded processor featured by LSI Logic Corp. (Milpitas, Calif.). The LR33000 CPU is a RISC-type processor compatible with the instruction sets of the MIPS R2000 or R3000,[2] designed by MIPS Computer Systems Co. (Palo Alto, CA). The MIPS follows the original design of the Stanford University MIPS, one of the first experimental RISC systems.[2] MIPS Computer Systems Co. completed the architectural design of the MIPS chips, but it does not manufacture them. It has licensed several companies to manufacture versions of MIPS chips. LSI Logic Corporation is one of these companies. It is also licensed to produce the Sun Microsystems SPARC.[2] Following the success of implementing RISC-type systems in a variety of workstations, personal computers, real-time boards, and others, LSI Logic decided to feature a new embedded controller, LR33000, based on the already successful MIPS CPU.

A block diagram of the LR33000 is shown in Fig. 7.26. It consists of the following subsystems:[16]

1. CPU based on the MIPS design, maintaining full binary compatibility with MIPS R2000 and R3000. It includes 32 CPU registers (32-bit) and a five-stage instruction pipeline.

2. On-chip instruction and data caches. The instruction cache is 8 kbytes, and the data cache is 1 kbyte. Both caches have the following identical parameters: direct mapped, line size of 16 bytes, write-through policy, and refill size of 4, 8, 16, 32, or 64 bytes. Software invalidate, bus snooping, and byte write (data cache only) are supported.

3. On-chip DRAM (dynamic RAM) control, featuring memory refresh circuitry, page-mode DRAM control signals, and DMA by separate I/O devices.

4. Bus interface unit (BIU), providing an interface to memory and I/O devices.

5. Three on-chip counter/timers. Two 24-bit general purpose timers and a 12-bit refresh timer that can support optional off-chip DRAM controllers.

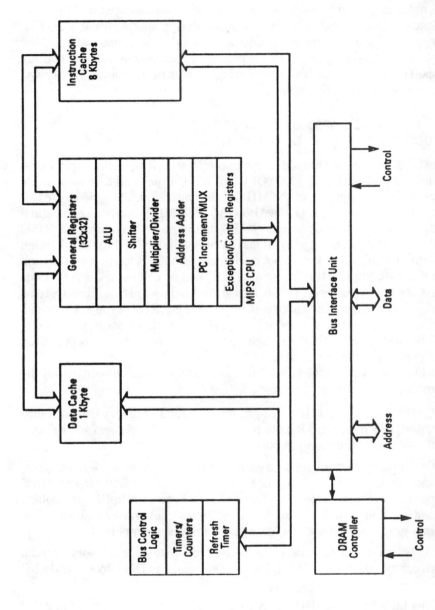

Figure 7.26 LR33000 processor features. (*Courtesy LSI Logic Corporation*)

The LR33000 CPU registers are illustrated in Fig. 7.27.[16] There are 32 general-purpose registers of 32 bits. Register r0 is hardwired to a zero value and r31 is defined as the link register for jump and link instructions. Otherwise, all general-purpose registers can be freely used by the programmer. The 32-bit program counter (PC) is separate. The two multiply/divide registers HI and LO store the double-word, 64-bit result of multiply and divide operations.

The data types recognized by the MIPS architecture implemented on the LR33000 are a 32-bit word, a 16-bit halfword, and an 8-bit byte.[16] The byte ordering within a word is configurable into either big-endian (lower byte has higher address) or little endian (lower byte has lower address) byte ordering. This configuration is controlled by an external input signal called BENDN (the pinout is described later in this section).

The LR33000 features the three instruction formats of the MIPS architecture, illustrated in Fig. 7.28:[16] Immediate, or I-type, Jump, or J-type, and Register, or R-type. The details of the format fields are listed in Fig. 7.28. All instructions have a standard 32-bit length. The LR33000 instruction set is listed in Table 7.5. These are essentially MIPS instructions; however, the LR33000 does not support MIPS instructions that relate to the translation lookaside buffer (TLB), and it does not support many of the MIPS coprocessor instructions, since they are not needed in embedded controller applications. The instruction types featured on the LR33000 are as follows:

1. Load/store

2. Arithmetic/logic, immediate, and three-operand register-to-register

3. Shift

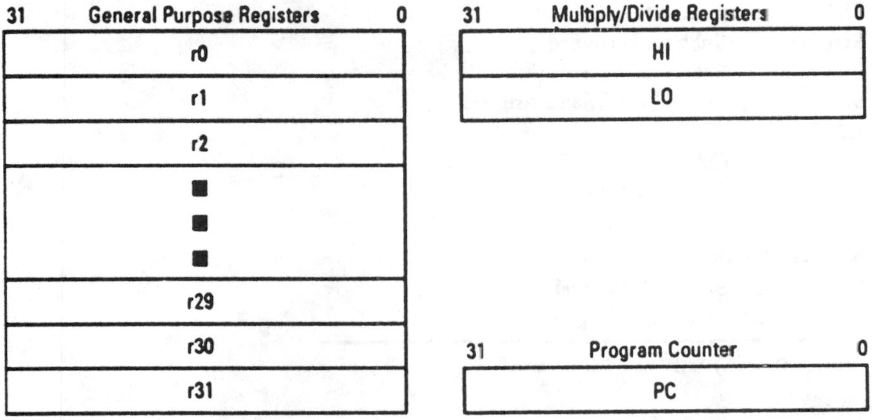

Figure 7.27 LR33000CPU registers. *(Courtesy LSI Logic Corporation)*

TABLE 7.5   LR33000 Instruction Summary

Op	Description
Load/Store Instructions	
LB	Load Byte
LBU	Load Byte Unsigned
LH	Load Halfword
LHU	Load Halfword Unsigned
LW	Load Word
LWL	Load Word Left
LWR	Load Word Right
SB	Store Byte
SH	Store Halfword
SW	Store Word
SWL	Store Word Left
SWR	Store Word Right
Arithmetic Instructions (ALU Immediate)	
ADDI	Add Immediate
ADDIU	Add Immediate Unsigned
SLTI	Set on Less Than Immediate
SLTIU	Set on Less Than Immediate Unsigned
ANDI	AND Immediate
ORI	OR Immediate
XORI	Exclusive OR Immediate
LUI	Load Upper Immediate
Arithmetic Instructions (3-Operand, Register-Type)	
ADD	Add
ADDU	Add Unsigned
SUB	Subtract
SUBU	Subtract Unsigned
SLT	Set on Less Than
SLTU	Set on Less Than Unsigned
AND	AND
OR	OR
XOR	Exclusive OR
NOR	NOR
Shift Instructions	
SLL	Shift Left Logical
SRL	Shift Right Logical

SOURCE: Courtesy LSI Logic Corporation.

TABLE 7.5   LR33000 Instruction Summary *(Continued)*

Op	Description
SRA	Shift Right Arithmetic
SLLV	Shift Left Logical Variable
SRLV	Shift Right Logical Variable
SRAV	Shift Right Arithmetic Variable

Multiply/Divide Instructions

MULT	Multiply
MULTU	Multiply Unsigned
DIV	Divide
DIVU	Divide Unsigned
MFHI	Move From HI
MTHI	Move To HI
MFLO	Move From LO
MTLO	Move To LO

Jump and Branch Instructions

J	Jump
JAL	Jump And Link
JR	Jump Register
JALR	Jump And Link Register
BEQ	Branch on Equal
BNE	Branch on Not Equal
BLEZ	Branch on Less than or Equal to Zero
BGTZ	Branch on Greater Than Zero
BLTZ	Branch on Less Than Zero
BGEZ	Branch on Greater than or Equal to Zero
BLTZAL	Branch on Less Than Zero And Link
BGEZAL	Branch on Greater than or Equal to Zero And Link

Special Instructions

SYSCALL	System Call
BREAK	Breakpoint

Coprocessor Instructions

BCzT	Branch on Coprocessor z True
BCzF	Branch on Coprocessor z False

System Control Coprocessor (CP0) Instructions

MTC0	Move To CP0
MFC0	Move From CP0
RFE	Restore From Exception

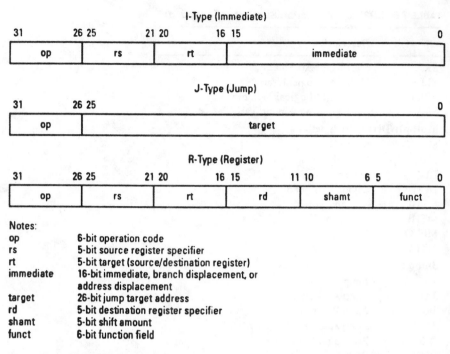

Figure 7.28 LR33000CPU instruction formats. *(Courtesy LSI Logic Corporation)*

4. Multiply/divide

5. Jump and branch

6. Special, coprocessor, and system control coprocessor (CP0)

The LR33000 has an on-chip system control coprocessor (CP0) that supports exception-handling functions. The CP0 includes 8 exception handling registers, illustrated in Fig. 7.29.[16] The exception handling registers are as follows:

1. *Status register*: Contains all major status bits for exception conditions.

2. *Cause register*: Describes the last occurring exception, indicating its cause.

3. *Bad address register (BadA)*: A read-only register that saves the address associated with an illegal access.

4. *Exception program counter (EPC) register*: Contains the address where processing resumes after an exception is serviced. In most cases it contains the address of the instruction that caused the exception.

5. *Processor revision identifier (PRId) register*: Contains information that identifies the implementation and revision level of the processor and CP0.

6. *Debug and cache invalidation control (DCIC) register*: Contains the enable and status bits for the breakpoint mechanism and the control bits for the cache controller's invalidate mechanism.

7. *Breakpoint program counter (BPC) register*: Used by software to specify a program counter breakpoint.

8. *Breakpoint data address (BDA) register*: Used by software to specify a data address breakpoint.

The LR33000 has two operating modes: *user* and *kernel* (called supervisor in other systems). Normally, the processor operates in user mode until an exception is detected, which forces it into the kernel mode.[16] It remains in the kernel mode until a restore from exception (RFE) instruction is executed. The user mode has a 2-Gbyte address space allocated to it, while the kernel mode can access 4 Gbytes, including the user space. The LR33000 address space is illustrated in Fig. 7.30. The LR33000 features the MIPS five-stage instruction pipeline, illustrated in Fig. 7.31. The pipeline stages are as follows:

1. *IF*: Fetch instruction from the instruction cache.

2. *RD*: Read any required operands from CPU registers while decoding the instruction.

3. *ALU*: Perform the required arithmetic or logical operation on the operands.

4. *MEM*: Access memory from the data cache.

5. *WB*: Write results back to register file.

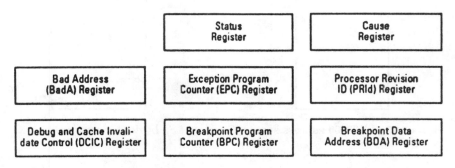

Figure 7.29   CPO exception-handling registers. *(Courtesy LSI Logic Corporation)*

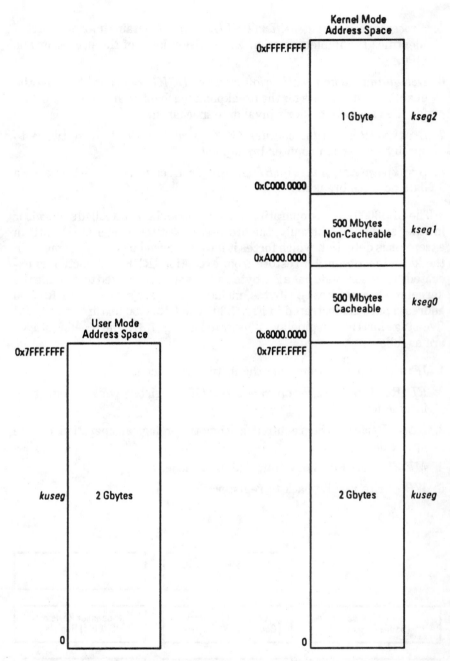

Figure 7.30 LR33000 address space. *(Courtesy LSI Logic Corporation)*

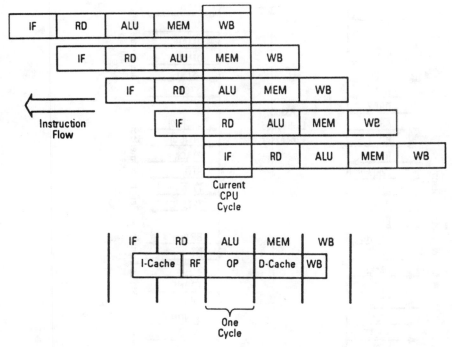

Figure 7.31 LR33000 instruction pipeline. *(Courtesy LSI Logic Corporation)*

The LR33000 interface input and output signals are illustrated in Fig. 7.32.[16] The interface signal functions are listed in Table 7.6.

The LR33000 on-chip DRAM controller provides the interface logic needed to address and control a DRAM array with very few additional components. It can support, but is not limited to, common DRAM configurations such as 256 Kwords by 4 bits (256K × 4), 1 Mword by 1 bit (1M × 1), and 1 Mword by 4 bits (1M × 4). The concept *word* in this case refers to the width of the DRAM chip (4 bits, 1 bit) and not to the processor word size. DRAM devices are organized as an array of rows and columns. To reduce the number of input address lines to DRAM chips, a multiplexed row and column address bus is used. During part of the address sequence for the DRAM, the row address is valid on the bus. Later in the sequence, the column address is valid. Two control strobes, row address strobe (RAS) and column address strobe (CAS), drive internal address latches within the DRAM to latch the row and column addresses at the proper time. The data within a DRAM (dynamic RAM) is dynamic, meaning that it decays over time. To prevent loss of data, DRAMs must be refreshed periodically (every few milliseconds). During a refresh operation, the electric charge representing the data at each bit is restored to its nominal value. DRAMs are

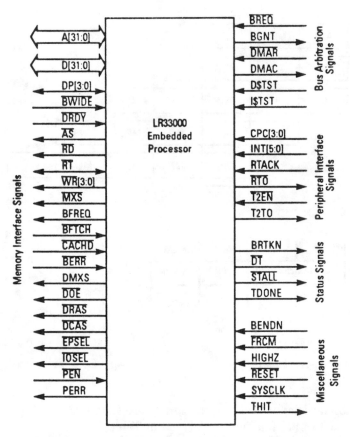

Figure 7.32 LR33000 interface signals. *(Courtesy LSI Logic Corporation)*

refreshed one row at a time. The LR33000 DRAM controller implements an implicit refresh mechanism. An implicit refresh is called a CAS before RAS refresh, since an external controller initiates the refresh by asserting DCAS# before asserting DRAS#, the reverse of the normal order of the two signals. During an implicit refresh, logic within the DRAM device generates the address of the next row to be refreshed.[16] A number of DRAM interfaces to the LR33000 are described in the following examples.

Example   4-Mbyte DRAM array. The interface to a 4 Mbyte DRAM, composed of 1M × 4 chips, is illustrated in Fig. 7.33.[16] Aside from the DRAMs, the only external component required to implement this configuration is an address multiplexer ('157) to multiplex the row and column addresses onto the DRAM address bus. Each byte of memory is composed of two 1M × 4 DRAM chips to store the data and one 1M × 1 DRAM chip to store the associated parity bit (optional). If parity is used, the PEN# pin should be asserted during transactions. If parity is not used, PEN# should be kept deactivated (high).

TABLE 7.6 LR33000 Interface Signals

A[31:0] **Address Bus [31:0]** Bidirectional
When it is bus master, the LR33000 uses A [31:0] to transmit instruction and data addresses to the memory and peripherals. When another device owns the bus, the LR33000 snoops addresses generated by the bus master (if $\overline{\text{CACHD}}$ is asserted).

$\overline{\text{AS}}$ **Address Strobe** Bidirectional
$\overline{\text{AS}}$ indicates that a valid address is present on the Address Bus. During memory transactions as bus master, the LR33000 asserts $\overline{\text{AS}}$ one half cycle after it asserts $\overline{\text{MXS}}$.

When the LR33000 is not bus master, it monitors $\overline{\text{AS}}$ to determine when a valid address is present on the Address Bus. The LR33000 uses the addresses generated by other bus masters for cache snooping and invalidation, and for DRAM Controller operations.

$\overline{\text{BERR}}$ **Bus Error** Input
External logic asserts this signal to indicate that an exceptional condition has occurred. When $\overline{\text{BERR}}$ is asserted, the LR33000 terminates the current memory transaction and takes a Bus Error Exception.

BFREQ **Block Fetch Request** Bidirectional
The LR33000 asserts this signal to indicate that it wants to perform a block-fetch transaction. The LR33000 asserts BFREQ whenever a cache miss occurs.

BFTCH **Block Fetch** Input
In response to the LR33000's assertion of BFREQ, external memory control logic asserts this signal to indicate that the LR33000 can expect a block transfer from memory.

When the LR33000's DRAM Controller is used to manage memory, the DRAM Controller acknowledges the BFREQ signal internally and automatically. Unless there is an external device that is capable of block-fetch transactions, BFTCH should be tied to VDD when using the internal DRAM Controller.

$\overline{\text{BWIDE}}$ **Byte-Wide Port** Input
External logic asserts $\overline{\text{BWIDE}}$ to indicate that the memory interface is eight bits wide. When $\overline{\text{BWIDE}}$ is asserted, the LR33000 executes four memory cycles with sequential byte addresses, beginning with the effective address. (The byte-wide feature supports partial-word accesses, but always performs four memory cycles.) If the effective address is not modulo four, the LR33000 wraps around to get all of the bytes in the word in which the effective address falls.

For byte-wide transactions, the LR33000 always takes data from bits D[7:0].

$\overline{\text{CACHD}}$ **Cacheable Data** Input
External logic asserts this signal to indicate that the instruction or data fetched during the current cycle may be stored in the LR33000's on-chip cache. If $\overline{\text{CACHD}}$ is not asserted, the LR33000 does not cache the instruction or data fetched during the current transaction.

D[31:0] **Data Bus [31:0]** Bidirectional
The LR33000 uses D[31:0] to transfer data and instructions to and from memory and peripherals. When another device owns the bus, it may drive or sample these signals.

TABLE 7.6   LR33000 Interface Signals *(Continued)*

$\overline{\text{DCAS}}$ **DRAM Column Address Strobe**   **Output**
During a DRAM memory transaction, the LR33000's DRAM Controller asserts this signal to indicate that the address presented to the DRAM array is the column address.

DMXS   **DRAM Mux Select**   **Output**
The LR33000's DRAM controller outputs DMXS for the Data Select input of the multiplexers required to multiplex the row and column address from the Address Bus on to the DRAMs' address bus. When HIGH, DMXS indicates that the row address should be selected. When LOW, DMXS indicates that the column address should be selected.

$\overline{\text{DOE}}$ **DRAM Output Enable**   **Output**
The LR33000's DRAM Controller asserts this signal to enable DRAM data outputs.

DP[3:0]   **Data Parity [3:0]**   **Bidirectional**
DP[3:0] allow the LR33000 to check and transmit parity bits for the four data bus bytes   The LR33000 uses even parity.

$\overline{\text{DRAS}}$ **DRAM Row Address Strobe**   **Output**
During a DRAM memory transaction, the LR33000's DRAM Controller asserts this signal to indicate that the address presented to the DRAM array is the row address.

$\overline{\text{DRDY}}$ **Data Ready**   **Input**
External logic asserts this signal to indicate that it can accept or is providing data. The assertion of DRDY terminates the memory transaction. Because the LR33000 must wait for external logic to assert DRDY before it can complete a memory transaction, slow devices can use DRDY to stall the processor. However, when the automatic wait state generator is enabled, the LR33000 asserts an internal version of DRDY to end transactions.

$\overline{\text{EPSEL}}$ **EPROM Select**   **Output**
The LR33000 asserts this signal to indicate that it is accessing the EPSEL address space (0x1E00.000 to 0x1EFF.FFFF). EPSEL can be used as a chip select. If automatic wait state generation is enabled, the LR33000 generates an internal DRDY signal a programmable number of cycles after the transaction begins. The external PROM logic may override the internal wait-state generator by asserting the DRDY signal before the programmed number of wait states have passed.

$\overline{\text{IOSEL}}$ **I/O Select**   **Output**
The LR33000 asserts this signal to indicate that it is accessing the IOSEL address space (0x1F00.000 to 0x1FFF.FFFF). IOSEL can be used as a chip select. If automatic wait state generation is enabled, the LR33000 generates its own DRDY signal a programmable number of cycles after the transaction begins. The external I/O logic may override the internal wait-state generator and use the DRDY signal to control the number of wait states.

$\overline{\text{MXS}}$ **Memory Transaction Start**   **Output**
The LR33000 asserts MXS for one clock cycle at the beginning of a memory transaction to indicate the start of the transaction.

TABLE 7.6   LR33000 Interface Signals (Continued)

$\overline{\text{PEN}}$ **Parity Enable**   **Input**
When asserted by external logic during a read transaction, the
LR33000 performs byte-wise checking for even parity on the Data
and Data Parity Buses. When deasserted, the LR33000 does not
check parity. External logic that is connected to the LR33000's byte-
wide port should never assert $\overline{\text{PEN}}$.

PERR   **Parity Error**   **Output**
The LR33000 asserts PERR to indicate that it has detected a parity
error during the just-completed memory transaction. PERR can be
tied to a interrupt input to cause a parity error exception.

$\overline{\text{RD}}$ **Read Strobe**   **Output**
The bus master asserts $\overline{\text{RD}}$ to indicate that the current transaction is a
read transaction and that memory may drive data onto the D[31:0].
When deasserted, only the bus master may drive the D[31:0].

$\overline{\text{RT}}$ **Read Transaction**   **Bidirectional**
The LR33000 asserts $\overline{\text{RT}}$ to indicate that the current memory cycle is
a read transaction. The LR33000 deasserts $\overline{\text{RT}}$ during write transactions. When the LR33000 is not bus master, the LR33000 monitors
this signal for cache snooping and invalidation.

When the LR33000 is in cache access mode, external logic uses this
signal to read and write the LR33000's internal cache.

$\overline{\text{WR}}$[3:0]   **Byte Write Strobes [3:0]**   **Output**
The LR33000 asserts these signals to indicate that there is valid data
on the corresponding segment of the Data Bus.

$\overline{\text{BREQ}}$ **Bus Request**   **Input**
External bus masters assert this signal to request control of the
Address and Data Buses and the relevant control signals.

BGNT   **Bus Grant**   **Output**
The LR33000 asserts this signal to grant control of the memory buses
to the requesting device. When BGNT is asserted, the LR33000
3-states the following signals:

D[31:0]	DP[3:0]
A[31:0]	$\overline{\text{MXS}}$
$\overline{\text{AS}}$	$\overline{\text{RT}}$
$\overline{\text{RD}}$	$\overline{\text{D}}$
$\overline{\text{WR}}$[3:0]	BFREQ

When BGNT is asserted, the requesting bus master can control the
3-state signals.

$\overline{\text{DMAR}}$ **DMA Request**   **Input**
External logic asserts this signal to request use of the LR33000's
DRAM Controller. The $\overline{\text{DMAR}}$ and DMAC signals are not related to
the $\overline{\text{BREQ}}$ and BGNT signals.

DMAC   **DMA Cycle**   **Output**
The LR33000 asserts this signal to acknowledge the $\overline{\text{DMAR}}$ and to
indicate that the DRAM Controller has started the access for the
requesting device. During such DMA accesses, the DRAM Control-
ler manages the $\overline{\text{DRAS}}$, $\overline{\text{DCAS}}$, and DMXS signals. The DMA device
must drive the RAM's write strobes, address bus, and data bus; a con-
figuration that requires multiport RAMs.

TABLE 7.6   LR33000 Interface Signals *(Continued)*

During DMA operations by other devices, the LR33000 continues to control and use the A[31:0], D[31:0], and DP[3:0]. If both external logic and the LR33000 need to use the DRAM Controller, they take turns, switching control at the end of each memory transaction.

D$TST   **Data Cache Access**   **Input**
When this signal is asserted, external logic can read and write the LR33000's data cache. In data cache access mode the external device controls A[31:0], $\overline{AS}$ and $\overline{RT}$ signals.

I$TST   **Instruction Cache Access**   **Input**
When this signal is asserted, external logic can read and write the LR33000's instruction cache. In instruction cache access mode the external device controls the A[31:0], $\overline{AS}$ and $\overline{RT}$ signals.

CPC[3:0] **Coprocessor Condition [3:0]**   **Input**
The four Coprocessor Condition signals, CPC[3:0], carry condition inputs that the LR33000 can separately test using coprocessor branch instructions. Note that the corresponding coprocessor usable bit (Cu[3:0]) in the Status Register must be set in order to test a condition input.

INT[5:0] **Interrupt [5:0]**   **Input**
External logic asserts these signals to cause the LR33000 to take an interrupt exception. Note that interrupts are not sampled in any other type of stall cycles. An instruction sees the interrupt only during the ALU pipestage. After the assertion of an interrupt and the occurrence of an interrupt exception, the interrupts continue to be sampled to provide a level-sensitive indication of the active interrupt or interrupts. The interrupts are not latched within the processor when an interrupt exception occurs.

$\overline{RTO}$ **Refresh Timeout**   **Output**
The LR33000 asserts this signal to indicate that the Refresh Timer has counted down from its preset value to zero. The LR33000 continues to assert $\overline{RTO}$ until external logic asserts RTACK. *When the DRAM Controller is enabled, this signal indicates that a refresh cycle is in progress.*

RTACK   **Refresh Timer Acknowledge**   **Input**
External logic asserts RTACK to indicate that it has completed the refresh operation initiated by the LR33000's assertion of $\overline{RTO}$. The assertion of RTACK causes the LR33000 to deassert $\overline{RTO}$, reload the Refresh Timer to the preset value, and begin counting. External logic must assert RTACK for two cycles to guarantee the Refresh Timer reset.

$\overline{T2EN}$ **Timer 2 Enable**   **Input**
External logic asserts this signal to enable the operation of Timer 2. Timer 2 counts as long as $\overline{T2EN}$ is asserted.

T2TO   **Timer 2 Timeout**   **Output**
The LR33000 toggles T2TO to indicate that the LR33000's Timer 2 has counted down from its preset value to zero. T2TO is set LOW on cold and warm starts.

BRTKN   **Branch Taken**   **Output**
When asserted, this signal indicates that the condition has been met for a branch instruction in the LR33000's pipeline. The LR33000 signals branches during the ALU stage of the pipeline, which is when the control transfer occurs.

TABLE 7.6   LR33000 Interface Signals (Continued)

$\overline{\text{DT}}$ **Data Transaction**   Output
The LR33000 asserts this signal to indicate that the current bus transaction is a data read or write transaction. The LR33000 deasserts this signal to indicate that the current bus transaction is an instruction fetch.

$\overline{\text{STALL}}$ **Stall**   Output
The LR33000 asserts this signal to indicate that it is in a stall state.

TDONE   **Test Mode Transaction Done**   Output
The LR33000 asserts this signal to indicate that a cache access mode read or write operation is complete.

BENDN   **Big Endian**   Input
This signal controls the byte ordering of the LR33000. When BENDN is tied HIGH, the LR33000 uses the IBM byte order convention (big endian). When it is tied LOW, the LR33000 uses the DEC byte order convention (little endian).

$\overline{\text{FRCM}}$ **Force Cache Miss**   Input
When external logic asserts $\overline{\text{FRCM}}$, the LR33000 detects a cache miss on the current instruction fetch. This signal is intended for test and in-circuit emulation purposes.

HIGHZ   **High Impedance**   Input
When external logic asserts HIGHZ, the LR33000 3-states all of its outputs. SYSCLK need not be active for HIGHZ to 3-state the LR33000 outputs. In addition, HIGHZ selects between warm and cold processor starts after reset. When HIGHZ and RESET are asserted together, the LR33000 executes a cold start after RESET is deasserted. When RESET is asserted alone, the LR33000 executes a warm start. The DRAM Controller is *not* initialized on a warm start and the contents of DRAM remain valid.

$\overline{\text{RESET}}$ **Reset**   Input
A low-to-high transition of $\overline{\text{RESET}}$ initializes the processor and initiates a non-maskable RESET exception. Driving $\overline{\text{RESET}}$ low does not halt the processor or cause any other action. Only a low-to-high transition of $\overline{\text{RESET}}$ initializes the processor.

The RESET Exception should vector the processor to a pre-defined bootstrap routine (typically, PROM-resident) that initializes the system. The RESET jump address is different from that used by interrupts, and $\overline{\text{RESET}}$ (unlike interrupts) forces the LR33000 out of any wait loops. (Refer to Chapter 4, Exception Processing, for a detailed description of the LR33000's response to the RESET Exception.) $\overline{\text{RESET}}$ need not be synchronized with SYSCLK, but SYSCLK must be active for $\overline{\text{RESET}}$ to take effect.

SYSCLK   **System Clock**   Input
This signal is the clock input to the processor. It determines the instruction cycle time of the processor.

THIT   **Test Hit**   Output
The LR33000 asserts THIT during a read of its instruction or data caches to indicate that the cache tags matched the address of the read. This signal is intended for test purposes.

SOURCE: Courtesy LSI Logic Corporation.

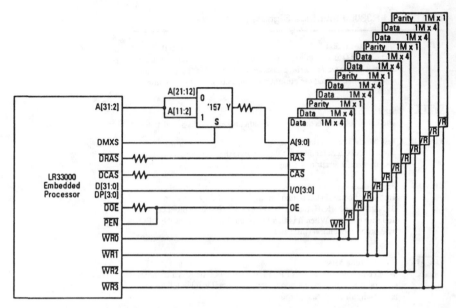

Figure 7.33   4-Mbyte memory subsystem with 1M x 4 DRAMs. *(Courtesy LSI Logic Corporation)*

The write strobes are connected to one of the WR#3 to WR#0 pins. Twenty address bits are required to address the 1 Mwords in this array. These 20 bits (A21 to A2) are multiplexed onto the DRAM's 10-line address bus (A9 to A0) with 10 two-into-one multiplexers.

Example   16-Mbyte DRAM array. A 16-Mbyte DRAM built with four 4-Mbyte banks is shown in Fig. 7.34.[16] The DRAM modules and the row/column multiplexer logic are identical to the ones in the preceding example. Because DRAS# must be asserted to enable a DRAM access, DRAS# is used to select between the four banks. A series of four AND gates combined with an address decoder routes DRAS# to the bank selected by A23 and A22. Four OR gates and RTO# (refresh timeout) are used to enable all four DRAS# signals simultaneously to accommodate refresh signals.[16]

Example   Interleaved 8-Mbyte array. An 8-Mbyte interleaved memory, built with two banks of 1M × 4 DRAMs, is shown in Fig. 7.35.[16] The two banks are interleaved to decrease overall memory delay on block-fetch operations. The design uses the same DRAM banks and address multiplexing logic as described in the first example. A block of control logic implemented with an FPLA (field programmed logic array) or ASIC (application-specific integrated circuit) manages the DCAS# signals and generates the block address (BLK_ADDR). The control block is also responsible for generating the BFTCH# and DRDY# signals.

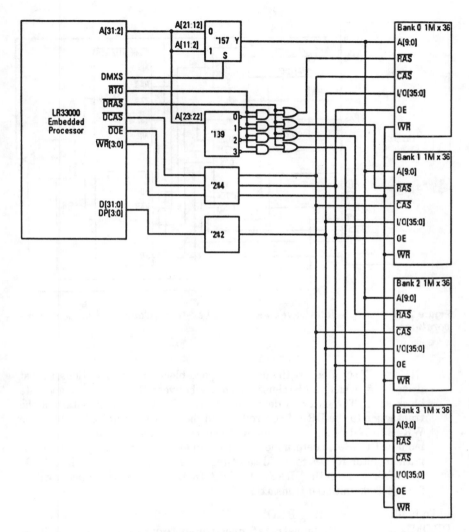

Figure 7.34 16-Mbyte memory subsystem with 1M x 4 DRAMs. *(Courtesy LSI Logic Corporation)*

Example   4-Mbyte DRAM with a DMA-capable external device. A system that includes an external device with DMA capability and a 4-Mbyte DRAM is shown in Fig. 7.36.[16] The external device can be any DMA-capable device, and modern 32-bit devices such as Ethernet controllers should require only minimal interface logic. Eight-bit devices will require more complicated circuitry and pack/unpack logic to manage the interface to the 32-bit LR33000. In all applications, the interface must control many of the memory interface signals for DMA transactions. BREQ# and BGNT are required to arbitrate the memory bus, and RD# and RT# are needed to signal the direction of the transaction (both are required because the DRAM controller uses RT# and the

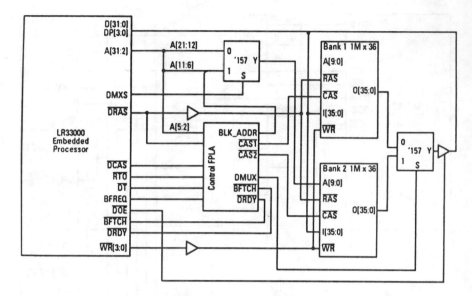

Figure 7.35   8-Mbyte interleaved memory subsystem. *(Courtesy LSI Logic Corporation)*

BIU uses RD#). Because this device supports block-fetch transactions, it must drive DT#, which the BIU uses to choose between the instruction and data block sizes. The external device must use AS# to signal the start of the transaction to the DRAM controller, and the device is required to drive the WR#3 to WR#0 signals during write operations. The external device must monitor DCAS# to determine when to sample D31 to D0 during reads, and it must monitor RTO# to avoid sampling during refresh cycles. The DRAM controller operates the DRAS#, DCAS#, DOE#, and DMXS signals as it does for LR33000 initiated transactions.[16]

To facilitate connecting PROMs to the LR33000, the design provides a PROM select signal that it asserts for read or write operations in the address range from 1F00 0000H to 1FFF FFFFH, and the LR33000 supports byte-wide (8-bit) devices by performing byte gathering when the BWIDE input pin is asserted. One of the system's two wait-state generators further simplifies the connection of PROMs within this address range. When enabled, the wait-state generator stalls the processor for a programmable number of cycles (from 1 to 15). During the last wait state, the wait-state generator asserts the data ready signal, terminating the memory transaction. To accommodate PROMs of various speeds, software can reprogram the wait-state generator for a different number of wait cycles before each access. In other systems, additional off-chip logic must be added to achieve the same performance features.

LSI Logic features an evaluation board for the LR33000 called *pocket rocket,* shown in Fig. 7.37. The board's CPU is a 25 MHz LR33000. The board includes the following:

1 Mbyte of DRAM, expandable to 4 Mbytes.

128 kbytes of PROM (expandable to 256 kbytes). The PROM houses a monitor and debugger for downloading and debugging user programs.

Two RS-232C serial channels providing interconnections to a host system and a terminal.

Two status indicators that can be directly controlled by user programs.

A 96-pin DIN connector providing easy access to the pocket rocket signals, many of which connect directly to the LR33000 CPU.

The board operates with a single 5-volt power supply capable of sourcing a minimum of 3 A.

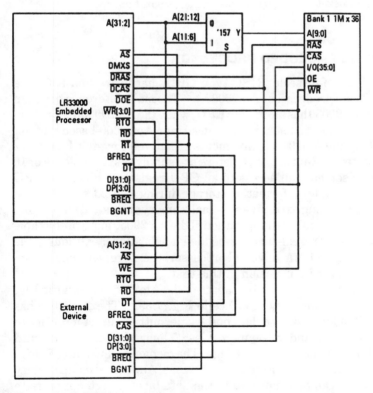

Figure 7.36 DMA-capable memory subsystem. *(Courtesy LSI Logic Corporation)*

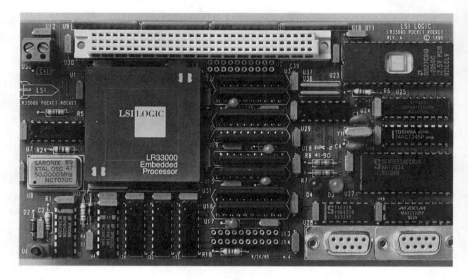

Figure 7.37 The Pocket Rocket. *(Courtesy LSI Corporation)*

7.3 AMD 29050 EMBEDDED PROCESSOR

The Am29050 is a full scale RISC-type CPU, hardware and software compatible with its predecessor Am29000.[2,17] The main difference between the two is that Am29050 has an on-chip FPU, while the Am29000 does not; an arithmetic accelerator Am29027 must be used in an Am29000-based system. The Am29050, as the Am29000, implements a relatively large CPU register file of 192 registers (32-bit). The registers are subdivided into blocks of 16 registers. The user has full access to all CPU registers. Blocks of CPU registers can be used for information storage during context switches and interrupts, thus avoiding excessive external memory access and minimizing response delays. This makes the operation of the Am29050 more predictable and better suited for real-time embedded applications. Although the Am29050 was designed as a universal processor, the aforementioned property is the reason for its inclusion in this text.

A block diagram of the internal datapath of the Am29050 is shown in Fig. 7.38 and its external memory interface in Fig. 7.39.[17] The Am29050 has separate, 32-bit instruction and data busses, permitting direct interface to separate instruction and data memories. Thus, a genuine Harvard architecture is implemented in this design. The memories illustrated in Fig. 7.39 can be either a part of main memory or caches. Additional main memory and I/O interfaces can be connected to the data bus through appropriate interface devices.

The Am29050 is a CMOS, 169-pin (pin grid array), up to 40 MHz (in 1990) microprocessor. Its main subsystems, illustrated in Fig. 7.38, are as follows:

Instruction fetch unit. This unit fetches instructions from the instruction memory and supplies instructions to other functional units. It incorporates the following parts:

1. Instruction prefetch buffer, consisting of 4 words (4 × 32 bits) to store four prefetched instructions (all instructions have a fixed 32-bit size).

2. Branch target cache, containing 256 instructions, subdivided into 64 branch targets of four instructions each. It is a two-way, set-associative cache containing the first four target instructions of up to 64 recently taken branches. The availability of this cache strongly reduces the delay

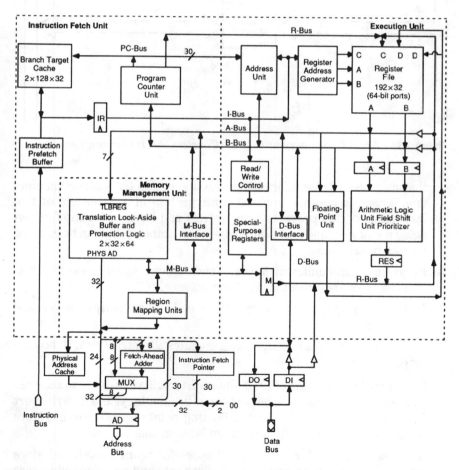

Figure 7.38 Am29050 microprocessor data flow. *(Courtesy AMD Corporation)*

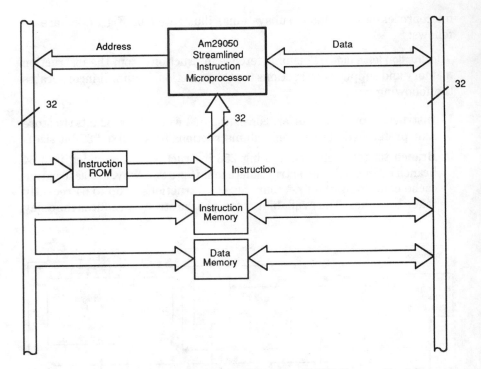

Figure 7.39 Am29050 memory interface. *(Courtesy AMD Corporation)*

that may be incurred in branch processing within the instruction pipe-line. There is another optional configuration of 128 targets of two instructions. This provides more entries and higher hit ratio for low-delay memories (one wait state) where the branch delay can be covered by two instructions.

3. Program counter unit, which creates and sequences addresses of instructions as they are executed by the processor.

4. Instruction register (IR), containing an instruction ready to be decoded and executed.

   Execution unit. The execution unit executes all instructions. It incorporates the following:

1. Register file, consisting of 192 32-bit registers. Two 64-bit read accesses and two write accesses can be performed in a single cycle. The write port for integer results is 32 bit, and for floating-point results it is 64 bit. More details on the register file are given later in this section.

2. Address unit, which evaluates addresses for branch, load, and store instructions. It also assembles immediate data and computes addresses for load-multiple and store-multiple sequences.

3. Arithmetic logic unit (ALU), which performs all integer arithmetic (including multiply step and divide step, included for compatibility, since the FPU supports multiply/divide), logical, and compare operations.

4. Floating-point unit (FPU), which performs single- and double-precision floating-point operations in accordance with the IEEE 754-1985 standard, as well as 32- and 64-bit integer multiply.

5. Field shift unit, which performs N-bit shifts, byte and halfword extract and insert operations, and extracts words from doublewords.

6. Prioritizer, which provides a count of the number of leading zero bits in a 32-bit word. This is useful for performing prioritization in a multilevel interrupt handler, for example.

7. Special-purpose registers (discussed later in this section).

Memory management unit (MMU). The MMU performs address translation and memory protection functions for all branches, loads, and stores. It includes a 64-entry translation lookaside buffer (TLB) for virtual to physical address translation. There are also two region mapping units for translating large areas of contiguous memory, such as a shared library or a frame buffer in graphics applications.

The Am29050 implements a four-stage integer instruction pipeline:

1. Fetch

2. Decode

3. Execute

4. Write back.

The pipeline is organized so the effective instruction execution rate is as high as one instruction per cycle. Data forwarding and pipeline interlocks to deal with pipeline hazards are handled by processor hardware. The execute stage of floating-point operations is further pipelined to a depth determined by the length of the operation. Most floating-point operations can be issued at a rate of one operation per cycle, though most operations take more than one cycle to complete.[17]

Data types recognized by the Am29050 architecture are 8-bit bytes, 16-bit halfwords, 32-bit words, and 64-bit doublewords. It also supports word and halfword integers (signed and unsigned) and single- and double-precision floating-point numbers (IEEE 754–1985 standard). Boolean and character data are also supported. The byte ordering within a word can be either little-endian or big-endian, depending on the setting of the appropriate bit in the configuration register.[17]

The pinout diagram of the Am29050 is illustrated in Fig. 7.40, and the pin description is given in Table 7.7.[17] The Am29050 has 192 general-purpose registers available for any program use. Most Am29050 instructions are

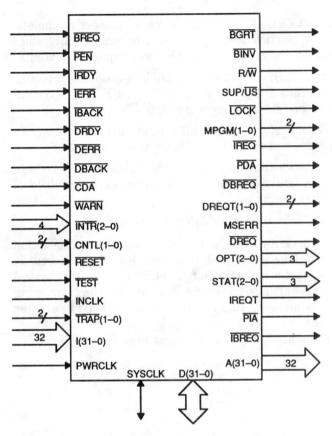

Figure 7.40 Am29050 pinout. *(Courtesy AMD Corporation)*

three-operand instructions. An instruction can specify any three of the 192 registers as operand holders; two for source operands, and one for destination. The 192 registers are divided into 64 global and 128 local registers. Global registers are addressed with absolute register numbers, while local registers are addressed relative to an internal stack pointer (register 1). The general-purpose register organization is illustrated in Fig. 7.41.[17] The following terminology is used to describe the addressing of general-purpose registers:

Register number: A software level number contained in an instruction field, ranging from 0 to 255.

Global register number: A software level number for a global register, ranging from 0 to 127. Registers 4 to 63 are not implemented (see Fig. 7.41).

TABLE 7.7   Am29060 Pin Description

PIN DESCRIPTION

Although certain outputs are described as being three-state or bidirectional outputs, all outputs (except MSERR) may be placed in a high-impedance state by the Test mode. The three-state and bidirectional terminology in this section is for those outputs (except SYSCLK) that are disabled when the processor grants the channel to another master.

A(31–0)
Address Bus (Three-state Outputs, Synchronous)

The Address Bus transfers the byte address for all accesses except burst-mode accesses. For burst-mode accesses, it transfers the address for the first access in the sequence.

BGRT
Bus Grant (Output, Synchronous)

This output signals to an external master that the processor is relinquishing control of the channel in response to BREQ.

BINV
Bus Invalid (Output, Synchronous)

This output indicates that the address bus and related controls are invalid. It defines an idle cycle for the channel.

BREQ
Bus Request (Input, Synchronous)

This input allows other masters to arbitrate for control of the processor channel.

CDA
Coprocessor Data Accept (Input, Synchronous)

This signal allows the coprocessor to indicate the acceptance of operands or operation codes. For transfers to the coprocessor, the processor does not expect a DRDY response; an active level on CDA performs the function normally performed by DRDY. CDA may be active whenever the coprocessor is able to accept transfers.

CNTL(1–0)
CPU Control (Inputs, Asynchronous)

These inputs control the processor mode:

CNTL1	CNTL0	Mode
0	0	Load Test Instruction
0	1	Step
1	0	Halt
1	1	Normal

D(31–0)
Data Bus (Bidirectionals, Synchronous)

The Data Bus transfers data to and from the processor for load and store operations.

DBACK
Data Burst Acknowledge (Input, Synchronous)

This input is active whenever a burst-mode data access has been established. It may be active even though no data are currently being accessed.

DBREQ
Data Burst Request (Three-state Output, Synchronous)

This signal is used to establish a burst-mode data access and to request data transfers during a burst-mode data access. DBREQ may be active even though the address bus is being used for an instruction access. This signal becomes valid late in the cycle, with respect to DREQ.

DERR
Data Error (Input, Synchronous)

This input indicates that an error occurred during the current data access. For a load, the processor ignores the content of the data bus. For a store, the access is terminated. In either case, a Data Access Exception trap occurs. The processor ignores this signal if there is no pending data access.

DRDY
Data Ready (Input, Synchronous)

For loads, this input indicates that valid data is on the data bus. For stores, it indicates that the access is complete, and that data need no longer be driven on the data bus. The processor ignores this signal if there is no pending data access.

DREQ
Data Request (Three-state Output, Synchronous)

This signal requests a data access. When it is active, the address for the access appears on the address bus.

DREQT(1–0)
Data Request Type (Three-state Outputs, Synchronous)

These signals specify the address space of a data access, as follows (the value "x" is a "don't care"):

DREQT1	DREQT0	Meaning
0	0	Data memory access
0	1	Input/output access
1	x	Coprocessor transfer

An interrupt/trap vector request is indicated as a data memory read. If required, the system can identify the vector fetch by the STAT(2–0) outputs. DREQT(1–0) are valid only when DREQ is active.

I(31–0)
Instruction Bus (Inputs, Synchronous)

The Instruction Bus transfers instructions to the processor.

IBACK
Instruction Burst Acknowledge (Input, Synchronous)

This input is active whenever a burst-mode instruction access has been established. It may be active even though no instructions are currently being accessed.

TABLE 7.7   Am29060 Pin Description *(Continued)*

IBREQ
Instruction Burst Request (Three-state Output, Synchronous)

This signal is used to establish a burst-mode instruction access and to request instruction transfers during a burst-mode instruction access. IBREQ may be active even though the address bus is being used for a data access. This signal becomes valid late in the cycle with respect to IREQ.

IERR
Instruction Error (Input, Synchronous)

This input indicates that an error occurred during the current instruction access. The processor ignores the content of the instruction bus, and an Instruction Access Exception trap occurs if the processor attempts to execute the invalid instruction. The processor ignores this signal if there is no pending instruction access.

INCLK
Input Clock (Input)

When the processor generates the clock for the system, this is an oscillator input to the processor at twice the processor's operating frequency. In systems where the clock is not generated by the processor, this signal must be tied High or Low, except in certain master/slave configurations.

INTR(3–0)
Interrupt Request (Inputs, Asynchronous)

These inputs generate prioritized interrupt requests. The interrupt caused by INTR0 has the highest priority, and the interrupt caused by INTR3 has the lowest priority. The interrupt requests are masked in prioritized order by the Interrupt Mask field in the Current Processor Status Register.

IRDY
Instruction Ready (Input, Synchronous)

This input indicates that a valid instruction is on the instruction bus. The processor ignores this signal if there is no pending instruction access.

IREQ
Instruction Request (Three-state Output, Synchronous)

This signal requests an instruction access. When it is active, the address for the access appears on the address bus.

IREQT
Instruction Request Type (Three-state Output, Synchronous)

This signal specifies the address space of an instruction request when IREQ is active:

IREQT	Meaning
0	Instruction random-access memory access
1	Instruction read-only memory access

LOCK
Lock (Three-state Output, Synchronous)

This output allows the implementation of various channel and device interlocks. It may be active only for the duration of an access, or active for an extended period of time under control of the Lock bit in the Current Processor Status.

The processor does not relinquish the channel (in response to BREQ) when LOCK is active.

MPGM(1–0)
MMU Programmable (Three-state Outputs, Synchronous)

These outputs reflect the value of two PGM bits in the Translation Look-Aside Buffer or Region Mapping Unit entry associated with the access. If no address translation is performed, these signals are both Low.

MSERR
Master/Slave Error (Output, Synchronous)

This output shows the result of the comparison of processor outputs with the signals provided internally to the off-chip drivers. If there is a difference for any enabled driver, this line is asserted.

OPT(2–0)
Option Control (Three-state Outputs, Synchronous)

These outputs reflect the value of bits 18–16 of the load or store instruction that begins an access. Bit 18 of the instruction is reflected on OPT2, bit 17 on OPT1, and bit 16 on OPT0.

The standard definitions of these signals (based on DREQT) are as follows (the value "x" is a "don't care"):

DREQT1	DREQT0	OPT2	OPT1	OPT0	Meaning
0	x	0	0	0	Word-length access
0	x	0	0	1	Byte access
0	x	0	1	0	Half-word access
0	0	1	0	0	Instruction ROM access (as data)
0	0	1	0	1	Cache control
0	0	1	1	0	Emulator accesses
—All Others—					Reserved

During an interrupt/trap vector fetch, the OPT(2–0) signals indicate a word-length access (000). Also, the system should return an entire aligned word for a read, regardless of the indicated data length.

The Am29050 microprocessor does not explicitly prevent a store to the instruction ROM. OPT(3–0) are valid only when DREQ is active.

PDA
Pipelined Data Access (Three-state Output, Synchronous)

If DREQ is not active, this output indicates that a data access is pipelined with another in-progress data access. The indicated access cannot be completed until the first access is complete. The completion of the first access is signaled by the assertion of DREQ.

PEN
Pipeline Enable (Input, Synchronous)

This signal allows devices that can support pipelined accesses (i.e., that have input latches for the address and required controls) to signal that a second access may begin while the first is being completed.

TABLE 7.7   Am29060 Pin Description *(Continued)*

PIA
Pipelined Instruction Access
(Three-state Output, Synchronous)

If $\overline{IREQ}$ is not active, this output indicates that an instruction access is pipelined with another in-progress instruction access. The indicated access cannot be completed until the first access is complete. The completion of the first access is signaled by the assertion of $\overline{IREQ}$.

PWRCLK
Power Supply for SYSCLK Driver

This pin is a power supply for the SYSCLK output driver. It isolates the SYSCLK driver, and is used to determine whether or not the Am29050 microprocessor generates the clock for the system. If power (+5 V) is applied to this pin, the Am29050 microprocessor generates a clock on the SYSCLK output. If this pin is grounded, the Am29050 microprocessor accepts a clock generated by the system on the SYSCLK input.

R/W̄
Read/Write (Three-state Output, Synchronous)

This signal indicates whether data is being transferred from the processor to the system, or from the system to the processor. R/W̄ is valid only when the address bus is valid. R/W̄ will be High when $\overline{IREQ}$ is active.

RESET
Reset (Input, Asynchronous)

This input places the processor in the Reset mode.

STAT(2–0)
CPU Status (Outputs, Synchronous)

These outputs indicate the state of the processor's execution stage on the previous cycle. They are encoded as follows:

STAT2	STAT1	STAT0	Condition
0	0	0	Halt or Step Modes
0	0	1	Pipeline Hold Mode
0	1	0	Load Test Instruction Mode, Synchronize
0	1	1	Wait Mode
1	0	0	Interrupt Return
1	0	1	Taking Interrupt or Trap
1	1	0	Non-sequential Instruction Fetch
1	1	1	Executing Mode

SUP/ŪS
Supervisor/User Mode
(Three-state Output, Synchronous)

This output indicates the program mode for an access.

SYSCLK
System Clock (Bidirectional)

This is either a clock output with a frequency that is half that of INCLK, or an input from an external clock generator at the processor's operating frequency.

TEST
Test Mode (Input, Asynchronous)

When this input is active, the processor is in Test mode. All outputs and bidirectional lines, except MSERR, are forced to the state.

TRAP̄(1–0)
Trap Request (Inputs, Asynchronous)

These inputs generate prioritized trap requests. The trap caused by $\overline{TRAP0}$ has the highest priority. These trap requests are disabled by the DA bit of the Current Processor Status Register.

WARN
Warn (Input, Asynchronous, Edge-sensitive)

A high-to-low transition on this input causes a non-maskable $\overline{WARN}$ trap to occur. This trap bypasses the normal trap vector fetch sequence, and is useful in situations where the vector fetch may not work (e.g., when data memory is faulty).

PIN169
Alignment pin

In the PGA package, this pin is used to indicate proper pin-alignment of the Am29050 microprocessor and is used by an in-circuit emulator to communicate its presence to the system.

SOURCE: Courtesy AMD Corporation.

Local register number: A software level number for a local register, ranging from 0 to 127.

Absolute register number: A hardware level number (in the left column of Fig. 7.41) used to select a general-purpose register, ranging from 0 to 255.

Register numbers are expressed by 8-bit binary numbers. When the most significant bit of a register number is 0, a global register is selected. The

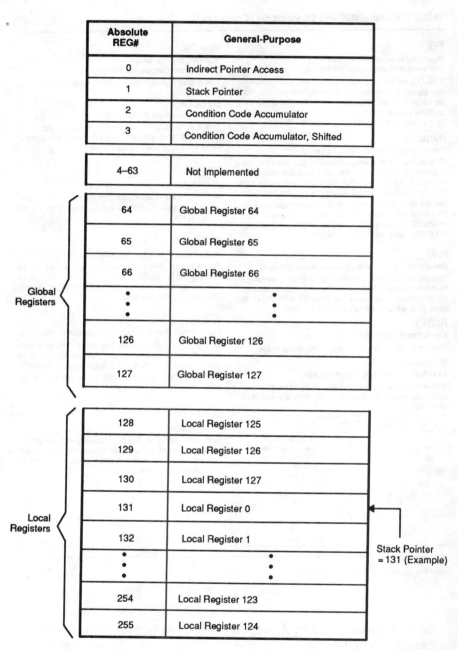

Absolute REG#	General-Purpose
0	Indirect Pointer Access
1	Stack Pointer
2	Condition Code Accumulator
3	Condition Code Accumulator, Shifted
4–63	Not Implemented
64	Global Register 64
65	Global Register 65
66	Global Register 66
• • •	• • •
126	Global Register 126
127	Global Register 127
128	Local Register 125
129	Local Register 126
130	Local Register 127
131	Local Register 0
132	Local Register 1
• • •	• • •
254	Local Register 123
255	Local Register 124

Global Registers (rows 64–127)

Local Registers (rows 128–255)

Stack Pointer = 131 (Example)

Figure 7.41   General-purpose register organization. *(Courtesy AMD Corporation)*

seven least significant bits of the register number give the global register number. For global registers, the absolute register number is equivalent to the register number. The register numbers associated with global registers 0 to 3 have a special meaning:

Register 0 specifies that an indirect pointer is to be used as the source of the register number; there is an indirect pointer for each of the instruction operand registers.

Register 1 contains the stack pointer, which is used in the addressing of local registers discussed later.

Registers 2 and 3 allow access to the condition code accumulator, which is used to concatenate boolean results from one or more operations into a single condition code.

When the most significant bit of a register number is 1, a local register is selected. The seven least significant bits of the register number give the local register number. For local registers, the absolute register number is obtained by adding the local register number to bits 8 to 2 of the stack pointer (register 1) and truncating the result to seven bits; the most significant bit of the original register number is unchanged (it remains a 1).

Example Local register 0 (Fig. 7.41). The content of the stack pointer (register 1) is 131. Thus, the absolute register number of local register 0 is: 0 + 131 = 131. Similarly, the absolute register number for local register 124 is: 124 + 131 = 255.

The stack pointer addition applied to local register numbers provides a limited form of base plus offset addressing within the local registers. The stack pointer contains the base address. This assists run-time storage management of variables for dynamically nested procedures.

For the purpose of access restriction and information protection, the general-purpose registers are divided into register banks, consisting of 16 registers (except for bank 0, which contains registers 4 to 15). Register banks are partitioned according to absolute register numbers, as illustrated in Fig. 7.42.[17] There is a register bank protect register (special register 7) that contains 16 protection bits, where each bit controls user mode accesses (read or write) to a bank of registers. Bits 0 to 15 of the register bank protect register protect register banks 0 to 15, respectively. When a bit in the register bank protect register is 1, and a register in the corresponding bank is specified as an operand register by a user mode instruction, a protection violation trap occurs. When the processor is in supervisor or monitor mode, the register bank protect register has no effect on general-purpose register accesses. The Am29050 contains 39 special-purpose registers, illustrated in Fig. 7.43.[17] A brief description of the tasks of the special purpose registers is listed in Table 7.8.[17]

TABLE 7.8 Special-Purpose Registers

0. **VAB**—Vector Area Base Address—Defines the beginning of the interrupt/trap Vector Area.

1. **OPS**—Old Processor Status—Receives a copy of the Current Processor Status when an interrupt or trap is taken. It is later used to restore the Current Processor Status on an interrupt return.

2. **CPS**—Current Processor Status—Contains control information associated with the currently executing process, such as interrupt disables and the Supervisor Mode bit.

3. **CFG**—Configuration—Contains control information which normally varies only from system to system, and usually is set only during system initialization.

4. **CHA**—Channel Address—Contains the address associated with an external access, and retains the address if the access does not complete successfully. The Channel Address Register, in conjunction with the Channel Data and Channel Control registers described below, allow the restarting of unsuccessful external accesses. This might be necessary for an access encountering a page fault in a demand-paged environment, for example.

5. **CHD**—Channel Data—Contains data associated with a store operation, and retains the data if the operation does not complete successfully.

6. **CHC**—Channel Control—Contains control information associated with a channel operation, and retains this information if the operation does not complete successfully.

7. **RBP**—Register Bank Protect—Restricts access of User-mode programs to specified groups of 16 registers. This protects operating-system parameters kept in the global registers from corruption by User-mode programs.

8. **TMC**—Timer Counter—Supports real-time control and other timing-related functions.

9. **TMR**-Timer Reload—Maintains synchronization of the Timer Counter. It includes control bits for the Timer Facility.

10. **PC0**—Program Counter 0—Contains the address of the instruction being decoded when an interrupt or trap is taken. The processor restarts this instruction upon interrupt return.

11. **PC1**—Program Counter 1—Contains the address of the instruction being executed when an interrupt or trap is taken. The processor restarts this instruction upon interrupt return.

12. **PC2**—Program Counter 2—Contains the address of the instruction just completed when an interrupt or trap is taken. This address is provided for information only, and does not participate in an interrupt return.

13. **MMU**—MMU Configuration—Allows selection of various memory-management options, such as page size.

14. **LRU**—LRU Recommendation—Simplifies the reload of entries in the Translation Look-Aside Buffer (TLB) by providing information on the least-recently used entry of the TLB when a TLB miss occurs (see Section Memory Management).

15. **RSN**—Reason Vector—Contains the vector number of the synchronous trap which caused entry into the Monitor mode.

16. **RMA0**—Region Mapping Address 0—Specifies a mapping from a region of virtual address space to physical address space; contains the Virtual Base Address (VBA) and the corresponding Physical Base Address (PBA). (See Section Address Translation.)

17. **RMC0**—Region Mapping Control 0—Contains control information associated with the region mapping specified by the Region Mapping Address Register 0.

18. **RMA1**—Region Mapping Address 1—Specifies a mapping from a region of virtual address space to physical address space; contains the Virtual Base Address (VBA) and the corresponding Physical Base Address (PBA).

19. **RMC1**—Region Mapping Control 1—Contains control information associated with the region mapping specified by the Region Mapping Address Register 1.

20. **SPC0**—Shadow Program Counter 0—Contains the address of the instruction being decoded when the processor enters Monitor mode. The processor restarts this instruction upon return from Monitor mode.

21. **SPC1**—Shadow Program Counter 1—Contains the address of the instruction being executed when the processor enters Monitor mode. The processor restarts this instruction upon return from Monitor mode.

22. **SPC2**—Shadow Program Counter 2—Contains the address of the instruction just completed when the processor enters Monitor mode. This address is provided for information only, and does not participate in the return from Monitor mode.

23. **IBA0**—Instruction Breakpoint Address 0—Contains the address of an instruction breakpoint.

24. **IBC0**—Instruction Breakpoint Control 0—Contains control and status information for the breakpoint comparison specified by the Instruction Breakpoint Address Register 0.

25. **IBA1**—Instruction Breakpoint Address 1—Contains the address of an instruction breakpoint.

26. **IBC1**—Instruction Breakpoint Control 1—Contains control and status information for the breakpoint comparison specified by the Instruction Breakpoint Address Register 1.

The unprotected special-purpose registers are defined as follows:

128. **IPC**—Indirect Pointer C—Allows the indirect access of a general-purpose register.

129. **IPA**—Indirect Pointer A—Allows the indirect access of a general-purpose register.

130. **IPB**—Indirect Pointer B—Allows the indirect access of a general-purpose register.

131. **Q**—Q—Provides additional operand bits for multiply step, divide step, and divide operations.

TABLE 7.8 Special-Purpose Registers *(Continued)*

132. **ALU**—ALU Status—Contains information about the outcome of integer arithmetic and logical operations, and holds residual control for certain instruction operations.

133. **BP**—Byte Pointer—Contains an index of a byte or half-word within a word. This register is also accessible via the ALU Status Register.

134. **FC**—Funnel Shift Count—Provides a bit offset for the extraction of word-length fields from double-word operands. This register is also accessible via the ALU Status Register.

135. **CR**—Load/Store Count Remaining—Maintains a count of the number of loads and stores remaining for load-multiple and store-multiple operations. The count is initialized to the total number of loads or stores to be performed before the operation is initiated. This register is also accessible via the Channel Control Register.

160. **FPE**—Floating-Point Environment—Controls the operation of floating-point arithmetic, such as rounding modes and exception reporting.

161. **INTE**—Integer Environment—Enables and disables the reporting of exceptions which occur during integer multiply and divide operations.

162. **FPS**—Floating-Point Status—Contains information about the outcome of floating-point operations.

164. **EXOP**—Exception Opcode—Reports the operation code of an instruction causing a trap. This register is provided primarily for recovery from floating-point exceptions, but is also set for other instructions that cause traps.

SOURCE: Courtesy AMD Corporation.

Register Bank Protect Register Bit	Absolute-Register Numbers	General-Purpose Registers
0	4 through 15	Bank 0 (not implemented)
1	16 through 31	Bank 1 (not implemented)
2	32 through 47	Bank 2 (not implemented)
3	48 through 63	Bank 3 (not implemented)
4	64 through 79	Bank 4
5	80 through 95	Bank 5
6	96 through 111	Bank 6
7	112 through 127	Bank 7
8	128 through 143	Bank 8
9	144 through 159	Bank 9
10	160 through 175	Bank 10
11	176 through 191	Bank 11
12	192 through 207	Bank 12
13	208 through 223	Bank 13
14	224 through 239	Bank 14
15	240 through 255	Bank 15

Figure 7.42 Register bank organization. *(Courtesy AMD Corporation)*

Register Number	Protected Registers	Mnemonic
0	Vector Area Base Address	VAB
1	Old Processor Status	OPS
2	Current Processor Status	CPS
3	Configuration	CFG
4	Channel Address	CHA
5	Channel Data	CHD
6	Channel Control	CHC
7	Register Bank Protect	RBP
8	Timer Counter	TMC
9	Timer Reload	TMR
10	Program Counter 0	PC0
11	Program Counter 1	PC1
12	Program Counter 2	PC2
13	MMU Configuration	MMU
14	LRU Recommendation	LRU
15	Reason Vector	RSN
16	Region Mapping Address 0	RMA0
17	Region Mapping Control 0	RMC0
18	Region Mapping Address 1	RMA1
19	Region Mapping Control 1	RMC1
20	Shadow Program Counter 0	SPC0
21	Shadow Program Counter 1	SPC1
22	Shadow Program Counter 2	SPC2
23	Instruction Breakpoint Address 0	IBA0
24	Instruction Breakpoint Control 0	IBC0
25	Instruction Breakpoint Address 1	IBA1
26	Instruction Breakpoint Control 1	IBC1

	Unprotected Registers	
128	Indirect Pointer C	IPC
129	Indirect Pointer A	IPA
130	Indirect Pointer B	IPB
131	Q	Q
132	ALU Status	ALU
133	Byte Pointer	BP
134	Funnel Shift Count	FC
135	Load/Store Count Remaining	CR
.		.
.		.
.		.
160	Floating-Point Environment	FPE
161	Integer Environment	INTE
162	Floating-Point Status	FPS
.		.
.		.
.		.
164	Exception Opcode	EXOP

Figure 7.43 Special-purpose registers. *(Courtesy AMD Corporation)*

The Am29050 implements 125 instructions. All instructions execute in a single cycle, except for IRET, IRETINV, LOADM, STOREM, floating-point, and integer multiply/divide instructions. The Am29050 instructions are listed in Table 7.9.[17] Most instructions deal with general-purpose registers for operands. A typical instruction format will contain an 8-bit opcode in the most significant 8 bits, and then three 8-bit fields for register operand specification. Specification of global register 0 as an instruction operand register causes an indirect access to the general-purpose registers. In this case, the absolute register number is provided by an indirect pointer contained in one of the special-purpose registers 128, 129, or 130 (see Fig. 7.43), called *indirect pointer C, A,* or *B,* respectively. The Am29050 instruction fields (8 bits each) are labeled as follows: opcode, C, A, B (C is the destination, and A and B are the source operands). For instance, if field A contains a zero, the operand register number will be fetched from special register 129 (indirect pointer A; see Fig. 7.43).

Interrupts and traps cause the Am29050 to suspend the execution of an instruction sequence and to begin the execution of an interrupt- or trap-handling routine. Interrupts allow external devices and the timer to control processor execution. They are always asynchronous to program execution. Traps are intended to be used for certain exceptional events that occur during instruction execution, and are generally synchronous to program execution. Interrupts are caused by signals applied to any of the external inputs INTR#3 to 0, or by the timer. The processor may be disabled from taking certain interrupts by the masking capability provided by certain bits in the current processor status register (special register 2). The vector number assignments for the Am29050 interrupts and traps are listed in Table 7.10.[17]

7.4 NATIONAL SEMICONDUCTOR EMBEDDED PROCESSORS

National Semiconductor has recently announced a whole family of embedded processors called the Series 32000/EP family of National Semiconductor Embedded System Processors.[18,19] It has features similar to those of the National Semiconductor NS32000 family of microprocessors.[20] The top current model of the 32000/EP family is the superscalar, RISC-type, integrated system processor NS32SF641, also known as Swordfish.[19] There is also a reduced model (without an on-chip FPU) NS32SF640 with an essentially identical architecture. It was designed especially for computation-intensive embedded applications. The software of the NS32SF641 is assembly compatible with other processors in the family. In addition, it provides new features that support graphics and digital signal processing. For the sake of simplicity, the NS32SF641 is called 641 in the rest of this section (and 640 for NS32SF640).

TABLE 7.9   Am29050 Instruction Set

Integer Arithmetic Instructions	
Mnemonic	**Operation Description**
ADD	DEST ← SRCA + SRCB
ADDS	DEST ← SRCA + SRCB IF signed overflow THEN Trap (Out of Range)
ADDU	DEST ← SRCA + SRCB IF unsigned overflow THEN Trap (Out of Range)
ADDC	DEST ← SRCA + SRCB + C
ADDCS	DEST ← SRCA + SRCB + C IF signed overflow THEN Trap (Out of Range)
ADDCU	DEST ← SRCA + SRCB + C IF unsigned overflow THEN Trap (Out of Range)
SUB	DEST ← SRCA − SRCB
SUBS	DEST ← SRCA − SRCB IF signed overflow THEN Trap (Out of Range)
SUBU	DEST ← SRCA − SRCB IF unsigned underflow THEN Trap (Out of Range)
SUBC	DEST ← SRCA − SRCB − 1 + C
SUBCS	DEST ← SRCA − SRCB − 1 + C IF signed overflow THEN Trap (Out of Range)
SUBCU	DEST ← SRCA − SRCB − 1 + C IF unsigned underflow THEN Trap (Out of Range)
SUBR	DEST ← SRCB − SRCA
SUBRS	DEST ← SRCB − SRCA IF signed overflow THEN Trap (Out of Range)
SUBRU	DEST ← SRCB − SRCA IF unsigned underflow THEN Trap (Out of Range)
SUBRC	DEST ← SRCB − SRCA − 1 + C
SUBRCS	DEST ← SRCB − SRCA − 1 + C IF signed overflow THEN Trap (Out of Range)
SUBRCU	DEST ← SRCB − SRCA − 1 + C IF unsigned underflow THEN Trap (Out of Range)
MULTIPLU	DEST ← SRCA · SRCB (unsigned)
MULTIPLY	DEST ← SRCA · SRCB (signed)
MUL	Perform one-bit step of a multiply operation (signed)
MULL	Complete a sequence of multiply steps
MULTM	DEST ← SRCA · SRCB (signed), most-significant bits
MULTMU	DEST ← SRCA · SRCB (unsigned), most-significant bits
MULU	Perform one-bit step of a multiply operation (unsigned)
DIVIDE	DEST ← (Q//SRCA)/(SRCB) (signed) Q ← Remainder
DIVIDU	DEST ← (Q//SRCA)/SRCB (unsigned) Q ← Remainder
DIV0	Initialize for a sequence of divide steps (unsigned)
DIV	Perform one-bit step of a divide operation (unsigned)

TABLE 7.9 Am29050 Instruction Set *(Continued)*

DIVL	Complete a sequence of divide steps (unsigned)
DIVREM	Generate remainder for divide operation (unsigned)

Compare Instructions

Mnemonic	Operation Description
CPEQ	IF SRCA = SRCB THEN DEST ← TRUE ELSE DEST ← FALSE
CPNEQ	IF SRCA <> SRCB THEN DEST ← TRUE ELSE DEST ← FALSE
CPLT	IF SRCA < SRCB THEN DEST ← TRUE ELSE DEST ← FALSE
CPLTU	IF SRCA < SRCB (unsigned) THEN DEST ← TRUE ELSE DEST ← FALSE
CPLE	IF SRCA < SRCB THEN DEST ← TRUE ELSE DEST ← FALSE
CPLEU	IF SRCA < SRCB (unsigned) THEN DEST ← TRUE ELSE DEST ← FALSE
CPGT	IF SRCA > SRCB THEN DEST ← TRUE ELSE DEST ← FALSE
CPGTU	IF SRCA > SRCB (unsigned) THEN DEST ← TRUE ELSE DEST ← FALSE
CPGE	If SRCA > SRCB THEN DEST ← TRUE ELSE DEST ← FALSE
CPGEU	IF SRCA > SRCB (unsigned) THEN DEST ← TRUE ELSE DEST ← FALSE
CPBYTE	IF (SRCA.BYTE0 = SRCB.BYTE0) OR (SRCA.BYTE1 = SRCB.BYTE1) OR (SRCA.BYTE2 = SRCB.BYTE2) OR (SRCA.BYTE3 = SRCB.BYTE3) THEN DEST ← TRUE ELSE DEST ← FALSE
ASEQ	IF SRCA = SRCB THEN Continue ELSE Trap (VN)
ASNEQ	IF SRCA <> SRCB THEN Continue ELSE Trap (VN)
ASLT	IF SRCA < SRCB THEN Continue ELSE Trap (VN)
ASLTU	IF SRCA < SRCB (unsigned) THEN Continue ELSE Trap (VN)
ASLE	IF SRCA < SRCB THEN Continue ELSE Trap (VN)
ASLEU	IF SRCA < SRCB (unsigned) THEN Continue ELSE Trap (VN)

TABLE 7.9   Am29050 Instruction Set *(Continued)*

ASGT	IF SRCA > SRCB THEN Continue ELSE Trap (VN)
ASGTU	IF SRCA > SRCB (unsigned) THEN Continue ELSE Trap (VN)
ASGE	IF SRCA > SRCB THEN Continue ELSE Trap (VN)
ASGEU	IF SRCA > SRCB (unsigned) THEN Continue ELSE Trap (VN)

Mnemonic	Operation Description
AND	DEST ← SRCA & SRCB
ANDN	DEST ← SRCA & ~ SRCB
NAND	DEST ← ~ (SRCA & SRCB)
OR	DEST ← SRCA | SRCB
ORN	DEST ← SRCA | ~ SRCB
NOR	DEST ← ~ (SRCA | SRCB)
XOR	DEST ← SRCA ^ SRCB
XNOR	DEST ← ~ (SRCA ^ SRCB)

Shift Instructions

Mnemonic	Operation Description
SLL	DEST ← SRCA << SRCB (zero fill)
SRL	DEST ← SRCA >> SRCB (zero fill)
SRA	DEST ← SRCA >> SRCB (sign fill)
EXTRACT	DEST ← high-order word of (SRCA//SRCB << FC)

Data Movement Instructions

Mnemonic	Operation Description
LOAD	DEST ← EXTERNAL WORD [SRCB]
LOADL	DEST ← EXTERNAL WORD [SRCB] assert LOCK output during access
LOADSET	DEST ← EXTERNAL WORD [SRCB] EXTERNAL WORD [SRCB] ← h'FFFFFFFF' assert LOCK output during access
LOADM	DEST .. DEST + COUNT ← EXTERNAL WORD [SRCB] .. EXTERNAL WORD [SRCB + COUNT · 4]
STORE	EXTERNAL WORD [SRCB] ← SRCA

TABLE 7.9 Am29050 Instruction Set *(Continued)*

STOREL	EXTERNAL WORD [SRCB] ← SRCA assert LOCK output during access
STOREM	EXTERNAL WORD [SRCB] .. EXTERNAL WORD [SRCB + COUNT · 4] ← SRCA .. SRCA + COUNT
EXBYTE	DEST ← SRCB, with low-order byte replaced by byte in SRCA selected by BP
EXHW	DEST ← SRCB, with low-order half-word replaced by half-word in SRCA selected by BP
EXHWS	DEST ← half-word in SRCA selected by BP, sign-extended to 32 bits
INBYTE	DEST ← SRCA, with byte selected by BP replaced by low-order byte of SRCB
INHW	DEST ← SRCA, with half-word selected by BP replaced by low-order half-word of SRCB
MFSR	DEST ← SPECIAL
MFTLB	DEST ← TLB [SRCA]
MTSR	SPDEST ← SRCB
MTSRIM	SPDEST ← 0I16
MTTLB	TLB [SRCA] ← SRCB

Constant Instructions	
Mnemonic	**Operation Description**
CONST	DEST ← 0I16
CONSTH	Replace high-order half-word of SRCA by I16
CONSTHZ	Replace high-order half-word of SRCA with I16, and replace low-order half-word of SRCA with zeros
CONSTN	DEST ← 1I16

Floating-Point Instructions		
Mnemonic	**Operation Description**	
FADD	DEST (single-precision)	← SRCA (single-precision) + SRCB (single-precision)
DADD	DEST (double-precision)	← SRCA (double-precision) + SRCB (double-precision)

TABLE 7.9   Am29050 Instruction Set *(Continued)*

FSUB	DEST (single-precision)	← SRCA (double-precision) – SRCB (single-precision)
DSUB	DEST (double-precision)	← SRCA (double-precision) – SRCB (double-precision)
FMUL	DEST (single-precision)	← SRCA (single-precision) · SRCB (single-precision)
FDMUL	DEST (double-precision)	← SRCA (single-precision) · SRCB (single-precision)
DMUL	DEST (double-precision)	← SRCA (double-precision) · SRCB (double-precision)
FDIV	DEST (single-precision)	← SRCA (single-precision) / SRCB (single-precision)
DDIV	DEST (double-precision)	← SRCA (double-precision) / SRCB (double-precision)
FMAC	ACC(ACN) (variable-precision)	← SRCA (single-precision) · SRCB (single-precision) + ACC(ACN) (variable precision)
DMAC	ACC(ACN) (double-precision)	← SRCA (double-precision) · SRCB (double-precision) + ACC(ACN) (double-precision)
FMSM	DEST (single-precision)	← SRCA (single-precision) · ACC(0) (single-precision) + SRCB (single-precision)
DMSM	DEST (double-precision)	← SRCA (double-precision) · ACC(0) (double-precision) + SRCB (double-precision)
MFACC	DEST ← ACC(ACN)	
MTACC	ACC(ACN) ← SRCA	
FEQ	IF SRCA (single-precision) = SRCB (single-precision) THEN DEST ← TRUE ELSE DEST ← FALSE	
DEQ	IF SRCA (double-precision) = SRCB (double-precision) THEN DEST ← TRUE ELSE DEST ← FALSE	
FGE	IF SRCA (single-precision) >= SRCB (single-precision) THEN DEST ← TRUE ELSE DEST ← FALSE	
DGE	IF SRCA (double-precision) >= SRCB (double-precision) THEN DEST ← TRUE ELSE DEST ← FALSE	
FGT	IF SRCA (single-precision) > SRCB (single-precision) THEN DEST ← TRUE ELSE DEST ← FALSE	

TABLE 7.9 Am29050 Instruction Set *(Continued)*

Floating-Point Instructions (continued)	
Mnemonic	**Operation Description**
DGT	IF SRCA (double-precision) > SRCB (double-precision) THEN DEST ← TRUE ELSE DEST ← FALSE
SQRT	DEST (single-precision, double-precision) ← SQRT[SRCA (single-precision, double-precision)]
CONVERT	DEST (integer, single-precision, double-precision) ← SRCA (integer, single-precision, double-precision)
CLASS	DEST ← CLASS[SRCA (single-precision, double-precision)]

Branch Instructions	
Mnemonic	**Operation Description**
CALL	DEST ← PC//00 + 8 PC ← TARGET Execute delay instruction
CALLI	DEST ← PC//00 + 8 PC ← SRCB Execute delay instruction
JMP	PC ← TARGET Execute delay instruction
JMPI	PC ← SRCB Execute delay instruction
JMPT	IF SRCA = TRUE THEN PC ← TARGET Execute delay instruction
JMPTI	IF SRCA = TRUE THEN PC ← SRCB Execute delay instruction
JMPF	IF SRCA = FALSE THEN PC ← TARGET Execute delay instruction
JMPFI	IF SRCA = FALSE THEN PC ← SRCB Execute delay instruction

JMPFDEC	IF SRCA = FALSE THEN
	  SRCA ← SRCA − 1
	  PC ← TARGET
	ELSE
	  SRCA ← SRCA − 1
	Execute delay instruction

Miscellaneous Instructions	
Mnemonic	**Operation Description**
CLZ	Determine number of leading zeros in a word
SETIP	Set IPA, IPB, and IPC with operand register numbers

TABLE 7.9   Am29050 Instruction Set *(Continued)*

EMULATE	Load IPA and IPB with operand register numbers, and Trap (VN)
INV	Reset all Valid bits in Branch Target Cache to zeros
IRET	Perform an interrupt return sequence
IRETINV	Perform an interrupt return sequence, and reset all Valid bits in Branch Target Cache to zeros
HALT	Enter Halt mode

SOURCE: Courtesy AMD Corporation.

The 641 is a 0.8-micron, double-metal CMOS, 223-pin, 50-MHz, over-1-million transistor chip. It can process up to 100 million instructions per second (MIPS). Its RISC CPU core incorporates two integer units, each of which has a four-stage pipeline, a floating-point unit (FPU) with a DSP (digital signal processing) array multiplier, an instruction and data caches, a two-channel DMA controller, a 15-level interrupt control unit, and a 16-bit timer. Such a composition of features on one chip make the 641 a very attractive option for high-performance embedded applications. It has an internal and external 64-bit data bus, which doubles the bandwidth of information transmission to and from the chip and significantly enhances overall performance. A block diagram of the 640 is shown in Fig. 7.44.[18]

The 641 has 105 internal registers grouped by function as follows:

Thirty-two general-purpose registers

Thirty-two single-precision (or 16 double-precision) floating-point data registers

Three dedicated address registers

One processor status register (PSR)

One configuration register (CFG)

One floating-point status register (FSR)

Two floating-point trapped result registers (FTRH and FTRL)

Six debug registers

Twenty-seven peripherals registers.

All registers, except six of the peripherals registers (interrupt control and timer registers), are 32 bits wide.

The 32 general-purpose registers are designated as R0 to R30 and C0. The C0 register is permanently wired to a zero value (similarly to r0 in many other RISC-type systems).[2] If a general purpose register is specified for an operand that is 8 or 16 bits long, then only the appropriate low part of the

TABLE 7.10 Vector Number Assignments

Number	Type of Trap or Interrupt	Cause
0	Illegal Opcode	Executing undefined instruction*
1	Unaligned Access	Access on unnatural boundary, TU = 1
2	Out of Range	Overflow or underflow
3	Coprocessor Not Present	Coprocessor access, CP = 0
4	Coprocessor Exception	Coprocessor $\overline{DERR}$ response
5	Protection Violation	Invalid User-mode operation
6	Instruction Access Exception	$\overline{IERR}$ response
7	Data Access Exception	$\overline{DERR}$ response, not coprocessor
8	User-Mode Instruction TLB Miss	No TLB entry for translation
9	User-Mode Data TLB Miss	No TLB entry for translation
10	Supervisor-Mode Instruction TLB Miss	No TLB entry for translation
11	Supervisor-Mode Data TLB Miss	No TLB entry for translation
12	Instruction MMU Protection Violation	TLB or RMU UE/SE = 0
13	Data MMU Protection Violation	TLB or RMU UR/SR = 0, UW/SW = 0 on write
14	Timer	Timer Facility
15	Trace	Trace Facility, breakpoint comparisons
16	$\overline{INTR0}$	$\overline{INTR0}$ input
17	$\overline{INTR1}$	$\overline{INTR1}$ input
18	$\overline{INTR2}$	$\overline{INTR2}$ input
19	$\overline{INTR3}$	$\overline{INTR3}$ input
20	$\overline{TRAP0}$	$\overline{TRAP0}$ input
21	$\overline{TRAP1}$	$\overline{TRAP1}$ input
22	Floating-Point Exception	Unmasked floating-point exception
23	Reserved	
24	FMAC exception	ACF in FPE Register = 00 or 11
25	DMAC exception	ACF in FPE Register = 00 or 11
26–27	Reserved	
28	Reserved for instruction emulation (opcodes BF, CF–D6, DC)	
29	Reserved for instruction emulation (opcode DD)	
30–32	Reserved	
33	DIVIDE	DIVIDE instruction
34	Reserved	
35	DIVIDU	DIVIDU instruction
36	CONVERT exception	FS = 00 or 11 or FD = 00 or 11
37	SQRT exception	FS = 00 or 11
38	CLASS exception	FS = 00 or 11
39	Reserved for instruction emulation (opcode E7)	
40	MTACC exception	FMT = 11 or FMT = 00 and ACF = 00 or 11
41	MFACC exception	FMT = 11 or FMT = 00 and ACF = 00 or 11
42–55	Reserved	
56	Reserved for instruction emulation (opcode F8)	
57	Reserved	
58–63	Reserved for instruction emulation (opcode FA–FF)	
64–255	ASSERT and EMULATE instruction traps (vector number specified by instruction)	

* This vector number also results if an external device removes $\overline{INTR}$(3–0) or $\overline{TRAP}$(1–0) before the corresponding interrupt or trap is taken by the processor.

SOURCE: Courtesy AMD Corporation.

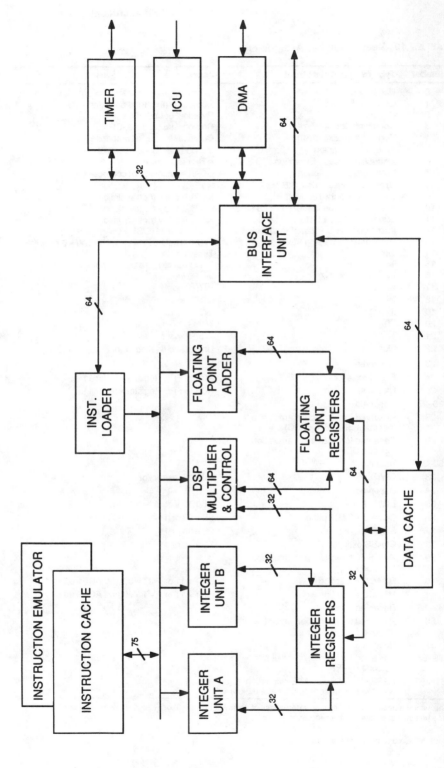

Figure 7.44 641 block diagram. *(Courtesy National Semiconductor Corporation)*

register is used; the high part is not affected. Access to registers R16 to R29 can be restricted to supervisor mode by setting the interrupt-service bit (ISR) of the configuration register (CFG) to 1. In this manner, 14 of the general-purpose registers can be dedicated to the handling of high-priority interrupts. This speeds interrupt service by eliminating the need to save and restore registers, making the 641 attractive for real-time applications. The floating-point data registers can be accessed as either 32 single-precision (32-bit) registers F0 to F31, or as 16 double-precision (64-bit) register pairs L0 to L30 (even numbers only).

The three dedicated address registers are as follows:

PC (program counter): The PC points to the first byte of the instruction currently being executed. It can be used as a base address.

SP (stack pointer): The SP points to the lowest address of the last item stored on the interrupt stack. This stack is normally used by the operating system (OS). It is used primarily for holding information by OS sub-routines and interrupt and trap service routines.

INTBASE (interrupt base register): The INTBASE holds the address of the dispatch table for interrupts and traps (usually called *vector table* in other systems).

The PSR holds status information and selects operating modes for the 641. Its least significant byte, called *user processor status register* (UPSR), is accessible to both user and supervisor mode programs. The upper 24 bits are accessible to supervisor mode programs only. The UPSR contains the flags of the CPU, such as carry (C), overflow (V), zero (Z), negative (N), and others. The CFG is used for enabling or disabling of various operating modes for the 641, execution of floating-point instructions, and control of the on-chip caches. The FSR controls the operating mode and records any exceptional conditions encountered during execution of floating-point instructions. The FTRH and FTRL registers are used by the FPU to store the results of operations that caused a floating-point trap. The six debug registers are dedicated to debugging functions.

The 27 peripherals registers are grouped according to functions as follows:

Nineteen DMA controller registers

Two interrupt control unit (ICU) registers

Four timer registers

Two debug counter registers.

The 641 implements full 32-bit addresses, directly accessing 4 Gbytes of memory. The memory is a uniform linear 4-Gbyte address space. Memory

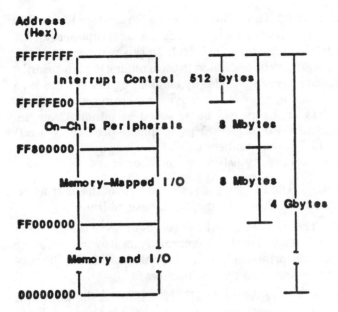

Figure 7.45 Memory organization. *(Courtesy National Semiconductor Corporation)*

locations are numbered sequentially starting at 0000 and ending at FFFF (hexadecimal). It is byte addressable. Instructions and data can be located at any byte address except for the top 16 Mbytes, which are dedicated, as illustrated in Fig. 7.45.[18] The 641 supports either little-endian or big-endian byte ordering, established by the B bit of PSR. When B = 0, we have little-endian ordering. When B = 1, we have big-endian ordering. The 641 architecture recognizes 8-bit bytes, 16-bit words, 32-bit doublewords, and 64-bit quad-words. A word, doubleword, or a quad-word is always referred to by the address of its lower-addressed byte (least significant for little-endian and most significant for big-endian byte ordering).

The 641 supports the use of memory-mapped I/O. Memory-mapped devices can be located arbitrarily in the address space except for the uppermost 8 Mbytes of memory, which are reserved for on-chip peripherals and interrupt control, as illustrated in Fig. 7.45. It is recommended, however, that peripheral devices be located in the specific 8-Mbyte region of memory (FF00 0000 to FF7F FFFF hexadecimal), as illustrated in Fig. 7.45. This is because the 641 detects references to the dedicated locations. It then applies special handling on its bus interface that simplifies system logic required for I/O references.

The 641 architecture features four addressing modes:

1. *Register*: The operand is contained in any integer or floating-point register.

2. *Immediate*: The operand is specified within the instruction.

3. *Relative*: The operand is located in memory, and its address is obtained by adding the content of a register in the range R0 to R30 or C0 to the value in the displacement field encoded in the instruction. If the C0 register is used (always zero), the displacement constitutes an absolute address to memory.

4. *Indexed*: The operand is located in memory. Its address is obtained by adding the content of a register (R0 to R30 or C0) to the content of an index register in the same range.

The instructions of 641 use one, two, or three registers as operands. Some instructions may also replace one register operand by an immediate value. Memory access is confined to the load/store instructions, as practiced in all RISC-type systems.[2] Memory addresses are specified either by the relative or indexed addressing modes. Instructions of 641 have either a 4- or 8-byte length. Since the data bus is 64 bits, it can always transfer one or two instructions at a time. Four byte instructions must be doubleword aligned in memory, and 8-byte instructions must be quad-word aligned. The two instruction formats are illustrated in Fig. 7.46.[18] The basic instruction contains an opcode and three specifier fields. Two of the specifier fields specify register numbers. The third specifies either a third register or an immediate value and may be encoded either in a short field of 11 bits or in a long field of 32 bits (in which case a 64-bit instruction is needed, as shown in Fig. 7.46). The k field (bits 8, 7) encodes the operation length for (byte, word, doubleword, or quad-word) integer instructions. The L bit (bit 9) specifies

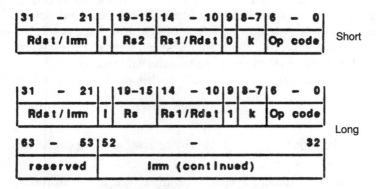

Figure 7.46   NS32SF641 instruction formats. *(Courtesy National Semiconductor Corporation)*

whether the instruction format is short (32-bit; L = 0) or long (64-bit; L = 1). The I bit (bit 20) specifies whether the third operand is immediate (I = 1 for immediate).

The 641 instruction cache (IC) is 4 kbytes in size, two-way, set associative, 8 bytes/line (for a total of 512 lines). The instructions stored in IC are partially decoded by the instruction loader (see Fig. 7.44). The contents of IC can be invalidated by software or hardware.

Software invalidation

1. By clearing bit IC (IC = 0) of CFG
2. By executing the CINV instruction

Hardware invalidation

By activating the INVIC# input signal

The 641 data cache (DC) is 1 kbyte in size, two-way, set associative, 16 bytes/line (for a total of 64 lines). The least recently used (LRU) replacement algorithm is implemented. A write-through policy is implemented. The DC can be invalidated by software or hardware.

Software invalidation

1. By clearing bit DC (DC = 0) of CFG
2. By executing the CINV instruction

Hardware invalidation

By activating INVDC# while INVBLK# is high

The on-chip DMA controller provides two channels for transferring blocks of data between memory and I/O devices. Source and destination addresses, as well as block length and type of operation, are set up in advance by loading the appropriate control registers. Actual transfers are handled by the DMA channels in response to external transfer requests. Upon receiving a transfer request from an I/O device, the DMA controller performs the following operations:

1. Acquires control of the bus.
2. Acknowledges the requesting I/O device by activating the appropriate DAK# signal.
3. Starts executing data transfer cycles according to the values stored in the control registers of the channel being serviced.
4. Terminates the data transfer operation whenever one of the following events occurs:

a. The scheduled number of bytes has been transferred.

b. The EOT# signal is activated during a data transfer cycle.

c. The software writes into the appropriate control registers.

The on-chip interrupt control unit (ICU) manages up to 15 levels of prioritized interrupt requests. Requests can be generated either externally or internally. External requests are binary encoded as a 4-bit value and are transmitted into the chip through the IR0# to IR3# pins. Internal requests are generated by the on-chip DMA controller and timer. Table 7.11 shows the possible interrupt sources and related priority levels. The ICU keeps track of the interrupt priority levels currently in service, and forwards to the CPU only those interrupt requests whose priority levels are higher than the highest-priority interrupt currently being serviced. In addition, the ICU monitors the system bus and responds to interrupt acknowledge and end-of-interrupt bus cycles by providing vector values to the CPU and updating the appropriate bits in the ISRV register. The 641 interface signals are illustrated in Fig. 7.47, and their brief description is given in Table 7.12.

Other members of the NS32000/EP family include two groups of controllers: the CG-Core and the GX-Core.[21,22] The CG-Core includes the following:

1. NS32CG16 processor intended for general imaging applications. It is fully programmable and optimized for graphics applications. A special graphics logic part is added to the regular NS32000 control logic.[20]

2. NS32CG160 processor. It is an extension of NS32CG16, permitting highly efficient operations on rectangular image areas. In addition to the NS32CG16 on-chip subsystems, the NS32CG160 also includes a (16 × 16) hardware multiplier, a two-channel DMA controller, a 15-level ICU, and three programmable 16-bit timers. It also operates at faster clock speeds.

3. NS32FX16 FAX (facsimile) processor, intended to control facsimile operations, data modems, voice mail systems, and laser printers. It has powerful on-chip digital signal processing capabilities and performs all the computations and control functions required for a stand-alone FAX system, a personal computer add-in FAX/modem card, or a laser/FAX system.

The GX-Core group includes the NS32GX32 and the NS32GX320 32-bit embedded system processors.[22] Both have the basic NS32000 architecture,[20] and they are particularly suited to be used as low-cost real-time controllers as well as highly sophisticated embedded systems.

TABLE 7.11   Interrupts and Priority Levels

Priority Level	Interrupt Source
INT15 (Highest)	External Only
INT14	External or DMA (DIP bit in IMSK is 1)
INT13	External or Timer IPFA or Capture Mode Underflow
INT12	External or Timer IPFB
INT11	External Only
INT10	External Only
INT9	External Only
INT8	External Only
INT7	External Only
INT6	External or DMA (DIP bit in IMSK is 0)
INT5	External Only
INT4	External Only
INT3	External Only
INT2	External Only
INT1 (Lowest)	External Only
—	No Interrupt

SOURCE: Courtesy National Semiconductor Corporation.

Both can run at up to 30 MHz. Both chips feature the basic subsystems found on the NS32532,[20] on-chip cache included. The NS32GX320 has the following additional features, not available on the NS32GX32: hardware multiplier, two-channel DMA controller, 15-level ICU, three 16-bit timers, complex number instructions, and 8-kbyte page mode (the regular NS32000 page is 4 kbytes in memory management operations) for DRAM access support. The NS32GX320 can be regarded as a forerunner of the NS32SF640.

7.5   COMPARISON AND EVALUATION OF 32-BIT MICROCONTROLLERS

Some of the properties of the four 32-bit microcontrollers (or embedded processors) described in the preceding sections of this chapter are listed in Table 7.13. A perfunctory glance at the table tells us that none of them excels in all areas. While one processor has certain advantages, another is endowed with other equally important features.

The availability of memory cache on chip, code or data, certainly enhances the overall performance of a processor. On that point the LR33000 takes the lead with its 9-kbyte total on-chip cache (8-kbyte code; 1-kbyte data),

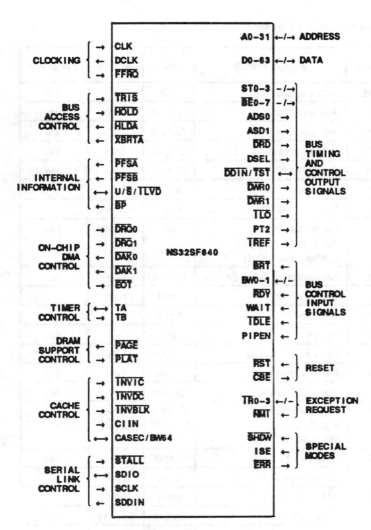

Figure 7.47   NS32SF640 interface signals. *(Courtesy National Semiconductor Corporation)*

followed by NS32SF641 with 5 kbytes (4-kbyte code; 1-kbyte data). The 80960CA is third with 1-kbyte code cache, and the Am29050 has no cache memory on chip, except for the branch target cache. However, the processors with less or no on-chip cache have other features that compensate for the absence (or small size) of on-chip cache. The 80960CA has 1 kbyte of main memory on chip (a property not shared by other processors) and a local register cache (LRC), containing a minimum of five local register sets (sixteen 32-bit registers each), extendable to 15 sets, also a unique property.

TABLE 7.12   Table of Interface Signals

Symbol	Pins	Signal Name	I/O	Active	Tri-State
A0-31	32	Address Bus	I/O	–	Yes
ADS0	1	Address Strobe 0	Output	High	No
ADS1	1	Address Strobe 1	Output	High	No
$\overline{BE}$0-7	8	Byte Enable (0-7)	Output	Low	Yes
$\overline{BP}$	1	Break Point	Output	Low	No
$\overline{BRT}$	1	Bus Retry	Input	Low	
BW0-1	2	Bus Width (0-1)	Input	Encode	
CASEC/ BW64	1	Cache Section/ 64-Bit Default Bus (Sampled at Reset)	I/O	Encode Encode	Yes
$\overline{CBE}$	1	Configuration Buffer Enable	Output	Low	No
CIIN	1	Cache Inhibit In	Input	High	
CLK	1	Bus Clock	Input	–	
D0-63	64	Data Bus (0-63)	I/O	–	Yes
$\overline{DAK0}$	1	DMA Acknowledge 0	Output	Low	No
$\overline{DAK1}$	1	DMA Acknowledge 1	Output	Low	No
DCLK	1	Delayed Bus Clock	Output	–	No
$\overline{DDIN}$/ $\overline{TST}$	1	Data Direction In/ Self Test Enable (Sampled at Reset)	I/O	Low Low	Yes
$\overline{DRD}$	1	Data Read	Output	Low	Yes
$\overline{DRQ0}$	1	DMA Request 0	Input	Low	
$\overline{DRQ1}$	1	DMA Request 1	Input	Low	
DSEL	1	Data Select	Output	Encode	Yes
$\overline{DWR0}$	1	Data Write Strobe 0	Output	Low	Yes
$\overline{DWR1}$	1	Data Write Strobe 1	Output	Low	Yes
$\overline{EOT}$	1	End Of DMA Transfer	Input	Low	
$\overline{ERR}$	1	Error	Output	Low	No
$\overline{FFRQ}$	1	Full Frequency Bus	Input	Encode	
$\overline{HLDA}$	1	Hold Acknowledge	Output	Low	No
$\overline{HOLD}$	1	Hold Request	Input	Low	
$\overline{IDLE}$	1	Idle Cycle Request	Input	Low	

TABLE 7.12 Table of Interface Signals *(Continued)*

Symbol	Pins	Signal Name	I/O	Active	Tri-State
$\overline{\text{ILO}}$	1	Interlocked Bus Cycle	Output	Low	No
$\overline{\text{INVBLK}}$	1	Invalidate Block	Input	Low	
$\overline{\text{INVDC}}$	1	Invalidate Data Cache	Input	Low	
$\overline{\text{INVIC}}$	1	Invalidate Instruction Cache	Input	Low	
$\overline{\text{IR}}$0-3	4	Maskable Interrupt (0-3)	Input	Encode	
$\overline{\text{IREF}}$	1	Internal Reference	Output	Low	No
ISE	1	In-System Emulation (Sampled at Reset)	Input	High	
$\overline{\text{NMI}}$	1	Non-Maskable Interrupt	Input	Edge	
$\overline{\text{PAGE}}$	1	Same Page	Output	Low	Yes
$\overline{\text{PFSA}}$	1	Program Flow Status (Pipe - A)	Output	Low	No
$\overline{\text{PFSB}}$	1	Program Flow Status (Pipe - B)	Output	Low	No
PIPEN	1	Pipelined Bus Enable	Input	High	
$\overline{\text{PLAT}}$	1	Page Latch Enable	Input	Low	
PT2	1	Possible T2 of the Bus Cycle	Output	High	Yes
$\overline{\text{RDY}}$	1	Ready	Input	Low	
$\overline{\text{RST}}$	1	Reset	Input	Low	
SCLK	1	Serial Link Clock	Input	Edge	
SDIO	1	Serial Data I/O	I/O	–	No
SDDIN	1	Serial Link Data Direction In	Output	Low	No
$\overline{\text{SHDW}}$	1	Shadow Mode	Input	Low	
ST0-3	4	Status (0-3)	Output	Encode	No
$\overline{\text{STALL}}$	1	Stall	Input	Low	
TXA	1	Timer Control	I/O	High	No
TXB	1	Timer Trigger	Input	High	
$\overline{\text{TRIS}}$	1	Tri-State All Outputs	Input	Low	
U/$\overline{\text{S}}$/ $\overline{\text{ILVD}}$	1	User/Supervisor Mode/ Interleaved Memory (Sampled at Reset)	I/O	Encode Encode	Yes
WAIT	1	Wait	Input	High	
$\overline{\text{XBRTA}}$	1	Extended Bus Retry Acknowledge	Output	Low	No

SOURCE: Courtesy National Semiconductor Corporation.

TABLE 7.13   Comparison of 32-Bit Embedded Processors

System:	80960CA	LR33000	Am29050	NS32SF641
Cache on-chip	1Kbyte code	8Kbyte code 1Kbyte data	no	4Kbyte code 1Kbyte data
Special cache on-chip	512 byte LRC	no	Branch target cache 1Kbyte	no
Main memory on-chip	1Kbyte	no	no	no
Number of Integer Units	1	1	1	2
Number of parallel pipelines	3 2: 3-stages 1: 2-stages	1 5-stages	1 4-stages	2 4-stages
FPU on-chip	no	no	yes	yes
CPU registers	32 (+LRC)	32	192	32 integer 32 floating-point
DMA controller on-chip	4-channel	no	no	2-channel
Interrupt controller on-chip	8 inputs + NMI	no	no	15-level, 4-input + NMI
DRAM controller on-chip	no	yes	no	no, some support signals
Timers,on-chip	no	3	1	1
Pins	168	155	169	223

LRC = local register cache; NMI = nonmaskable interrupt; FPU = floating-point unit; CPU = central processing unit; DMA = direct memory access; DRAM = dynamic RAM.

These features make the 80960CA competitive in fast handling of interrupts, since important interrupt data can be stored on chip, saving on CPU-memory traffic. This, in turn, makes it competitive for real-time embedded applications, even though it does not have a data cache on chip. The compensatory feature of the Am29050 is the 1-kbyte branch target cache, containing 64 branch target groups of instructions (4 instructions per target, or 128 groups with 2 instructions per target). This feature significantly enhances the performance of the instruction pipeline, particularly for control hazards caused by conditional branches, whose potential outcome is difficult to predict. Target instructions for nearly all possible branch outcomes can be stored in the branch target cache, and promptly retrieved from it, enhancing the overall performance. Another feature of the Am29050, compensating for the lack of an on-chip cache, is the 192-register file, accessible by the user in its entirety (despite of the 80960CA's LRC, any procedure running on it can access only 32 registers). The availability of such a large register file permits the storage of interrupt information and process context on chip, speeding up interrupt and context switch handling and making the Am29050 attractive for real-time embedded processing.

Many embedded applications need fast and massive floating-point computations. In this case, the availability of an on-chip FPU is very important. The Am29050 and NS32SF641 take the lead with their on-chip FPUs. No matter how fast the CPU is, if an external coprocessor is to be used, the floating-point operations will be slower compared to an on-chip FPU. This is particularly enhanced on the NS32SF641, which has a 64-bit data bus (in and out of chip) and a separate floating-point register file of 32-bit or sixteen 64-bit registers.

While the 80960CA has only one REG pipeline for register-to register operations, the NS32SF641 has two parallel execution units. This means that on the 960CA subsequent register operations can be pipelined, while on the 641 two register instructions can be executed in complete concurrency. This gives the 641 an edge over the 960CA in executing sequences of register-to-register operations.

The availability of I/O interface devices on chip has two major advantages for embedded applications:

1. It speeds up I/O and interrupt operations.
2. It saves on the overall chip count, allowing more compact (reduced volume) realization of embedded processors, and yields simplicity of design and reduced cost.

In this case, the 80960CA and NS32SF641 take the leading role among the others. The 80960CA has a four-channel DMA controller and an eight-input (plus NMI) interrupt controller. The NS32SF641 has a two-channel DMA controller, a four-input, 15-level interrupt controller (plus NMI), and a 16-bit timer. The LR33000, on the other hand, has a DRAM controller (a unique property) and three timers (two 24-bit, one 12-bit). The NS32SF641 has page-mode support to enable page access detection in DRAMs. The Am29050 does not have any on-chip I/O interface subsystems. However, the Am29050 was designed as a universal processor and not just as an embedded controller, for which it is also eminently suited.

LSI Logic performed a number of runs using the Dhrystone benchmark, comparing the LR33000 with the 80960CA. A 33-MHz LR33000 HTP board was used for the LR33000 runs, and a 33-MHz i960CA ASV960CA board was used for the 80960CA runs. The number of LR33000 Dhrystones for the same time interval was 53300, and for the 80960CA it was 41030.

None of the aforementioned processors can be declared an absolute winner. Each may prove to be more powerful and more advantageous for different applications. Only additional practical experience and extensive benchmark experimentation may provide more precise comparison and evaluation.

REFERENCES

1. G. J. Myers and D. L. Budde, *The 80960 Microprocessor Architecture*. New York: Wiley, 1988.
2. D. Tabak, *RISC Systems*. U.K.: Research Studies Press; and New York: Wiley, 1990.
3. *Intel's Superscalar i960CA Processor for Embedded Systems*. Technical background information, Intel Corporation, #270804-001, September 12, 1989.
4. S. McGready, "Inside Intel's i960CA Superscalar Processor," *Microprocessors and Microsystems*, 14(6):385–396, July-August, 1990
5. *80960CA User's Manual*. Intel Corporation, #270710, 1990.
6. *Solutions 960*. Intel Corporation, #270791-001, 1989.
7. *80960CA Product Overview*. Intel Corporation, #270669-001, 1989.
8. J. P. Hayes, *Computer Architecture and Organization*, 2nd ed. New York: McGraw-Hill, 1988.
9. V. C. Hamacher, Z. G. Vranesic, and S. G. Zaky, *Computer Organization*, 3rd ed. New York: McGraw-Hill, 1990.
10. A. J. Smith, "Line(Block) Size Choice for CPU Cache Memories," *IEEE Trans. on Computers*, C-36(9):1063–75, September 1987.
11. S. Przybylski, M. Horowitz and J. Hennessy, "Performance Tradeoffs in Cache Design," *Proc. 15th Annual International Symposium on Computer Architecture*, pp. 290-98, May, 1988.
12. *80960KB Programmer's Reference Manual*. Intel Corporation, #270567-001, 1988.
13. R. S. Sandige, *Modern Digital Design*. New York: McGraw-Hill, 1990.
14. D. Tabak, *Multiprocessors*. Englewood Cliffs, NJ: Prentice-Hall, 1990.
15. B. W. Kernighan and D. M. Ritchie, *The C Programming Language*, 2nd ed. Englewood Cliffs, NJ: Prentice-Hall, 1988.
16. *LR33000 MIPS Embedded Processor User's Manual*, #LR33000UM Rev.A, LSI Logic Corporation, Milpitas, CA, 1990.
17. *AM29050 Microprocessor User's Manual*. Advanced Micro Devices (AMD) Inc., Sunnyvale, CA, 1991.
18. NS32SF'640 Superscalar 64-Bit Intergrated System Processor. National Semiconductor, August, 1990.
19. W. Andrews, "Embedded Processor Hits 100 MIPS Mark," *Computer Design*, pp. 38–43, March 1, 1991.
20. D. Tabak, *Advanced Microprocessors*. New York: McGraw-Hill, 1991.
21. *Introduction to CG-Core Embedded Processors*. National Semiconductor, 1990.
22. *Introduction to GX-Core Embedded Processors*. National Semiconductor, 1990.

PROBLEMS

All problems are for the 80960CA.

1. A four-instruction sequence is fetched on the 80960CA:

```
addi    r1,r2,r3    ;add integer, (r1)+(r2)-->r3
cmpi    r5,r3       ;compare integer, (r3)-(r5), condition code affected
be      addr1       ;branch to addr1 if equal
st      r3,memr     ;store, (r3)-->memr
```

where addr1 is an address label and memr is a location in memory. Provide a complete analysis of the pipeline work involved and calculate, using an appropriate timing diagram, how many cycles will it take to complete the execution of all four instructions.

2. In a specific implementation of the 80960CA, 10 local register set spaces are allocated in the local register cache (LRC). Provide a complete analysis of the 80960CA on-chip RAM allocation.

3. Provide two examples of 80960CA instruction sequences, capable of keeping all three pipelines busy at the same time.

4. It is necessary to transfer every eighth byte from a 128-byte array, starting at location AA00H, into every fourth byte location of another 64-byte array, starting at location F000H. Using appropriate addressing modes, provide a short program performing this transfer.

5. Move a triple word located within the global register file starting at g0 to another set of registers starting at g8. Which registers are affected by this operation?

6. Explain in detail what happens when instruction lda 8(r5)[r6*4],r8 is executed. Initial values: (r5) = 2000H, (r6) = 4, (r8) = 0.

7. Given the following assembly code:

```
TARG EQU 128
b   TARG
bx TARG(g1)[g2*2]
```

and given (g1) = 1000H, (g2) = 8, explain in detail the difference between the two branch instructions.

8. Explain in detail the consequences to pipeline operation after using be.t and be.f options of the branch-if-equal instruction.

9. Interpret the following settings of the AC register (Fig. 7.8): 0000 1104H, 0000 8002H.

10. Set the AC register for the following conditions: no imprecise faults, integer overflow not masked, an integer overflow occurred, and the condition code is greater than. Present your answer in hexadecimal.

11. Interpret the following settings of the process control register (Fig.7.9): 000B 2401H, 0012 0000H.

12. Set the process control register for the following conditions: The processor, executing in supervisor mode, was interrupted and is executing an interrupt-handling routine at priority level 7, there is no trace fault pending; and the trace is disabled. Present your answer in hexadecimal.

13. There is a main program and three levels of nested procedures running on the 960CA. Each has eight optional 32-bit variables on stack. Draw a detailed stack frame map and linkage diagram for these conditions. Stress all appropriate address pointings.

14. Explain in detail, providing examples, the differences between local and system calls, *call* and *calls*.

15. Explain the difference between *call* and *callx*.

16. Explain the following instruction: *calls* 255.

17. Interpret the following procedure entry content: 00FF 8602H, 0010 FF00H.

18. Explain the differences between system-local calls and system-supervisor calls.

19. Interrupt vector 62H of priority 16 is pending. What will be the setting of the pending priorities and pending interrupt fields in the interrupt table? Where else can the interrupt vector be stored?

20. A 960CA-based microcomputer has eight interrupt sources. What is the most appropriate interrupt mode (see Fig.7.14) to be used and why?

21. Repeat Problem 20 for 128 interrupt sources. Draw an interconnection diagram.

22. Repeat Problem 21 for (a) 16, and (b) 32 interrupt sources.

23. Double the 27960CX-based memory in Fig.7.19. Draw a detailed interconnection diagram.

24. Convert Fig.7.20, featuring the 80960CA with the 82596CA LAN coprocessor, into a detailed interconnection diagram.

80960CA Pin Description*

The 80960CA pins are described in Tables A.2 to A.4. Table A.1 presents the notation used in these tables. The 80960CA external bus signals are described in Table A.2. The pins associated with basic processor configuration and control are described in Table A.3. The pins associated with the DMA controller and interrupt units are described in Table A.4.

---

*All tables of App. A are reproduced courtesy of Intel Corporation.

TABLE A.1 Pin Description Nomenclature

Symbol	Description
I	Input only pin
O	Output only pin
I/O	Pin can be either an input or output
—	Pins "must be" connected as described
S(...)	Synchronous. Inputs must meet setup and hold times relative to PCLK2:1 for proper operation of the processor. All outputs are synchronous to PCLK2:1. S(E) Edge sensitive input S(L) Level sensitive input
A(...)	Asynchronous. Inputs may be asynchronous to PCLK2:1. A(E) Edge sensitive input A(L) Level sensitive input
H(...)	While the processor's bus is in the Hold Acknowledge state, the pin: H(1) is driven to V_{CC} H(0) is driven to V_{SS} H(Z) floats H(Q) continues to be a valid output
R(...)	While the processor's RESET pin is low, the pin R(1) is driven to V_{CC} R(0) is driven to V_{SS} R(Z) floats R(Q) continues to be a valid output

TABLE A.2 80960CA Pin Description—External Bus Signals

Name	Type	Description
A31:2	O S H(Z) R(Z)	ADDRESS BUS carries the upper 30 bits of the physical address. A31 is the most significant address bit and A2 is the least significant. During a bus access, A31:2 identify all external addresses to word (4-byte) boundaries. The byte enable signals indicate the selected byte in each word. During burst accesses, A3 and A2 increment to indicate successive data cycles.
D31:0	I/O S(L) H(Z) R(Z)	DATA BUS carries 32, 16, or 8-bit data quantities depending on bus width configuration. The least significant bit of the data is carried on D0 and the most significant on D31. When the bus is configured for 8 bit data, the lower 8 data lines, D7:0 are used. For 16 bit bus widths, D15:0 are used. For 32 bit bus widths the full data bus is used.
$\overline{BE3}$ $\overline{BE2}$ $\overline{BE1}$ $\overline{BE0}$	O S H(Z) R(Z)	BYTE ENABLES select which of the four bytes addressed by A31:2 are active during an access to a memory region configured for a 32-bit data-bus width. $\overline{BE3}$ applies to D31:24; $\overline{BE2}$ applies to D23:16; $\overline{BE1}$ applies to D15:8; and $\overline{BE0}$ applies to D7:0. 32-bit bus: $\overline{BE3}$ –Byte Enable 3 –enable D31:24 $\overline{BE2}$ –Byte Enable 2 –enable D23:16 $\overline{BE1}$ –Byte Enable 1 –enable D15:8 $\overline{BE0}$ –Byte Enable 0 –enable D7:0 For accesses to a memory region configured for a 16-bit data-bus width, the processor directly encodes $\overline{BE3}$, $\overline{BE1}$ and $\overline{BE0}$ to provided $\overline{BHE}$, A1 and $\overline{BLE}$ respectively. 16-bit bus: $\overline{BE3}$ –Byte High Enable ($\overline{BHE}$) –enable D15:8 $\overline{BE2}$ –Not used (is driven high or low) $\overline{BE1}$ –Address Bit 1 (A1) $\overline{BE0}$ –Byte Low Enable ($\overline{BLE}$) –enable D7:0 For accesses to a memory region configured for an 8-bit data bus width, the processor directly encodes $\overline{BE1}$ and $\overline{BE0}$ to provide A1 and A0 respectively. 8-bit bus: $\overline{BE3}$ –Not used (is driven high or low) $\overline{BE2}$ –Not used (is driven high or low) $\overline{BE1}$ –Address Bit 1 (A1) $\overline{BE0}$ –Address Bit 0 (A0)
W/$\overline{R}$	O S H(Z) R(Z)	WRITE/READ is low (0) for read accesses and high (1) for write accesses. The W/$\overline{R}$ signal changes in the same clock cycle as $\overline{ADS}$. It remains valid for the entire access in non-pipelined regions. In pipelined regions, W/$\overline{R}$ may not be valid in the last cycle of a read accesses.
$\overline{ADS}$	O S H(Z) R(1)	ADDRESS STROBE indicates valid address and the start of a new bus access. $\overline{ADS}$ is asserted for the first clock of a bus access.

TABLE A.2   80960CA Pin Description—External Bus Signals (Continued)

Name	Type	Description
READY	I S(L) H(Z) R(Z)	READY is an input which signals the termination of a data transfer. READY is used to indicate that read data on the bus is valid, or that a write-data transfer has completed. The READY signal works in conjunction with the internally programmed wait-state generator. If READY is enabled in a region, the pin is sampled after the programmed number of wait-states has expired. If the READY pin is deasserted high, wait states will continue to be inserted until READY becomes asserted low. This is true for the N_{RAD}, N_{RDD}, N_{WAD}, and N_{WDD} wait states. The N_{XDA} wait states cannot be extended.
BTERM	I S(L) H(Z) R(Z)	BURST TERMINATE is an input which signals the termination of an access. The assertion of BTERM causes another address cycle to occur. The BTERM signal works in conjunction with the internally programmed wait-state generator. If READY and BTERM are enabled in a region, the BTERM pin is sampled after the programmed number of wait states has expired. When BTERM is asserted (low), READY is ignored.
WAIT	O S H(Z) R(1)	WAIT indicates the status of the internal wait state generator. WAIT is active when wait states are being caused by the internal wait state generator and not by the READY or BTERM inputs. WAIT can be used to derive a write-data strobe. WAIT can also be thought of as a READY output that the processor provides when it is inserting wait states.
BLAST	O S H(Z) R(0)	BURST LAST indicates the last transfer in a bus access. BLAST is asserted in the last data transfer of burst and non-burst accesses after the wait state counter reaches zero. BLAST remains active until the clock following the last cycle of the last data transfer of a bus access. If the READY or BTERM input is used to extend wait states, the BLAST signal remains active until READY or BTERM terminates the access.
DT/R̄	O S H(Z) R(1)	DATA TRANSMIT/RECEIVE indicates direction for data transceivers. DT/R̄ is used in conjunction with DEN to provide control for data transceivers attached to the external bus. When DT/R̄ is low (0), the signal indicates that the processor will receive data. Conversely, when high (1) the processor will send data. DT/R̄ will change only while DEN is high.
DEN	O S H(Z) H(Z)	DATA ENABLE indicates data cycles in a bus access. DEN is asserted (low) at the start of the first data cycle of a bus access and is deasserted (high) at the end of the last data cycle. DEN is used in conjunction with DT/R̄ to provide control for data transceivers attached to the external bus. DEN will remain asserted for sequential reads from pipelined memory regions. DEN will be high when DT/R̄ changes.
LOCK	O S H(Z) R(1)	BUS LOCK indicates that an atomic read-modify-write operation is in progress. LOCK may be used to prevent external agents from accessing memory which is currently involved in an atomic operation. LOCK is asserted (0) in the first clock of an atomic operation, and deasserted in the clock cycle following the last bus access for the atomic operation. To allow the most flexibility for a memory system enforcement of locked accesses, the processor will acknowledge a bus hold request when LOCK is asserted. The processor will perform DMA transfers while LOCK is active.
HOLD	I S(L) H(Z) R(Z)	HOLD REQUEST signals that an external agent requests access to the external bus. The processor asserts HOLDA after completing the current bus request. HOLD, HOLDA, and BREQ are used together to arbitrate access to the processor's external bus by external bus agents.

TABLE A.2 80960CA Pin Description—External Bus Signals (Continued)

Name	Type	Description
HOLDA	O S H(1) R(Q)	**HOLD ACKNOWLEDGE** indicates to a bus requestor that the processor has relinquished control of the external bus. When HOLDA is asserted, the external address bus, data bus, and bus control signals are floated. HOLD, HOLDA, and BREQ are used together to arbitrate access to the processor's external bus by external bus agents. Since the processor will grant HOLD requests and enter the Hold Acknowledge state even while RESET is active, the state of the HOLDA pin will be independent of the RESET pin.
BREQ	O S H(Q) R(0)	**BUS REQUEST** indicates that the processor wishes to perform a bus request. BREQ can be used by external bus arbitration logic in conjunction with HOLD and HOLDA to determine when to return mastership of the external bus to the processor.
D/C̄	O S H(Z) R(0)	**DATA OR CODE** indicates that a bus access is a data access (1) or a instruction access (0). D/C̄ has the same timing as W/R̄
DMA	O S H(Z) R(1)	**DMA ACCESS** indicates whether the bus access was initiated by the DMA controller. DMA will be asserted (low) for any DMA access. DMA will be deasserted (high) for all other accesses.
SUP	O S H(Z) R(0)	**SUPERVISOR ACCESS** indicates whether the bus access originates from a request issued while in supervisor mode. SUP will be asserted (low) when the access has supervisor privileges, and will be deasserted (high) otherwise. SUP can be used to isolate supervisor code and data structures from non-supervisor access.

TABLE A.3 80960CA Pin Description—Processor Control Signals

Name	Type	Description
RESET	I A(L) H(Z) R(Z) N(Z)	**RESET** causes the chip to reset. When RESET is asserted (low), all external signals return to the reset state. When RESET is deasserted, initialization begins. When the two-x clock mode is selected, RESET must remain asserted for 16 PCLK2:1 cycles before being deasserted in order to guarantee correct initialization of the processor. When the one-x clock mode is selected, RESET must remain asserted for 10,000 PCLK2:1 cycles before being deasserted in order to guarantee correct initialization of the processor. The CLKMODE pin selects one-x or two-x input clock division of the CLKIN pin. The processor's Hold Acknowledge bus state functions while the chip is reset. If the processor's bus is in the Hold Acknowledge state when RESET is activated, the processor will internally reset, but will maintain the Hold Acknowledge state on external pins until the Hold request is removed. If a hold request is made while the processor is in the reset state the processor bus will grant HOLDA and enter the Hold Acknowledge state.
FAIL	O S H(Q) R(0)	**FAIL** indicates failure of the processor's self-test performed at initialization. When RESET is deasserted and the processor begins initialization, the FAIL pin is asserted (0). An internal self-test is performed as part of the initialization process. If this self-test passes, the FAIL pin is deasserted (1) otherwise it remains asserted. The FAIL pin is reasserted while the processor performs and external bus self-confidence test. If this self-test passes, the processor deasserts the FAIL pin and branches to the users initialization routine, otherwise the FAIL pin remains asserted. Internal self-test and the use of the FAIL pin can be disabled with the STEST pin.

TABLE A.3 8096aCA Pin Description—Processor Control Signals (Continued)

Name	Type	Description
STEST	I S(L) H(Z) R(Z)	**SELF TEST** causes the processor's internal self-test feature to be enabled or disabled at initialization. STEST is read on the rising edge of $\overline{\text{RESET}}$. When asserted (high) the processor's internal self-test and external bus confidence tests are performed during processor initialization. When deasserted (low), no self-tests are performed during initialization.
$\overline{\text{ONCE}}$™	I A(L) H(Z) R(Z)	**ON CIRCUIT EMULATION** causes all outputs to be floated when asserted (low). $\overline{\text{ONCE}}$ is continuously sampled while $\overline{\text{RESET}}$ is low, and is latched on the rising edge of $\overline{\text{RESET}}$. To place the processor in the ONCE state: (1) assert $\overline{\text{RESET}}$ and $\overline{\text{ONCE}}$ (order does not matter) (2) wait for at least 16 clocks in one-x mode, or 10,000 clocks in two-x mode, after V_{CC} and CLKIN are within operating specifications (3) deassert $\overline{\text{RESET}}$ (4) wait at least 16 clocks (The processor will now be latched in the ONCE state as long as $\overline{\text{RESET}}$ is high.) To exit the ONCE state, bring V_{CC} and CLKIN to operating conditions, then assert $\overline{\text{RESET}}$ and bring $\overline{\text{ONCE}}$ high prior to deasserting $\overline{\text{RESET}}$. CLKIN must operate within the specified operating conditions of the processor until step 4 above has been completed. The CLKIN may then be changed to D.C. to achieve the lowest possible ONCE mode leakage current. $\overline{\text{ONCE}}$ can be used by emulator products or for board testers to effectively make an installed processor transparent in the board.
CLKIN	I A(E) H(Z) R(Z)	**CLOCK INPUT** is an input for the external clock needed to run the processor. The external clock is internally divided as prescribed by the CLKMODE pin to produce PCLK2:1.
CLKMODE	I A(L) H(Z) R(Z)	**CLOCK MODE** selects the division factor applied to the external clock input (CLKIN). When CLKMODE is high (1), CLKIN is divided by one to create PCLK2:1 and the processor's internal clock. When CLKMODE is low (0), CLKIN is divided by two to create PCLK2:1 and the processor's internal clock. CLKMODE should be tied high, or low in a system, as the clock mode is not latched by the processor. If left unconnected, the processor will internally pull the CLKMODE pin low (0), enabling the two-x clock mode.
PCLK2 PCLK1	O S H(Q) R(Q)	**PROCESSOR OUTPUT CLOCKS** provide a timing reference for all inputs and outputs of the processor. All inputs and output timings are specified in relation to PCLK2 and PCLK1. PCLK2 and PCLK1 are identical signals. Two output pins are provided to allow flexibility in the system's allocation of capacitive loading on the clock. PCLK2:1 may also be connected at the processor to form a single clock signal.
V_{SS}	–	**GROUND** connections consist of 24 pins which must be shorted externally to a V_{SS} board plane.
V_{CC}	–	**POWER** connections consist of 24 pins which must be shorted externally to a V_{CC} board plane.
N/C	–	**NO CONNECT** pins must not be connected in a system.

TABLE A.4 80960CA Pin Description—EMA and Interrupt Unit Control Signals

Name	Type	Description
DREQ3 DREQ2 DREQ1 DREQ0	I A(L) H(Z) R(Z)	DMA REQUEST causes a DMA transfer to be requested. Each of the four signals requests a transfer on a single channel. DREQ0 requests channel 0, DREQ1 requests channel 1, etc. When two or more channels are requested simultaneously, the channel with the highest priority is serviced first. The channel priority mode is programmable.
DACK3 DACK2 DACK1 DACK0	O S H(Z) R(1)	DMA ACKNOWLEDGE indicates that a DMA transfer is being executed. Each of the four signals acknowledges a transfer for a single channel. DACK0 acknowledges channel 0, DACK1 acknowledges channel 1, etc. DACK3:0 are active (0) when the requesting device of a DMA is accessed.
EOP3/TC3 EOP2/TC2 EOP1/TC1 EOP0/TC0	I / O A(L) H(Z) R(Z)	END OF PROCESS/TERMINAL COUNT can be programmed as either an input (EOP3:0) or as an output (TC3:0), but not both. Each pin is individually programmable. When programmed as an input, EOPx causes the termination of a current DMA transfer for the channel corresponding to the EOPx pin. EOP0 corresponds to channel 0, EOP1 corresponds to channel 1, etc. When a channel is configured for source *and* destination chaining, the EOP pin for that channel causes termination of only the current buffer transferred and causes the next buffer to be transferred. EOP3:0 are asynchronous inputs. When programmed as an output, the channel's TCx pin indicates that the channel byte count has reached 0 and a DMA has terminated. TCx is driven active (0) for a single clock cycle after the last DMA transfer is completed on the external bus. TC3:0 are synchronous outputs.
XINT7 XINT6 XINT5 XINT4 XINT3 XINT2 XINT1 XINT0	I A(E/L) H(Z) R(Z)	EXTERNAL INTERRUPT PINS cause interrupts to be requested. These pins can be configured in three modes. In the Dedicated Mode, each pin is a dedicated external interrupt source. Dedicated inputs can be individually programmed to be level (low) or edge (falling) activated. In the Expanded Mode, the 8 pins act together as an 8-bit vectored interrupt source. The interrupt pins in this mode are level activated. Since the interrupt pins are active low, the vector number requested is the one's complement of the positive logic value place on the port. This eliminates glue logic to interface to combinational priority encoders which output negative logic. In the Mixed Mode, XINT7:5 are dedicated sources and XINT4:0 act as the 5 most significant bits of an expanded mode vector. The least significant bits are set to 010 internally.
NMI	I A(E) H(Z) R(Z)	NON-MASKABLE INTERRUPT causes a non-maskable interrupt event to occur. NMI is the highest priority interrupt recognized. NMI is an edge (falling) activated source.

Design Examples

EXAMPLE B-1: 80960CA-BASED DISK DRIVE CONTROLLER*

A preliminary block diagram of the disk drive control system, showing its main parts, is given in Fig. B.1. The memory was selected according to the needs of the application software. The static RAM (SRAM) chosen for the system is Intel 5164S/L (8K × 8 bits). It was chosen for its low access time of 100 ns. A bank of four such chips is used to interface with the full 32-bit data bus of the 960CA. The 5164 has the following control pins: two chip select (CS1# and CS2), write enable (WE#), and output enable (OE#). When CS1# and OE# are low, and CS2 and WE# are high, the 5164 is set for a read operation to send data to the CPU. When CS1# and WE# are low, and CS2 and OE# are high, the 5164 is set for a write operation receiving data from the CPU on the data bus. When the chip is not accessed, CS1# should normally be kept high (inactivated) and CS2 should be kept low to reduce power consumption. The CS1# and CS2 signals are driven by a 74LS138 decoder. The WE# signal is controlled by a combination of signals BE0# to BE3#, W/R#, and WAIT#. This circuitry, illustrated in Fig. B.2, allows access to any of the 4 bytes of a 32-bit word, depending on the BE# signals. The OE# signal is controlled by a combination of BE0# to BE3#, and W/R# signals.

The DRAM chip selected for the system is Intel 21464 (64K × 4 bits). It was chosen because of its low access time (100 to 120 ns), ease of interfacing to a DRAM controller, its size (dictated by software needs), and its relatively low cost per bit. A bank of eight such chips is required to interface

---

*Worked out by graduate students of GMU William B. Bugert and Raymond Kaddissi.

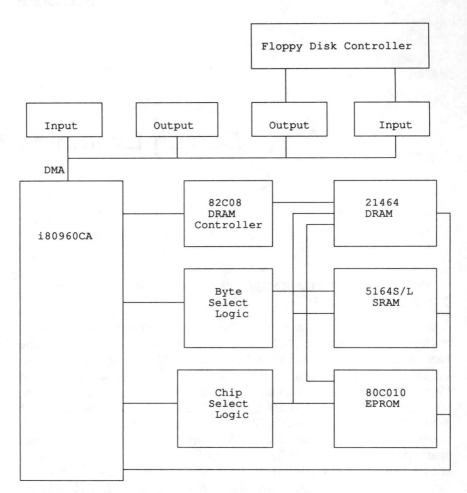

Figure B.1 i80960CA system major components block diagram.

to the 32-bit data bus of the 960CA. A diagram of the CPU and DRAM interface is shown in Fig. B.3. The standard Intel 82C08 DRAM controller chip is implemented. It provides the necessary signals to address, refresh rows and columns, and directly drive the DRAM chips.

The EPROM chip selected for the system is Intel 27C010 (128K × 8 bits). It was chosen because of its low access time (120 ns) and low price. The CPU interface to the four EPROM chips is shown in Fig. B.4. The chip is activated by the chip enable (CE#) and output enable (OE#) control inputs. The CE# signal is generated by the 74LS138 decoder. The OE# signal is generated by a logic combination of the W/R# and BE0# to BE3# signals. Although usually one would fetch a 4-byte (32-bit) instruction from an EPROM, the capability of accessing each separate byte was added for the convenience of the user, in case of storage of permanent data smaller than a word.

A block diagram of the I/O interface, showing the DMA control lines, is illustrated in Fig. B.5. A more detailed diagram, illustrating the interface to the floppy disk controller, is shown in Fig. B.6. A standard WD2793 floppy disk controller/formatter was selected. It was specifically designed by its manufacturer to be controlled by a microcontroller. The WD2793 accepts commands and data from the CPU via the data bus and generates all of the control signals required to read and write data with a floppy disk drive. The RE# and WE# control lines of the WD2793, activated by the W/R# line from the CPU, control the direction of the data flow. The WD2793 takes up four memory-mapped addresses selected by its A1 and A0 inputs, activated by the CPU's BE0# to BE3# lines and additional logic circuitry. Data is transferred to or from the WD2793 when A1, A0 = 11. The other three combinations of A1, A0 select internal control functions and allow status to be read. The chip select (CS#) pin is activated by the decoder.

When writing data to the floppy disk, the CPU writes sector commands to the floppy disk controller (FDC), and the FDC searches for the correct track number, sector number, and side number of the disk. When the correct position on the disk is found, the FDC asserts the data request (DRQ) signal. This signal indicates that the FDC data register is ready to accept a byte of data from the CPU. The DRQ signal requests a DMA transfer from channel 0 of the four-channel, on-chip DMA unit of the 80960CA. The CPU replies by activating the DACK0# (DMA acknowledge, channel 0) signal, accesses the FDC data port by asserting A1, A0 # 11. The DMA controller (DMAC) unit of the 80960CA fetches a data byte from a given address from memory and transfers it directly to the FDC on D0 to D7 lines of the data bus. The address is then incremented or decremented to access the next byte. The entire operation requires 0.8 μ. However, the floppy disk is ready for the next byte after 16 μ. For this reason the READY# input to the DMAC unit on the CPU chip must be controlled by the FDC to insert wait states until the next data byte can be read by the disk. This is taken care of by the D-type flip-flop, shown in Fig. B.6. The particular disk drive being controlled is the SA800. The WD2793 FDC is compatible with all of the essential control signals of the SA800 disk drive. The FDC head load (HLD) output signal is used to select the drive. The head load settling time can be programmed using a one-shot circuit (74LS121 chip) connected to the head load timing (HLT) input of the FDC. Once HLD has been activated, the disk head is assumed loaded when the HLT input goes high.

EXAMPLE B.2: 80960CA-BASED PACKETIZED AUDIO MIXER*

The advent of the 100 Mbit/s fiber data distribution interface (FDDI) packetized local area network (LAN) has provided an open systems inter-

---

*Worked out by graduate students of GMU Thomas J. Bova, Lorraine M. Brown, and Timothy J. Trapp.

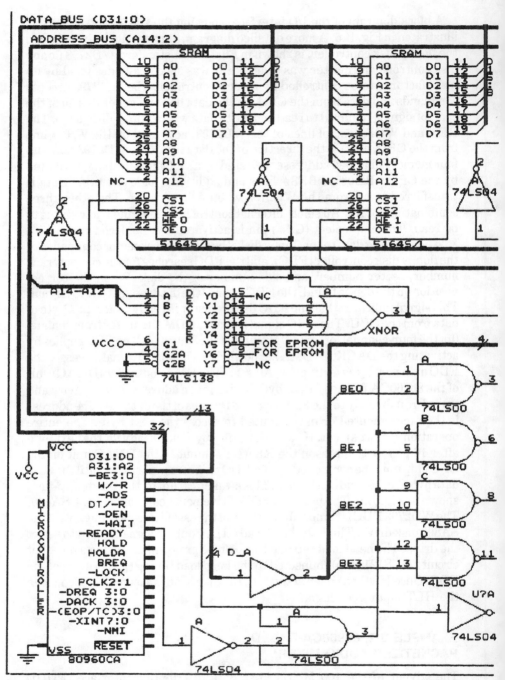

Figure B.2 SRAM interface.

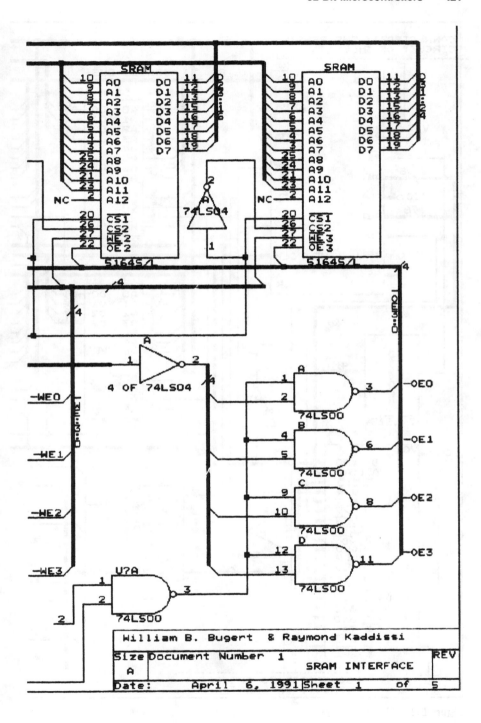

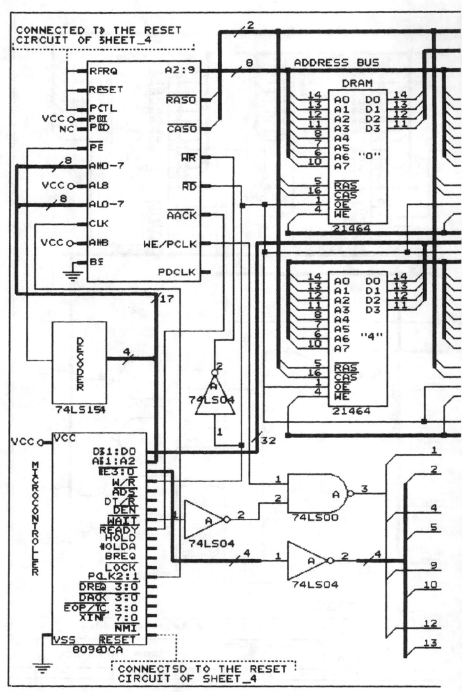

Figure B.3 DRAM interface.

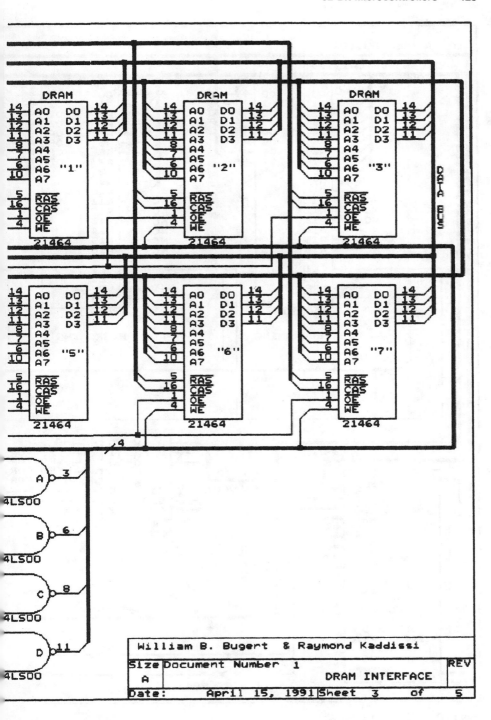

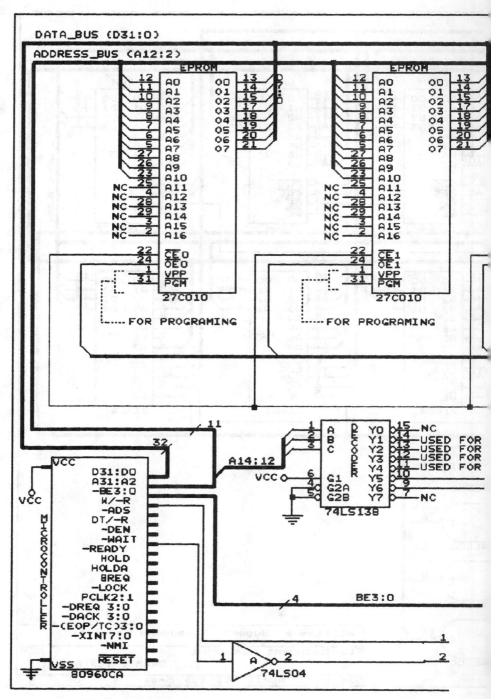

Figure B.4 EPROM interface.

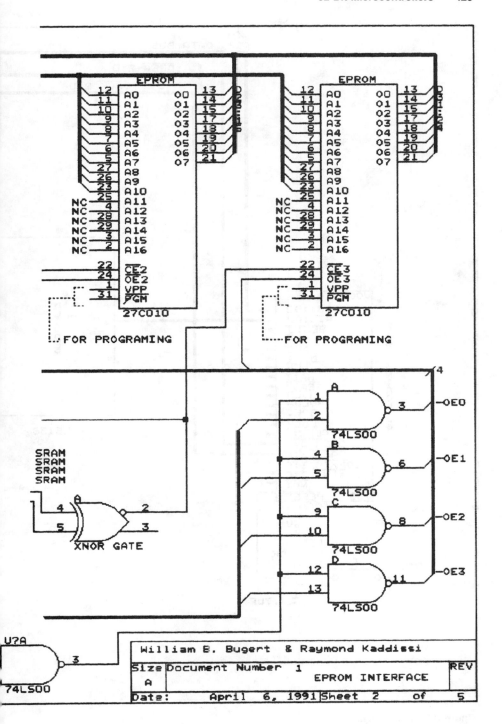

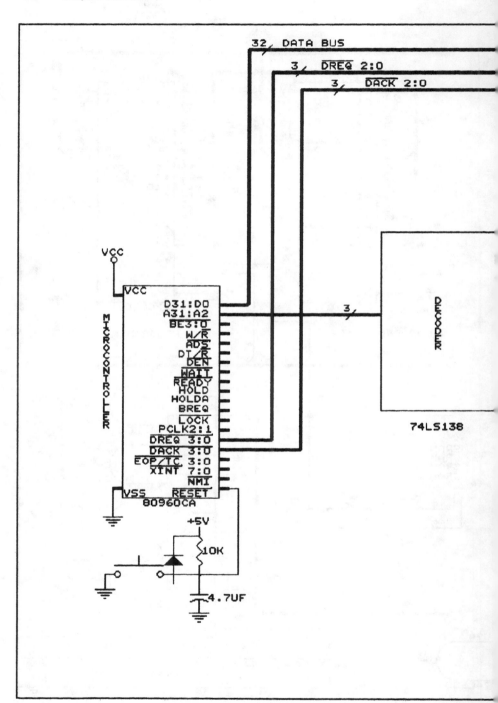

Figure B.5 I/O interface.

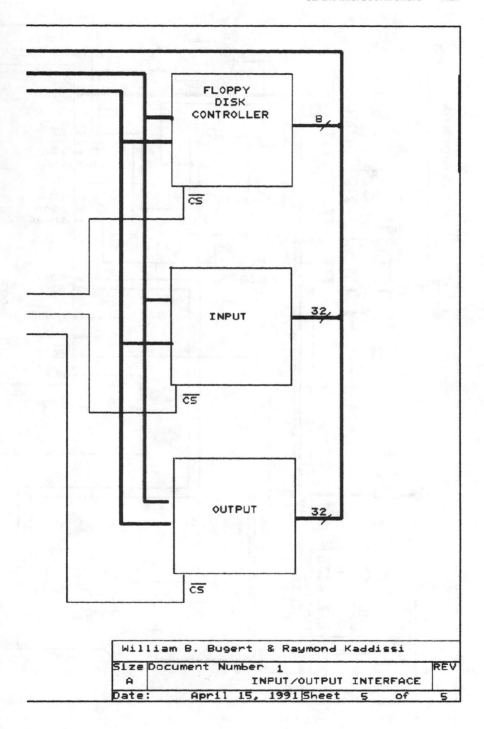

FLOPPY
DISK
CONTROLLER

8

$\overline{CS}$

INPUT

32

$\overline{CS}$

OUTPUT

32

$\overline{CS}$

William B. Bugert & Raymond Kaddissi		
Size Document Number 1		REV
A	INPUT/OUTPUT INTERFACE	
Date: April 15, 1991	Sheet 5 of 5	

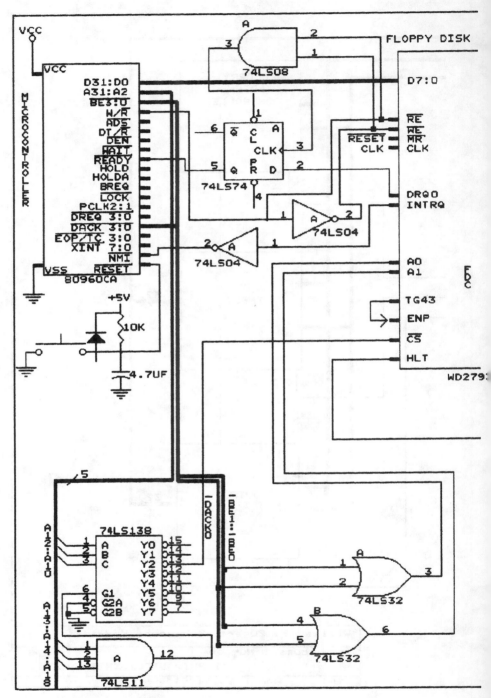

Figure B.6 Floppy disk interface.

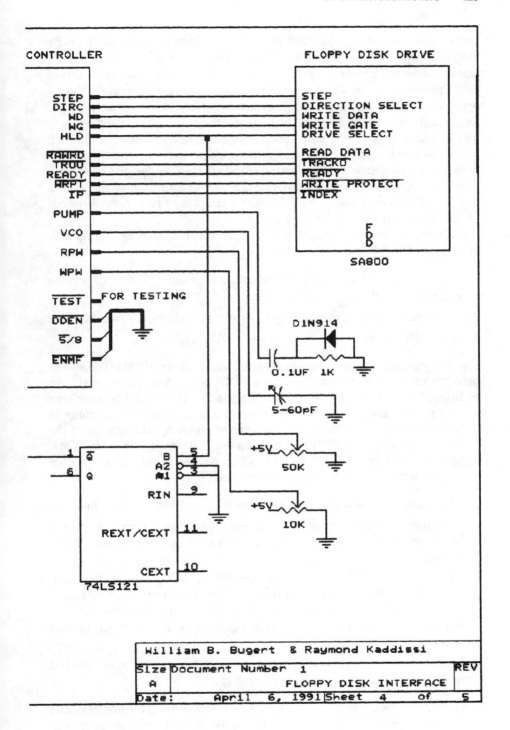

CONTROLLER

STEP
DIRC
WD
WG
HLD

RAWRD
TRU0
READY
WRPT
IP

PUMP

VCO

RPW

WPW

TEST
DDEN
5/8
ENMF

FOR TESTING

FLOPPY DISK DRIVE

STEP
DIRECTION SELECT
WRITE DATA
WRITE GATE
DRIVE SELECT

READ DATA
TRACK0
READY
WRITE PROTECT
INDEX

F
D
D

SA800

D1N914

0.1UF   1K

5-60pF

+5V
50K

+5V
10K

B
A2
A1
RIN
REXT/CEXT
CEXT

1
6
5
4
3
9
11
10

74LS121

	William B. Bugert & Raymond Kaddissi	
Size	Document Number 1	REV
A	FLOPPY DISK INTERFACE	
Date:	April 6, 1991	Sheet 4 of 5

connection technology that will simplify distributed systems. A single standardized interconnection, capable of supporting traditional LAN command and control protocol, as well as significant (tens of Mbit/s) amounts of data, provides a simplified interface for new processing nodes. Furthermore, a simple standardized interface removes dedicated data paths. Dedicated data paths are costly because they are often nonstandard, increase the cost of maintenance, and require dedicated connections for each node they service. In the past, audio data in a distributed system was often presented to consumers over a dedicated data path. This design focuses on a technique for a workstation audio data receiver to mix and route data between the packet based FDDI LAN and a time division multiplexed (TDM) standard format denoted by T1.

A minimum configuration of such a workstation is illustrated in Fig. B.7. It includes the following:

1. A packetized audio mixer (PAM) circuit card assembly (CCA) to provide audio data routing and audio mixing

2. A host command and control processor for the human-machine interface

3. An FDDI engine to interface to the FDDI LAN.

The design concentrates on the packetized audio mixer (PAM). The 80960-based PAM exchanges data between FDDI and T1 simultaneously. In addition, the PAM mixes 16-bit audio samples received from the interfaces and can present the results to any of these interfaces. The PAM can route 64 different signals. The 64 signals come from or go to any combination of the I/O ports: FDDI on VME (VERSA Module Eurocard), or VSB (VME secondary bus) bus, or T1. The mix function provides a 48-channel mixing capability with the following audio features:

1. The audio process is designed to support up to 16-bit-wide audio data.

2. The mixing of left and right ear volumes independently provides a stereo platform for presenting data to the operator in spatially divided audio format.

3. The 48 signal sources can be a mixture of linear and u-law encoded data. Any channel to be mixed will be converted to linear encoding before processing. The output of the mix task can be converted to either u-law or linear encoding. This provides the ability to adjust for incompatible data sources in the system.

4. The mixed output can be sent to one or more of the I/O interfaces (FDDI or T1), thus providing a flexible product.

The PAM operation involves three tasks and eight tables/buffers and channels for process initialization, audio mixing, and audio routing. The

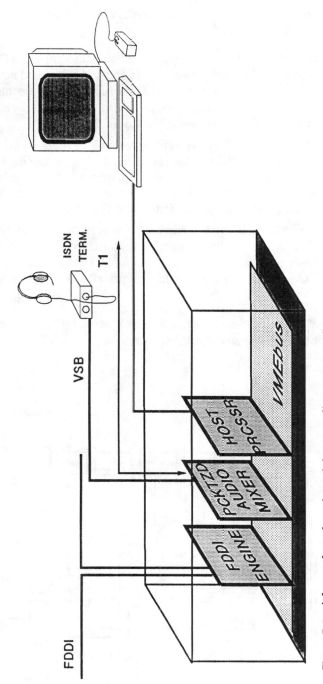

Figure B.7 Advanced workstation minimum audit support configuration.

interconnection between these tasks and buffers is illustrated in Fig. B.8. The three PAM tasks are as follows:

1. *Command and control task (CC task)*: The main function of the CC task is to interface with the host processor via the VME bus and act on its commands. The CC task's key commands are to set up a new connection or to disconnect an existing one. The CC task informs the traffic cop task recording which audio addresses it should be looking for.

2. *Traffic cop task*: The traffic cop's main function is to route audio data to and from the interfaces. The traffic cop also initiates the mixing task. It must operate in such a way as to keep up with all the incoming and outgoing data. This criterion makes it time critical. Therefore, the traffic cop task must run at a higher priority than the CC task.

3. *Mix task*: The mix task mixes a sample from each data queue that has been identified as being included in the mix algorithm. The mix task is the most processor intensive and time critical of all the tasks in the PAM. Therefore, the mix task runs at a higher priority than the other two tasks.

The eight PAM tables/buffers, shown in Fig. B.8, are listed and described in Table B.1. The main steps of the PAM mix task are as follows:

Initialization

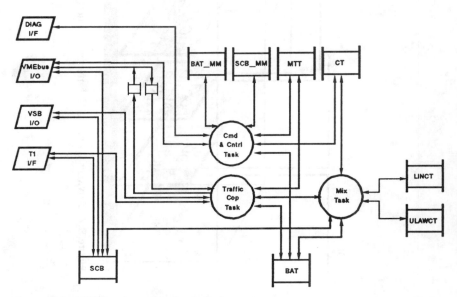

Figure B.8 PAM process blocks and elements.

Symbolic Name	Name	Description
BAI	Basic Attribute Table	Table used to hold attribute parameters for each active buffer.
BAT_MM	Basic Attribute Table Memory Manager	Management table used to allocate Basic Attribute Tables.
MTT	MAC Address to BAT_P Translation Table	Table used to map FDDI MAC address to internal Basic Attribute Tables
CT	Connection Table	Table used to identify which queues will be used in the mixing function.
LINCT	Linear Lookup Conversion Table	Lookup table used to convert linear encoded data to u–Law encoded data.
ULAWCT	u–Law Lookup Coversion Table	Lookup table used to convert u–Law encoded data to linear encoded data.
SCB	Signal Circular Buffer	Buffer space reserved for one signal or channel.
SCB_MM	Signal Circular Buffer Memory Manager	Management table used to allocate data buffers.

TABLE B.1 i960 PAM Tables and Buffers

433

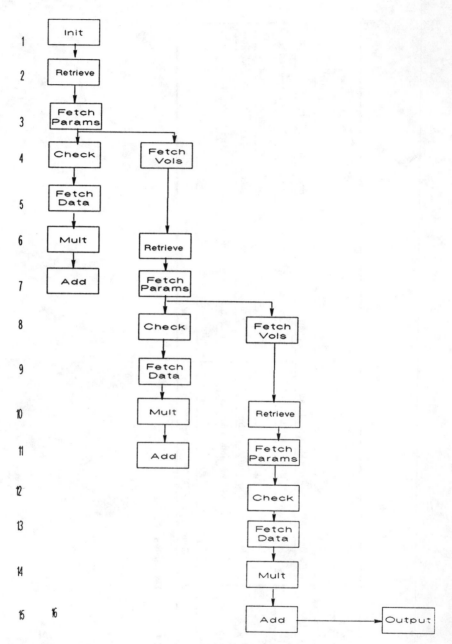

Figure B.9 Mix task parallel thread flow diagram.

Retrieve queue pointer

Fetch parameters

Check difference

Fetch data

Multiply

Add

Output

The design benefits from the superscalar property of the 960CA by executing steps of different threads simultaneously. This is illustrated in the flow diagram in Fig. B.9. In this example, three samples are to be mixed. Due to the availability of three concurrent pipelines on the 960CA, some operations are executed simultaneously. For instance, the add of the first sample (REG pipeline) and the fetch parameters of the second sample (MEM pipeline) are executed simultaneously. By performing these operations concurrently, the number of steps for a three-sample mix can be reduced from 26 to 16. When this scenario is expanded to 48 samples per mix, the reduction is from 386 to 196 steps (about 50 percent).

A block diagram of the PAM implementation is shown in Fig. B.10. The VME bus interface provides a high-bandwidth data path for data transfers between the PAM and other memory-mapped devices, such as the FDDI board. The VSB (VME secondary bus) provides a secondary path to memory-mapped devices that is independent of the VME bus, to provide both additional bandwidth and reduce delay. The T1 interface provides access to the outside world via a T1 line. The interrupt programming feature of the 960CA allows for the best-suited interrupt scheme to be used. In this case, a dedicated mode (see Fig. 7.14) provides the minimal delay for handling interrupts. This allows the PAM to meet the stringent processing requirements for the real-time mixing and routing. The burst-mode memory access capability of the 960CA allows more data to be transferred per clock cycle. The 960CA on-chip instruction cache speeds up the overall execution. The 960CA on-chip, four-channel DMA controller allows for the best mode (demand or block) to be specified for each DMA channel. In addition, the DMA process may take advantage of the burst mode capability. The variety in the DMA controller operation provides flexibility in presenting data to the various consumers. Memory may be sectioned by the 960CA. Each section may be programmed as to data width, burst access, and wait states. This enables the PAM to take advantage of the best mode of operation for each memory section. The diagnostic port provides a direct connection to processes running on the PAM to allow debugging of I/O and internal operations. The diagnostic port provides the capability to monitor built-in test procedures and on-line performance.

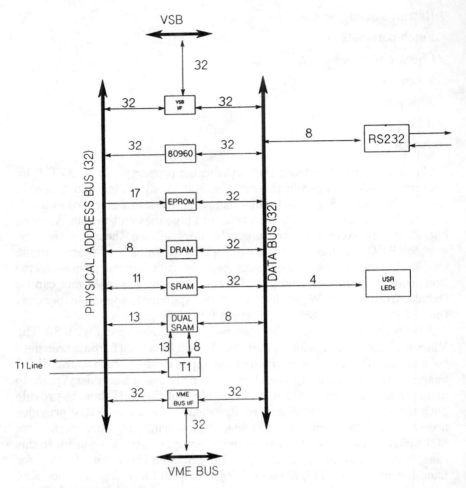

Figure B.10 PAM architecture.

TABLE B.2 Interrupt Table

INTERRUPT SIGNAL	PRIORITY	SOURCE
XINT0	3	T1
XINT1	5	RS232
XINT2	X	UNASSIGNED
XINT3	2	VME
XINT4	4	VSB
XINT5	1	CLK TICK
XINT6	X	UNASSIGNED
XINT7	X	UNASSIGNED
NMI		ABORT

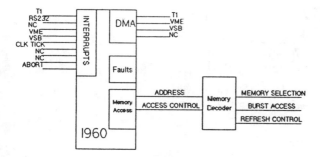

Figure B.11 Core overview.

The 960CA core interaction is illustrated in Fig. B.11. The interrupt priority assignments are listed in Table B.2. The CLK TICK is required for context switching and therefore has the highest assigned priority. The VME bus is crucial for getting FDDI audio packets to the PAM CCA, and thus has priority 2, freeing the FDDI to service consumers other than the PAM. Since the T1 and VSB also need to communicate with the PAM CCA for audio mixing, they have priorities 3 and 4, respectively. The RS232 port is used for diagnostics and has the lowest priority (5). The nonmaskable interrupt (NMI) is not necessary for the function of PAM, but will be available through the diagnostics port to abort all operations at any time for testing. Table B.3 shows the specifications for each DMA channel. Only three channels are implemented. All channels use the block mode to transfer data. In this mode, DMA transfers are initiated by software. The memory address decoder (MAD) is shown in Fig. B.12. The PAM memory map is shown in Fig. B.13.

The MAD accesses SRAM using 960CA signals ADS, BE, BLAST, and WAIT, and producing as outputs to the SRAM the A2 and A3 signals. The BLAST signal is used to terminate the burst access. The DRAM is accessed similarly; however, refresh signals RAS and CAS are generated by MAD in

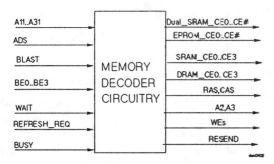

Figure B.12 Memory address decoder.

TABLE B.3   DMA Table

DMA CHANNEL	SOURCE	DESTINATION	MODE	PURPOSE
0	DUAL SRAM	DUAL SRAM	BLOCK	T1
0	DUAL SRAM	DUAL SRAM	BLOCK	T1
1	VME	DRAM	BLOCK	VME
1	DRAM	VME	BLOCK	VME
2	VSB	DRAM	BLOCK	VSB
2	DRAM	VSB	BLOCK	VSB
3				UNASSIGNED

```
FFFF FFFFH  ┌─────────────────┐
            │                 │
FFFF 1000H  │   VME CTRL      │
FFFF 0000H  ├─────────────────┤
FFFB 0000H  │   VSB CTRL      │
FFFA 0000H  ├─────────────────┤
            │                 │
4000 4000H  │                 │
            │   DUAL SRAM     │
4000 0000H  ├─────────────────┤
3000 2000H  │                 │
            │   SRAM          │
3000 0000H  ├─────────────────┤
200C 0000H  │                 │
            │   DRAM          │
2000 0000H  ├─────────────────┤
1004 0000H  │                 │
            │   EPROM         │
1000 0000H  ├─────────────────┤
            │                 │
            │   VMEbus        │
            │                 │
0800 0000H  ├─────────────────┤
            │                 │
            │   VSB           │
            │                 │
0000 0000H  └─────────────────┘
```

Figure B.13 PAM memory map.

addition. The dual-ported SRAM (DPS) is accessed in a manner similar to SRAM, however, in addition, MAD must monitor the BUSY signal from the DPS. This signal will be asserted if the processor attempts to access an address currently being accessed (through T1, for instance; see Fig. B.10).

The bootup code and all of the assembly code to run the PAM process is stored in the EPROM. The 27960CS EPROM was selected in this design because of its ease of connectivity to the 960CA and for its burst-mode capabilities. A total of four EPROM chips are implemented for a total of 512 kbytes. The SRAM bank contains the assembly code that has been downloaded from the EPROMs. This allows the code to run much faster than if it were accessed in the EPROM while running. Four 6116 SRAM chips are used for a total of 8 kbytes. The DRAM stores all of the data queues and tables required during the audio mixing. The DRAM implements the TMS44C256 chip (256K × 4 bits). A total of 8 DRAM chips are used (768 kbytes).

The dual-ported SRAM (DPS) is used to store T1 data. It is accessible by the T1 circuitry on one side, and by the PAM address and data busses on the other (see Fig. B.10). The CY7C144 chip was selected, providing 32 kbytes of storage.

The VME bus interface is used to support I/O from the FDDI and the host processor. The PAM circuitry supports DMA transfers in both directions with the 960CA or another VME bus device as the master. The VME bus interface is built around the VIC068 chip manufactured by Cypress Semiconductor. This chip provides a complete VME bus interface controller and arbiter, complete master/slave capability, interleaved block transfers over the VME bus, interrupt support, and interprocessor communications. A block diagram of the VME bus interface is shown in Fig. B.14.

The VME subsystem bus (VSB) is a local subsystem extension bus. It allows a processor board, such as the PAM, to access additional memory or I/O over a separate local bus, removing traffic from the main VME bus and improving the total throughput of the system. The VSB interface, illustrated in Fig. B.15, is built around the Motorola MVSB2400 chip. The MVSB2400 is a gate array designed to provide most of the I/O signals and control the data transfers and arbitration of the VSB.

The T1 circuitry provides the interface between the T1 line and the PAM process. The T1 circuitry receives data from the T1 line and places it in the DPS. The T1 circuitry also takes data from the DPS and transmits it back to the T1 line (see Fig. B.10). The data is available in the DPS, addressable by the 960CA CPU, for the purpose of processing T1 data and sending data out to the T1 line. The T1 circuitry interacts with the 960CA directly through control signals. The T1 circuit can process up to 32 transmit and 32 receive channels.

A block diagram of the serial interface is shown in Fig. B.16. The serial interface is built around the Zilog Z85C30 chip. The Z85C30 is a dual-channel

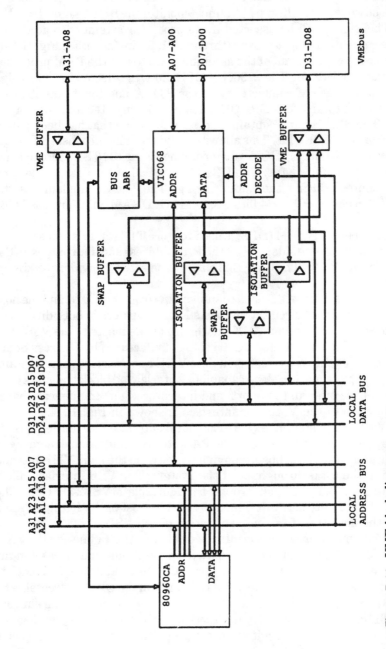

Figure B.14 VME block diagram.

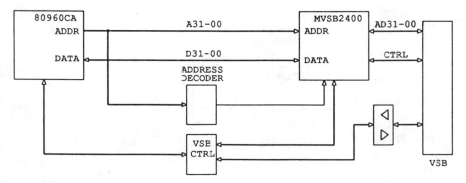

Figure B.15   VSB block diagram.

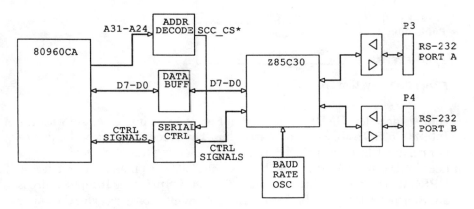

Figure B.16   Serial interface block diagram.

multiprotocol serial interface. The Z85C30 contains parallel-to-serial and serial-to-parallel conversion circuitry and baud rate generators.

EXAMPLE B.3: UNIVERSAL 80960CA-BASED PROCESSOR BOARD*

The 80960CA-based board includes the following subsystems: 512-kbyte EPROM, 32-kbyte SRAM, 512-kbyte DRAM, a high-speed serial communication device, two 16-bit timer/counters, one 24-bit timer/counter, a 16-bit event counter, a bus watchdog timer, and three 8-bit parallel I/O ports. The 80960CA memory space is subdivided into 16 regions, 0 to 15. Different memory subsystems are assigned to different regions. The memory map and region assignment are illustrated in Fig. B.17. The system

---

*Worked out by GMU graduate students Scott McNutt and Ronald Portee.

ADDRESS

0000 0000H	Region 0
	Unused
0FFF FFFFH	
1000 0000H	Region 1
	Input/Output
1FFF FFFFH	
2000 0000H	Region 2-12
CFFF FFFFH	Unused
D000 0000H	Region 13
	DRAM
DFFF FFFFH	
E000 0000H	Region 14
	SRAM
EFFF FFFFH	
F000 0000H	Region 15
	EPROM
FFFF FFFFH	

Figure B.17 System "region" memory map.

address decoder (SAD), illustrated in Fig. B.18, is implemented with the Intel 80C508 CHMOS decoder/latch chip. The decoding speed of this device is 7 ns, and this is the primary reason for selecting this chip. A high speed decoder is crucial for systems that use the 27960CX pipelined burst access EPROM, as is the case in this design. The 80C508 provides 16 input address and 8 output select pins. In the current implementation it receives the upper 14 address lines A18 to A31 from the CPU (see Fig. B.18). The SUP# signal indicates whether a bus access originated from a request issued in super-

TABLE B.4 Address Decoder Regions

Signal	SUP#	D/C#	A31:28	A27:24	A23:20	A19:18	Address Range
EPROM#	X	X	1111	1111	1111	10	FFF8000 - FFFFFFFF
SRAM#	X	X	1110	0000	0000	00	E000000 - E003FFFF
TIMER#	0	1	0001	0000	0000	00	1000000 - 1003FFFF
PARIO#	0	1	0001	0000	0000	01	1004000 - 1007FFFF
SERIO#	0	1	0001	0000	0000	10	1008000 - 100BFFFF
IO#	0	1	0001	0000	0000	XX	1000000 - 100FFFFF
EXPAND1#	X	X	XXXX	XXXX	XXXX	XX	Unused
EXPAND2#	X	X	XXXX	XXXX	XXXX	XX	Unused

EPROM#   EPROM module select.
SRAM#   Selects the SRAM module select.
TIMER#   82C54-2 Programmable Interval Timer select.
PARIO#   82C55A Programmable Peripheral Interface select.
SERIO#   MC2652 Multi-Protocol Communications Controller select.
IO#   IO Region select.
EXPAND1# Unused. For future expansion.
EXPAND2# Unused. For future expansion.

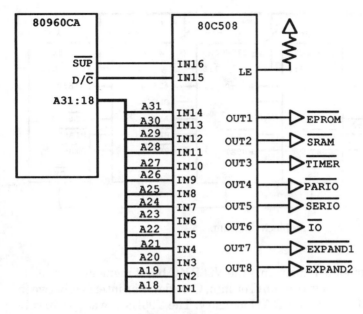

Figure B.18 System address decoder schematic diagram.

visor mode, and the D/C# signal indicates a bus access to code or data. These signals prevent the user from accessing supervisor memory areas. The D/C# signal also prevents code accesses to data-only areas (such as memory-mapped I/O). The address regions selected by the decoder are summarized in Table B.4.

The system EPROM module, illustrated in Fig. B.19, implements four 27960CX chips. The 27960CX was designed to be directly interfaced to the 80960CA CPU, without any additional logic circuitry (except the decoder, which provides the EPROM# chip select line, connected to the CS# pin of the 27960CX). The RESET# signal resets the internal state to its initial conditions. The BLAST# (burst last) signal indicates that the data transfers of an access are complete. It is asserted in the last data transfer of every bus access, burst and nonburst.

The SRAM module, illustrated in Fig. B.20, is implemented using the Intel 5164 high speed 8K × 8 SRAM device. The timing parameters of the 5164 are listed in Table B.5. The 5164 provides two key features that help simplify and enhance the design. The first is its low access time for both read and write operations, allowing the SRAM module to operate with zero wait states. The second is the availability of two chip select inputs (CS1# and CS2) that help reduce the required device select logic. The 5164 also provides a standby or reduced power consumption mode when the device is not activated (not selected).

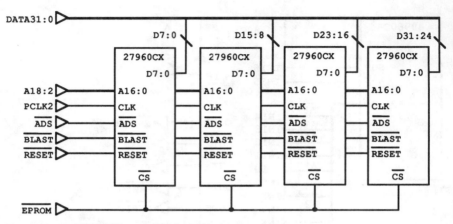

Figure B.19   EPROM module block diagram.

The DRAM module implements the V96BMC burst memory controller, developed by the V3 corporation (Toronto, Canada) as an interface between the 80960CA CPU and the DRAM memory. The V96BMC was specifically designed to interface with the 960CA directly, it supports burst accesses, and it is completely software programmable. In addition, its cost is comparable to that of Intel 82C08 DRAM controller. The V96BMC organizes the DRAM devices into two 32-bit banks and allows the 960CA to access DRAM at a rate of one memory cycle per processor clock cycle (no wait states) during burst accesses. This high transfer rate is accomplished by perform-

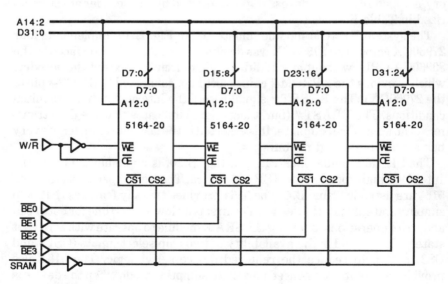

Figure B.20   SRAM module schematic.

TABLE B.5   Intel 5164 High-Speed Static RAM Timing Parameters

Symbol	Parameter	5164-20	5164-25	5164-30	5164-35	Unit
t_{RC}	Read Cycle Time	20	25	30	35	ns
t_{AA}	Address Access Time	20	25	30	35	ns
t_{ACS}	Chip Select Access Time	20	25	30	35	ns
t_{OE}	Output Enable to Output Valid	15	15	20	20	ns
t_{WC}	Write Cycle Time	20	25	30	35	ns

ing interleaved page-mode accesses to the two banks. Both banks can be configured for up to 64 Mbytes (128 Mbytes total) using 16M × 4 devices. The V96BMC supports burst write accesses via special buffer control signals. The signals are used to latch the data bus in high speed latching buffers to satisfy the data hold time of the DRAM.

The interface between the 960CA and the V96BMC is illustrated in Fig. B.21. The BERR# (bus error) output of the V96BMC is asserted when the V96BMC watchdog timer completes a time out. This signal can be used to generate an interrupt on one of the interrupt inputs of 960CA. The TINT#

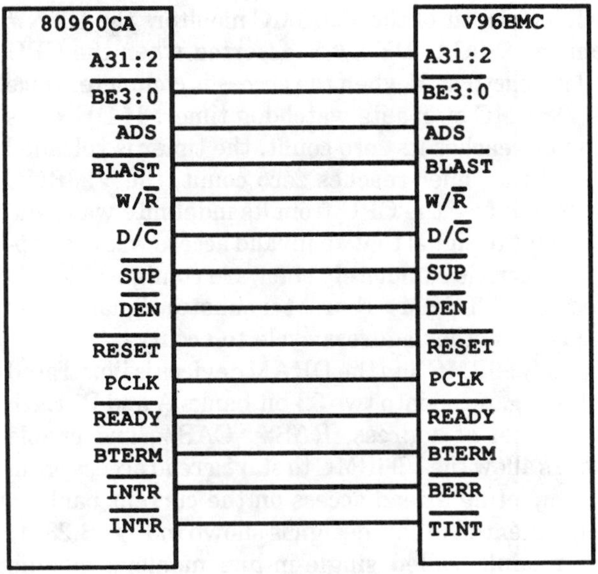

Figure B.21   Interface between V96BMC and the i960 CA.

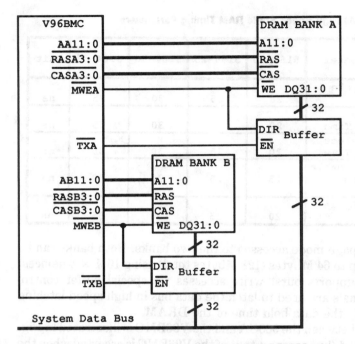

Figure B.22 Interface between V96BMC and DRAM.

(timer interrupt) output of the V96BMC indicates that the 24-bit internal timer/counter has "timed out." It can also be used to generate an interrupt on the 960CA. The watchdog circuit of the V96BMC monitors the DEN# (data enable) signal from the 960CA. DEN# is asserted when the CPU initiates a bus access and is deactivated when the access is complete. Once DEN# is asserted, the V96BMC starts its watchdog timer. If DEN# is deactivated before the timer reaches its zero count, the timer is reloaded and the process repeats. If the timer reaches zero count, the V96BMC asserts the READY# signal to free the CPU from its indefinite wait, and generates a BERR# interrupt to signal that an invalid access was attempted. The watchdog timer feature is completely software configurable and can be optionally disabled. The BTERM# (burst terminate) signal breaks up a burst access and causes another address cycle to occur.

The interface between the V96BMC and the DRAM devices is illustrated in Fig. B.22. The DRAM is organized into two 32-bit banks A and B. Each bank has its own dedicated set of address, RAS#, CAS#, and enable signals. The data bus buffers allow the V96BMC to start a read access on an opposing bank prior to completing a read access on the current bank.

The DRAM module implemented in this design is shown in Fig. B.23. It features an elementary module called single-in-line memory module (SIMM), shown in Fig. B.24. Each SIMM is driven by its own CAS# signal

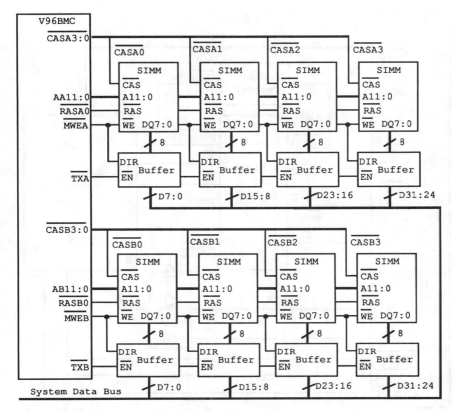

Figure B.23 DRAM module schematic diagram.

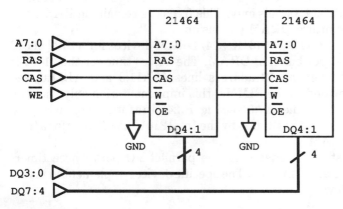

Figure B.24 Single-in-line memory module schematic.

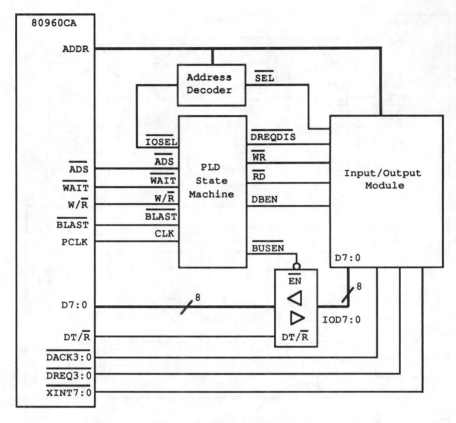

Figure B.25 I/O interface module schematic diagram.

from the V96BMC. The SIMM RAS# inputs are driven by the V96BMC RASA0# and RASB0# signals only. This frees the remaining RASA and RASB signals for future DRAM expansion.

This design uses Intel 21464 (64K × 4) DRAM devices for a total of 512 kbytes (256 kbytes per bank) of DRAM. The access time for the 21464 is about 60 ns. Although all 12 address lines (AA11:0 and AB11:0; see Fig. B.23) are provided to the SIMMs, this implementation only uses the eight low-order address lines A7:0 (see Fig. B.24). In a future expansion, the 21464 can be replaced in the SIMMs by the Intel 21014 (256K × 4) for a total of 2 Mbytes of DRAM.

The I/O subsystem includes a 24-line parallel I/O port, three timer/counters, and a serial I/O module. The specific devices implemented in this design are as follows:

82C54-2 programmable interval timer (PIT)

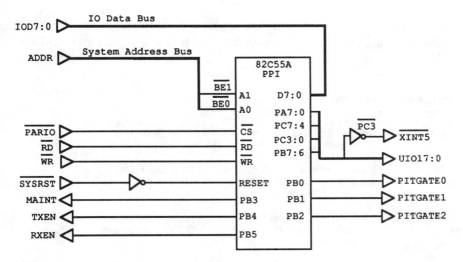

Figure B.26 Parallel I/O module schematic diagram.

82C55A programmable peripheral interface (PPI)

MC2652 multiprotocol communications controller (MPCC).

The system also implements a programmed logic device (PLD) subsystem to generate all of the required control signals. A diagram of the I/O interface module is shown in Fig. B.25, and the main signals are listed in Table B.6. The parallel I/O module, implementing the 82C55A PPI, is illustrated in Fig. B.26. The timer/counter module, implementing the 82C54 PIT, is

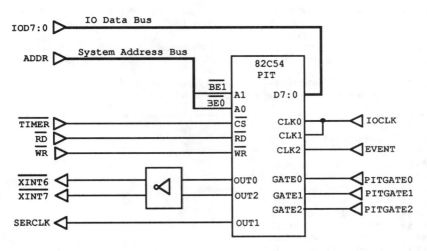

Figure B.27 Timer/counter module schematic diagram.

shown in Fig. B.27. The serial I/O module, implementing the MC2652, is illustrated in Fig. B.28. The system list of interrupt sources is given in Table B.7.

TABLE B.6   I/O Module Interface Signals

ADDR	**System address bus (I).** This includes BE0# and BE1#.
SEL#	**Module select (I).** This signal is from the system address decoder and indicates a bus access to the specified I/O module.
DREQDIS#	**DMA Request Disable (I).** This input is used by the I/O module to negate the DREQ3# - DREQ0# outputs to the processor.
WR#	**Write(I).** This signal is from the PLD state machine and indicates a write access to an iAPX standard peripheral device.
RD#	**Read(I).** This signal is from the PLD state machine and indicates a read access to an iAPX standard peripheral device.
DBEN	**Data Bus Enable(I).** This signal is from the PLD state machine and is used to latch input data during write cycles or drive the data bus during read cycles. The actual type of cycle is qualified by the W/R# input.
IOD7:0	**I/O data bus (I/O).** The (buffered) 8-bit input/output data bus.
DACK3:0#	**DMA Acknowledge (I).** These inputs are tied directly to the processors DMA Acknowledge pins.
DREQ3:0#	**DMA Request (O).** These outputs are tied directly to the processors DMA Request pins.
XINT7:0#	**External Interrupt (O).** These outputs are tied directly to the processors External Interrupt pins.

TABLE B.7   System Interrupts

XINT0#	Serial Data Available
XINT1#	Serial Transmit Buffer Empty
XINT2#	Serial Port Receive Status Available
XINT3#	Serial Transmitter Underrun
XINT4#	Serial Sync/Flag Detected
XINT5#	Parallel I/O Interrupt
XINT6#	Operating System Clock Timeout
XINT7#	Event Counter Interrupt

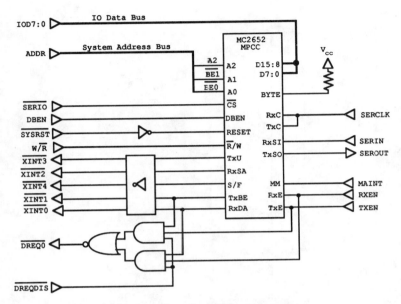

Figure B.28   Serial I/O module schematic diagram.

Concluding Comments

Microcontrollers have come a long way since their inception in mid-seventies to the nineties. They were started as simple 4- or 8-bit chips with modest capabilities. They started with less than 200 bytes of RAM and just 1 Kbyte of ROM or EPROM (if at all). Progress was made through 16-bit microcontrollers, gradually increasing the memory and other on-chip features. This development culminated in modern, 32-bit microcontrollers, with large on-chip caches (9 Kbyte on LR33000, 5 Kbyte on NS32SF640), larger on-chip main memory RAM (1 Kbyte on 80960CA), and up to 16 Kbyte on-chip ROM or EPROM (83C196KC, 87C196KC). In addition to a superscalar CPU (80960CA, NS32SF640) and memory, modern microcontroller chips contain many other subsystems necessary for their efficient implementation. One can find today a 4-channel DMA controller (80960CA), a 15-level interrupt controller (NS32SF640), a DRAM controller (LR33000), 3 timers (LR33000), an A/D converter (80C196KC), and a serial port (80C196KC). A typical microcontroller has about 48 I/O pins, some of which may have an additional use as address and data lines.

The performance of some of the microcontrollers becomes comparable to the performance of leading microprocessors, and sometimes it may surpass it. All of the 32-bit microcontrollers are endowed with one or more 3- to 5-stage instruction pipelines. Some have a superscalar design: the 80960CA with 3 concurrent pipelines and 4 instructions fetched and decoded simultaneously, and the NS32SF640 with 2 concurrent integer units. Some have an on-chip FPU (NS32SF640, Am29050, 80960KB,

80960MC), significantly enhancing their floating-point performance. All 32-bit microcontrollers have adopted notable features of RISC, such as memory access by load or store instructions only, register-to-register operations, and relatively large CPU register files (192 registers on Am29050, 32 integer and 32 floating-point registers on NS32SF640, and 32 on 80960CA and LR33000). The 80960CA has an additional 512 byte local register cache (not counted as a part of cache or main memory). The Am29050 has an additional 1 Kbyte branch target cache. All of the RISC type features help streamline and speed up the handling of the instructions on the microcontrollers and enhance the overall performance significantly. The LR33000 has a RISC-type CPU, the MIPS (microprocessor without interlocked pipeline stages), designed by MIPS Computer (Palo Alto,CA) and originated at Stanford University. The Am29050, preceded by Am29000, was designed as a RISC-type system to begin with, by AMD (Advanced Micro Devices).

The applications of microcontrollers have penetrated into practically every area of industry, government and academia. Microcontrollers can be found embedded into robots, machine tools, automobiles, aircraft, aerospace vehicles, medical electronics and many other installations. They are used as secondary control systems on larger computing installations to control printers and other peripherals. The extent of microcontroller applications in various fields is expected to grow significantly in the future. In anticipation of such growth, an attempt was made in this text to present to the reader some basic principles underlying the microcontrollers and their implementation, along with detailed examples of some leading specimens of such systems.

Appendices

APPENDIX I: SMART COMPASS

Assembly Language Source Code for Smart Compass in MC68H11

```
*************************************************************************
* NAME:   Smart Compass                                                *
*                                                                      *
* BY:  Marc Gilstead                                                   *
* DATE BEGUN:  March 24, 1989                                          *
* LAST UPDATE:  March 31, 1991KJH: INTROL assembler compatibility      *
*                                                                      *
* This program reads in two binary value representing the orthogonal   *
* components of the earth's magnetic field and calculates the heading  *
* angle of the sensor.  It then outputs the angle to an LED display and *
* a latch.                                                             *
*                                                                      *
* This program is based on the original Smartcompass program written  *
* by Doug Garner for an Intel 8748.  This version is for a Motorola    *
* 68HC811A2.                                                           *
*************************************************************************
*************************************************************************
*                           EQUATES                                    *
*************************************************************************

ADCTL      EQU   $1030      *A/D Control register
ADRTRAN    EQU   $1031      *A/D Result -Transverse
ADRLONG    EQU   $1032      *A/D Result -Longitudinal
ADRREF     EQU   $1033      *A/D Result -Reference
ADCTLWD    EQU   $10        *A/D Control word
CCF        EQU   $80        *Conversion complete mask
```

```
PORTB      EQU  $1004     *O/P Port B -to display
PORTC      EQU  $1003     *O/P Port C -to robot
DDRC       EQU  $1007     *Data direction register for C
OPTION     EQU  $1039     *Option register
OPTIONWD   EQU  $90       *Choose start up options
RAMSTART   EQU  $0000     *Start of RAM
EPROMSTART EQU  $F800     *Start of EEPROM
TRANLONG   EQU  $99       *Transverse > Longitudinal Mask
PERIOD     EQU  $2FF      *Display period
OCT1       EQU  $01       *
OCT2       EQU  $02       * OCTANT
OCT3       EQU  $04       *
OCT4       EQU  $08       * CODE
OCT5       EQU  $10       *
OCT6       EQU  $20       * MASKS
OCT7       EQU  $40       *
OCT8       EQU  $80       *

*****************************************************************************
*                             VARIABLES                                    *
*****************************************************************************

           ORG  RAMSTART

STACK      RMB  20        *Define stack
STACKSTART RMB  1         *Initial stack bottom
OCTANT     RMB  1         *8 bit octant code
ABSTRAN    RMB  1         *Magnitude of Transverse vector
ABSLONG    RMB  1         *Magnitude of Longitudinal vector
PHI        RMB  1         *Angle to nearest axis
THETA      RMB  2         *Heading angle
HUNDREDS   RMB  1         *BCD hundreds digit
TENS       RMB  1         *BCD tens digit
ONES       RMB  1         *BCD ones digit
COUNT      RMB  2         *Display freq. counter

*****************************************************************************
* MAIN PROGRAM - Takes two words representing magnetic field vectors       *
* and outputs a magnetic compass heading                                   *
*****************************************************************************

           ORG  EPROMSTART   *Set address for compilation
           LDS  #STACKSTART  *Initialize stack
           JSR  INIT         *Initialize vars
main:      JSR  IPVECTORS    *Read in field component
           JSR  CALCOCTANT   *Calculate which octant
           JSR  CALCPHI      *Calculate angle to axis
           JSR  CALCTHETA    *Calculate heading angle
           JSR  DISPLAY      *Output BCD to display
           JSR  ROBOT        *Output angle to latch
           JMP  main         *Continue

*****************************************************************************
* INIT - Initialize variables and hardware.                               *
*****************************************************************************

INIT       LDAA OPTIONWD  *Power up the A/D converter
           STAA OPTION
           CLR  DDRC      *Initialize port C for output
           COM  DDRC
           RTS

*****************************************************************************
* IPVECTORS - Initiates and reads results of A/D conversion of the 3      *
* channels.  Returns with values in memory at ADRTRAN, ADRLONG, ADRREF,   *
* ABSTRAN,  ABSLONG.                                                       *
*****************************************************************************
```

```
IPVECTORS   LDAA  #ADCTLWD    *Put control word in control register
            STAA  ADCTL       *start A/D
            LDX   #ADCTL      *Wait until conversion is complete
NOTRDY      BRCLR 0,X,CCF,NOTRDY
            LDAA  ADRTRAN     *Put Transverse in ACCA
            SUBA  ADRREF      *Subtract Reference
            BHS   TPOSITIVE   *If positive, jump
            NEGA              *Else, take magnitude
TPOSITIVE   STAA  ABSTRAN     *Store |Transverse|
            LDAB  ADRLONG     *Put Longitudinal in ACCB
            SUBB  ADRREF      *Subtract Reference
            BHS   LPOSITIVE   *If positive, jump
            NEGB              *Else, Take magnitude
LPOSITIVE   STAB  ABSLONG     *Store |Longitudinal|
            RTS


*****************************************************************************
* CALCOCTANT - Determines which of the 8 octants the vector is in.        *
* Returns with the corrosponding bit in OCTANT set.                       *
*****************************************************************************

CALCOCTANT  CLR   OCTANT      *Clear octant code
            LDAA  ADRTRAN     *Put Transverse in ACCA
            LDAB  ADRLONG     *Put Longitudinal in ACCB
            LDX   #OCTANT     *X gets OCTANT adx
            CMPA  ADRREF      *Compare Transverse to Reference
            BLO   WESTHALF    *If Transverse < Reference, westhalf
EASTHALF    CMPB  ADRREF      *Compare Longitudinal to Reference
            BLO   SEQUAD      *If Longitudinal < Reference, southeast
NEQUAD      LDAA  ABSTRAN     *Else, northeast
            CMPA  ABSLONG     *Comp. |Transverse| to |Longitudinal|
            BLS   OCTANT1     *If |Transverse| <= |Longitudinal|, jump
OCTANT2     BSET  0,X,OCT2    *Else set bit 2 in code
            RTS
OCTANT1     BSET  0,X,OCT1    *Set bit 1 in code
            RTS
SEQUAD      LDAA  ABSTRAN     *ACCA gets |Transverse|
            CMPA  ABSLONG     *Compare |Transverse| to |Longitudinal|
            BLS   OCTANT4     *If |Transverse|<=|Longitudinal|, jump
OCTANT3     BSET  0,X,OCT3    *Else set bit 3 in code
            RTS
OCTANT4     BSET  0,X,OCT4    *Set bit 4 in code
            RTS
WESTHALF    CMPB  ADRREF      *Comp. Longitudinal to Reference
            BLO   SWQUAD      *If Longitudinal < Reference, southwest
NWQUAD      LDAA  ABSTRAN     *Else Northeast
            CMPA  ABSLONG     *compare |Transverse| to |Longitudinal|
            BLS   OCTANT8     *If |Transverse| < |Longitudinal|, jump
OCTANT7     BSET  0,X,OCT7    *Else set bit 7 in code
            RTS
OCTANT8     BSET  0,X,OCT8    *Set bit 8 in code
            RTS
SWQUAD      LDAA  ABSTRAN     *ACCA gets |Transverse|
            CMPA  ABSLONG     *Compare |Transverse| to |Longitudinal|
            BLS   OCTANT5     *If |Transverse| < |Longitudinal|, jump
OCTANT6     BSET  0,X,OCT6    *Else set bit 6 in code
            RTS
OCTANT5     BSET  0,X,OCT5    *Set bit 5 in code
            RTS


*****************************************************************************
* CALCPHI - Divides vector components and takes the arctan to yield the   *
* angle PHI.                                                              *
*****************************************************************************

CALCPHI     LDX   #OCTANT              *If |Transverse| >
            BRCLR 0,X,TRANLONG,ATANLT  *|Longitudinal| then jump
ATANTL      LDAA  ABSTRAN              *Else ACCA gets |Transverse|
            LDAB  ABSLONG              *ACCB gets |Longitudinal|
```

```
                BSR    DIVIDE              *Divide them
                BSR    ARCTAN              *Take Arctangent
                RTS
ATANLT          LDAA   ABSLONG            *ACCA gets |Longitudinal|
                LDAB   ABSTRAN            *ACCB gets |Transverse|
                BSR    DIVIDE              *Divide them
                BSR    ARCTAN              *Take Arctangent
                RTS

*******************************************************************************
* DIVIDE - Divides ACCA by ACCB and returns with result in ACCA that is  *
* 64 times the quotient.                                                  *
*******************************************************************************

DIVIDE          PSHA               *Push ACCA onto stack
                CLRA               *Clear ACCA
                PSHA               *Push $00 onto stack
                PSHB               *Push ACCB onto stack
                CLRB               *Clear ACCB
                PSHB               *Push $00 onto stack
                PULX               *Pull $00:B into X
                PULA               *Pull $00:A into D (D=A:B)
                PULB
                FDIV               *X gets D/X
                XGDX               *D gets X, ACCA = 256 * quotient
                CLC                *Clear carry
                RORA               *ACCA gets ACCA/2
                CLC                *Clear carry
                RORA               *ACCA gets ACCA/2
                RTS

*******************************************************************************
* ARCTAN - Takes a value in ACCA which is 64 times quot. and looks up  *
* the arctan of the quotient in table.  Returns  with angle in PHI     *
*******************************************************************************

ARCTAN          LDX    #TABLESTART *X gets lookup table address
                TAB                *Transfer ACCA to ACCB
                ABX                *Add 8bit B to 16bit X
                LDAA   0,X         *ACCA gets phi from tbl
                STAA   PHI         *Store angle in PHI
                RTS

*******************************************************************************
* CALCTHETA - Takes the variable PHI and subtracts or adds 90, 180,  *
* 270, or 360 degrees to yield the  heading angle THETA.             *
*******************************************************************************

CALCTHETA       CLRA                      *Clear ACCA
                LDAB   PHI                *ACCB gets PHI
                LDX    #OCTANT            *X gets OCTANT code
                BRSET  0,X,OCT1,ADD0          *Octant 1, add 0
                BRSET  0,X,OCT2,SUB90         *Octant 2, sub 90
                BRSET  0,X,OCT3,ADD90         *Octant 3, add 90
                BRSET  0,X,OCT4,SUB180   *Octant 4, sub 180
                BRSET  0,X,OCT5,ADD180   *Octant 5, add 180
                BRSET  0,X,OCT6,SUB270   *Octant 6, sub 270
                BRSET  0,X,OCT7,ADD270   *Octant 7, add 270
                BRSET  0,X,OCT8,SUB360   *Octant 8, sub 360
ADD0            BRA    STORE
ADD90           ADDD   #$005A             *Add 90 to PHI
                BRA    STORE
ADD180          ADDD   #$00B4             *Add 180 to PHI
                BRA    STORE
ADD270          ADDD   #$010E             *Add 270 to PHI
                BRA    STORE
SUB90           SUBD   #$005A             *Subtract 90
                BRA    MAG                *Take magnitude
SUB180          SUBD   #$00B4             *Subtract 180
```

```
               BRA   MAG                 *Take magnitude
SUB270         SUBD  #$010E              *Subtract 270
               BRA   MAG                 *Take magnitude
SUB360         SUBD  #$0168              *Subtract 360
MAG            COMA                      *1's complement ACCA
               COMB                      *1's complement ACCB
               ADDD  #$0001              *Add 1 to ACCD=ACCA:ACCB
STORE          STD   THETA               *Store result in THETA
               RTS
```

```
*****************************************************************
* DISPLAY - Takes angle in THETA, converts it to BCD, and outputs it to *
* display at port B.                                            *
*****************************************************************
```

```
DISPLAY        LDX   COUNT               *X gets counter
               DEX                       *Decrement counter
               STX   COUNT               *Store counter
               CPX   #$00                *Counter equal zero?
               BNE   SKIP                *No.. do not display
               LDX   #PERIOD             *X gets display period
               STX   COUNT               *Store counter
               LDD   THETA               *ACCD gets THETA
               BSR   BIN_BCD             *Convert to BCD
               LDX   #PORTB              *X points at port B
               LDAA  HUNDREDS            *ACCA gets Hundreds
               STAA  PORTB               *Write to port B
               BSET  0,X,$40             *Pulse B6
               BCLR  0,X,$40
               LDAA  TENS                *ACCA gets Tens
               STAA  PORTB               *Write to port B
               BSET  0,X,$20             *Pulse B5
               BCLR  0,X,$20
               LDAA  ONES                *ACCA gets Ones
               STAA  PORTB               *Write to port B
               BSET  0,X,$10             *Pulse B4
               BCLR  0,X,$10
SKIP           RTS
```

```
*****************************************************************
* BIN_BCD - Takes a 16-bit number in ACCD and converts it to 3 bytes  *
* whose least significant nibble is the BCD code for the hundreds,    *
* tens, and ones place of THETA.  Returns with BCD digits in HUNDREDS,*
* TENS, and ONES.                                               *
*****************************************************************
```

```
BIN_BCD        LDY   #$0003              *Initialize counter
NEXT           DEY                       *Decrement counter
               LDX   #$000A              *Load X with 10d
               IDIV                      *Divide ACCD by X
               PSHB                      *Push BCD digit to stack
               XGDX                      *Swap ACCD and X
               CPY   #$0000              *Test counter for zero
               BNE   NEXT                *If not $00, do next
               PULA                      *Pull hundreds off stack
               STAA  HUNDREDS            *Store in memory
               PULA                      *Pull tens off stack
               STAA  TENS                *Store in memory
               PULA                      *Pull ones off stack
               STAA  ONES                *Store in memory
               RTS
```

```
*****************************************************************
* ROBOT - Takes the heading angle in THETA, converts it to an 8 bit   *
* approximation of the angle, and outputs it to Port C.         *
*****************************************************************
```

```
ROBOT          LDD   THETA               *ACCD gets heading angle
               LDX   #$168               *X gets 360 degrees
               FDIV                      *X gets ACCD / X
```

```
            XGDX              *D gets quotient
            STAA  PORTC       *Port C gets 256 * quotient
            LDX   #PORTB      *Pulse B7 to enable latch inputs
            BSET  0,X,$80
            BCLR  0,X,$80
            RTS

***********************************************************************
* ARCTAN LOOKUP TABLE - Data is the Arctangent of 1/64 of the one byte *
* address offset                                                      *
***********************************************************************

TABLESTART  FCB   $00,$01,$02,$03,$04,$05,$05,$06,$07
            FCD   $08,$09,$0A,$0B,$0B,$0C,$0D,$0E,$0F
            FCB   $10,$11,$11,$12,$13,$14,$15,$15,$16
            FCB   $17,$18,$18,$19,$1A,$1B,$1B,$1C,$1D
            FCB   $1D,$1E,$1F,$1F,$20,$21,$21,$22,$23
            FCB   $23,$24,$24,$25,$25,$26,$27,$27,$28
            FCB   $28,$29,$29,$2A,$2A,$2B,$2B,$2C,$2C
            FCB   $2D,$2D
```

C Language Source Code for smart Compass in MC68HC11

```c
/*~~~~~~~~~~~~~~~~~~~~~~~~~~~~~~~~~~~~~~~~~~~~~~~~~~~~~~~~~~~~~~~~~~~~~~~~~*/
/* SmrtComp.C                                                          */
/*                                                                     */
/* Original: KJH, 3/31/91                                              */
/* Last Update:                                                        */
/*                                                                     */
/* C language implementation of MC68HC11 SmartCompass Program          */
/* Derived from Assembly language program by M. Gilstead               */
/* A/D addresses changed to work with Motorola M68HC11EVB              */
/*~~~~~~~~~~~~~~~~~~~~~~~~~~~~~~~~~~~~~~~~~~~~~~~~~~~~~~~~~~~~~~~~~~~~~~~~~*/

#include "c:\introl\include\math.h"

/*#define ADinAssembly*/

#ifndef ADinAssembly
#define ADinC
#endif

/*~~~~~~~~~~~~~~~~~~~~~~~~~~~~~~~~~~~~~~~~~~~~~~~~~~~~~~~~~~~~~~~~~~~~~~~~~*/
/*                  Function Prototypes                                */
/*~~~~~~~~~~~~~~~~~~~~~~~~~~~~~~~~~~~~~~~~~~~~~~~~~~~~~~~~~~~~~~~~~~~~~~~~~ */
#ifdef ADinAssembly
void InputVectors(void);
#endif


/*~~~~~~~~~~~~~~~~~~~~~~~~~~~~~~~~~~~~~~~~~~~~~~~~~~~~~~~~~~~~~~~~~~~~~~~~~*/
/*                  Variables                                          */
/*~~~~~~~~~~~~~~~~~~~~~~~~~~~~~~~~~~~~~~~~~~~~~~~~~~~~~~~~~~~~~~~~~~~~~~~~~*/
#ifdef ADinAssembly
extern unsigned char AbsTran; /* variables defined in Assy             */
extern unsigned char AbsLong; /* language module                       */
extern unsigned char MagRef;
#endif

int    Transverse;          /* values read from coils by               */
int    Longitudinal;             /* A/D converter                       */
int    MagneticReference;
int    AbsoluteTransverse;       /* absolute values of readings         */
int    AbsoluteLongitudinal;
```

```
float Phi;                  /* computed values                    */
float Theta;
int   Octant;
float Heading;

unsigned char     *OptionRegisterAddress; /* Hardware Addresses    */
unsigned char     *DataDirectionRegisterC;
unsigned char     *ADAddress;
unsigned char     *LongitudinalAddress;
unsigned char     *TransverseAddress;
unsigned char     *ReferenceAddress;


#define ADRTRAN    0x1032     /*A/D Result -Transverse          */
#define ADRLONG    0x1033     /*A/D Result -Longitudinal        */
#define ADRREF     0x1034     /*A/D Result -Reference           */
#define OPTIONWD   0x90       /* Choose start up opts.          */
#define ADCTLWD    0x10       /* A/D Control word               */
#define CCF    0x80           /* Conversion complete            */
#define H11OPTION 0x1039
#define H11DDRC   0x1007
#define H11ADCTL  0x1030
#define DEGREES_PER_RADIAN 57.29577


/*~~~~~~~~~~~~~~~~~~~~~~~~~~~~~~~~~~~~~~~~~~~~~~~~~~~~~~~~~~~~~~~~~~~~~~~~~~~*/
void InitializeSmartCompass(void)

{/* initialize pointers */
 OptionRegisterAddress   = (unsigned char *)H11OPTION;
 DataDirectionRegisterC  = (unsigned char *)H11DDRC;
 ADAddress               = (unsigned char *)H11ADCTL;
 LongitudinalAddress     = (unsigned char *)ADRLONG;
 TransverseAddress       = (unsigned char *)ADRTRAN;
 ReferenceAddress        = (unsigned char *)ADRREF;
 *OptionRegisterAddress = OPTIONWD;        /* Power up A/D          */
 *DataDirectionRegisterC= 0xFF;            /* Set Port C for Output  */
}

/*~~~~~~~~~~~~~~~~~~~~~~~~~~~~~~~~~~~~~~~~~~~~~~~~~~~~~~~~~~~~~~~~~~~~~~~~~~~*/
int CalculateOctant(void)

/* Determines which of the 8 octants the vector is in. Returns      */
/* the octant                                                       */

{
 if(Transverse < MagneticReference)
  {                         /* WestHalf = TRUE;                      */
   if (Longitudinal < MagneticReference)
    {                       /* SouthWestQuad = TRUE;                 */
     if (AbsoluteTransverse <= AbsoluteLongitudinal)
      return (5);
     else
      return (6);
    }
   else
    {                       /* NorthWestQuad = TRUE;                 */
     if (AbsoluteTransverse <= AbsoluteLongitudinal)
      return (8);
     else
      return (7);
    }
  }
 else
  {                         /* EastHalf = TRUE;                      */
   if (Longitudinal < MagneticReference)
    {                       /* SouthEastQuad = TRUE;                 */
     if (AbsoluteTransverse <= AbsoluteLongitudinal)
      return (4);
```

```
    else
     return (3);
    }
   else
   {                            /* NorthEastQuad = TRUE;                        */
    if (AbsoluteTransverse <= AbsoluteLongitudinal)
     return (1);
    else
     return (2);
   }
  }
}


/*~~~~~~~~~~~~~~~~~~~~~~~~~~~~~~~~~~~~~~~~~~~~~~~~~~~~~~~~~~~~~~~~~~~~~~~~~~~~~*/
float CalculatePhi(void)

/* Divides vector components and takes the arctan and returns               */
/* the angle PHI.                                                           */

{
float Ratio;

 if (AbsoluteTransverse > AbsoluteLongitudinal)
  Ratio = (float)AbsoluteLongitudinal / (float)AbsoluteTransverse;
 else
  Ratio = (float)AbsoluteTransverse / (float)AbsoluteLongitudinal;
 return (atan(Ratio)*DEGREES_PER_RADIAN);
}


/*~~~~~~~~~~~~~~~~~~~~~~~~~~~~~~~~~~~~~~~~~~~~~~~~~~~~~~~~~~~~~~~~~~~~~~~~~~~~~*/
float CalculateTheta(float Phi, int Octant)

{
 switch (Octant)
   {
   case 1:    return(Phi);
   case 2:    return(Phi - 90.0);
   case 3:    return(Phi + 90.0);
   case 4:    return(Phi - 180.0);
   case 5:    return(Phi + 180.0);
   case 6:    return(Phi - 270.0);
   case 7:    return(Phi + 270.0);
   case 8:    return(Phi - 360.0);
   default:   {
                printf("error in theta\n, Octant = %d\n",Theta,Octant);
                return(-1);
                }
   }
}


/*~~~~~~~~~~~~~~~~~~~~~~~~~~~~~~~~~~~~~~~~~~~~~~~~~~~~~~~~~~~~~~~~~~~~~~~~~~~~~*/
float ComputeHeading(void)

{
 Octant = CalculateOctant();
 Phi    = CalculatePhi();
 Theta  = CalculateTheta(Phi,Octant);
 return(Theta);
}


/*~~~~~~~~~~~~~~~~~~~~~~~~~~~~~~~~~~~~~~~~~~~~~~~~~~~~~~~~~~~~~~~~~~~~~~~~~~~~~*/
void DisplayHeading(void)

{
 printf("The heading is %d, the octant is %d.\n",(int)Heading,Octant);
}
```

```
#ifdef ADinC
/*~~~~~~~~~~~~~~~~~~~~~~~~~~~~~~~~~~~~~~~~~~~~~~~~~~~~~~~~~~~~~~~~~~~~~~~~~~~~*/
void InputVectorsC(void)

/* reads the values of the coils using the internal A/D             */
/* stores the values in global variables                           */

{
int   Difference;

  *ADAddress = ADCTLWD;          /* initialize A/D converters      */
  while (!(*ADAddress & CCF)); /*Wait for conversion complete      */
  MagneticReference = (int)*ReferenceAddress;    /* read           */
  Transverse    = (int)*TransverseAddress;       /* coils          */
  Longitudinal  = (int)*LongitudinalAddress;
  Difference    = Transverse - MagneticReference; /* abs           */
  AbsoluteTransverse  = fabs(Difference);         /* value         */
  Difference          = Longitudinal - MagneticReference;
  AbsoluteLongitudinal = fabs(Difference);
}
#endif             /*ADinC*/


/*~~~~~~~~~~~~~~~~~~~~~~~~~~~~~~~~~~~~~~~~~~~~~~~~~~~~~~~~~~~~~~~~~~~~~~~~~~~~*/
int main()

{
 InitializeSmartCompass();
 for (;;)                    /* do forever                        */
  {
#ifdef ADinAssembly
  InputVectors();
  AbsoluteTransverse    = AbsTran;
  AbsoluteLongitudinal  = AbsLong;
  MagneticReference     = MagRef;
#endif
#ifdef ADinC
  InputVectorsC();
#endif
  Heading = ComputeHeading();
  DisplayHeading();
  }
}
```

Assembly Language Source Code for A/D
Input to Link with C Language Smart
Compass in MC68HC11

```
*************************************************************************
* NAME:   SC.s11                                                       *
*                                                                      *
* BY:   K. J. Hintz                                                    *
* DATE BEGUN:   March 31, 1991                                         *
* LAST UPDATE: April 1, 1991                                           *
*                                                                      *
* This program reads in two binary value representing the orthogonal   *
* components of the earth's magnetic field as well as a reference      *
* voltage.  This is an assembly language program which is meant to be  *
* called by a C program to do the rest of the calculations.           *
*                                                                      *
* This program is based on the original Smartcompass program written  *
* by Doug Garner for an Intel 8748, and M. Gilstead's version for      *
* a Motorola 68HC811A2.                                                *
* A/D addresses changed to work with Motorola M68HC11EVB              *
*************************************************************************
```

```
ADCTL        EQU     $1030 *A/D Control register
ADRTRAN              EQU     $1032 *A/D Result -Transverse
ADRLONG      EQU     $1033 *A/D Result -Longitudinal
ADRREF       EQU     $1034 *A/D Result -Reference
ADCTLWD      EQU     $10   *A/D Control word
CCF          EQU     $80   *Conversion complete mask

      section .text
AbsTran:     RMB     1       *exportable Magnitude of Transverse vec.
AbsLong:     RMB     1       *exportable Magnitude of Longitud. vec.
MagRef:              RMB     1       *exportable magnetic reference value

************************************************************************
* InputVectors - Initiates and reads results of A/D conversion of the 3 *
* channels.  Puts values read in AbsTran, AbsLong, and MagRef.          *
************************************************************************

InputVectors:        LDAA  #ADCTLWD     *initialize A/D converters
             STAA  ADCTL          *start A/D
             LDX   #ADCTL         *Wait for conversion complete
NOTRDY               BRCLR 0,X,CCF,NOTRDY
             LDAA  ADRREF         *get magnetic reference
             STAA  MagRef         *save it
             LDAA  ADRTRAN        *Put Transverse in ACCA
             SUBA  MagRef         *Subtract Reference
             BHS   TPOSITIVE      *If positive, jump
             NEGA                 *Else, take magnitude
TPOSITIVE    STAA  AbsTran        *Store |Transverse|
             LDAB  ADRLONG        *Put Longitudinal in ACCB
             SUBB  MagRef         *Subtract Reference
             BHS   LPOSITIVE      *If positive, jump
             NEGB                 *Else, Take magnitude
LPOSITIVE    STAB  AbsLong        *Store |Longitudinal|
             RTS
```

APPENDIX II: UAV CODE

```
/*~~~~~~~~~~~~~~~~~~~~~~~~~~~~~~~~~~~~~~~~~~~~~~~~~~~~~~~~~~~~~~~~~~~~~~~~~~*/
/* Partial C coding UAV Autopilot                                       */
/* for MC68HC11                                                         */
/*                                                                      */
/*~~~~~~~~~~~~~~~~~~~~~~~~~~~~~~~~~~~~~~~~~~~~~~~~~~~~~~~~~~~~~~~~~~~~~~~~~~*/
/*~~~~~~~~~~~~~~~~~~~~~~~~~~~~~~~~~~~~~~~~~~~~~~~~~~~~~~~~~~~~~~~~~~~~~~~~~~*/
/*                     Defines                                          */
/*~~~~~~~~~~~~~~~~~~~~~~~~~~~~~~~~~~~~~~~~~~~~~~~~~~~~~~~~~~~~~~~~~~~~~~~~~~*/
#ifndef MAX_HISTORY
#define MAX_HISTORY 10
#endif

#ifndef FALSE
#define FALSE 0
#endif

#define TRUE !FALSE

#define ONE_MILLISECOND   _____       /* about counter mode assumptions    */
#define TWO_MILLISECOND   _____
#define THREE_MILLISECOND _____
#define COUNTER_PHYSICAL_ADDRESS  ____       /* Address of Master Counter  */
#define MAX_COUNTER_VALUE 65535    /* 16 bit counter overflow value     */
#define TCTL2            1021H     /* address of Timer Control Reg2      */
#define TMSK2            1024H     /* Address of Timer Mask 2            */
#define TIC1LOW          1011H     /* Low byte of Timer Input           */
                                   /* capture register One              */
#define TIC1HIGH 1010H        /* High byte of Timer Input              */
                              /* capture register One                  */

#define PASS_LOW_BYTE    0FFH      /* AND with int to keep low byte     */
#define PASS_HIGH_BYTE   0FF00H    /* AND with int to keep hi byte      */
#define BYTE_LENGTH      8         /* length of byte for shifting       */
#define FOREVER          TRUE      /* Conditon for infinite while       */
```

```
#define TMSK_TWO_INIT    0              /* prescale by 1 and no interrupt    */
                                        /* assumes _____ MHz clock           */
                                        /* which yields _____ ns/clock       */


/*~~~~~~~~~~~~~~~~~~~~~~~~~~~~~~~~~~~~~~~~~~~~~~~~~~~~~~~~~~~~~~~~~~~~~~~~~~~~~*/
/*                    Global Variables                                      */
/*~~~~~~~~~~~~~~~~~~~~~~~~~~~~~~~~~~~~~~~~~~~~~~~~~~~~~~~~~~~~~~~~~~~~~~~~~~~~~*/
int    PassedSelfTest;     /* True after self test passed, initially FALSE */
int    Synchronized;       /* True if program synchronized to              */
                           /* incoming pulse train                         */
int    ChannelOne;         /* True if Channel 1 pulse width has            */
                           /* been received                                */
int    ChannelTwo;         /* True if Channel 2 pulse width has            */
                           /* been received                                */
int    PreviousPlaces[MAXHISTORY];  /*Array to store history               */
                                    /*of places, 0 is most                 */
                                    /*recent place visited                 */
unsigned int ChannelOnePulseWidth;    /* measured duration of Channel      */
                                      /* one pulse width                   */
unsigned int ChannelTwoPulseWidth;    /* measured duration of Channel      */
                                      /* two pulse width                   */
unsigned int ChannelThreePulseWidth;   /* measured duration of Channel     */
                                       /* three pulse width                */
unsigned int TemporaryPulseWidth;     /* Pulse Width as read but not       */
                                      /* yet associated with channel       */
unsigned int CounterBeginValue;       /* Value of Counter when timer       */
                                      /* is started                        */
unsigned int CounterEndValue;         /* Value of Counter when edge        */
                                      /* received                          */
unsigned int  *TimerOneAddress;       /* pointer to hardware counter       */
unsigned char *TimerControlRegisterTwoAddress; /* pointer to Timer         */
                                      /*control register 2                 */
unsigned char *TimerInterruptMaskRegisterTwo;


/*~~~~~~~~~~~~~~~~~~~~~~~~~~~~~~~~~~~~~~~~~~~~~~~~~~~~~~~~~~~~~~~~~~~~~~~~~~~~~*/
/*                    Function Prototypes                                   */
/*~~~~~~~~~~~~~~~~~~~~~~~~~~~~~~~~~~~~~~~~~~~~~~~~~~~~~~~~~~~~~~~~~~~~~~~~~~~~~*/
unsigned int ReadPulseWidth(void);
void             TransitionThree(void);
int              TransitionSix(void);
void             TransitionFifteen(void);
unsigned int     ReadAndSwapEndianBytes(unsigned int *HardwareAddress);


/*~~~~~~~~~~~~~~~~~~~~~~~~~~~~~~~~~~~~~~~~~~~~~~~~~~~~~~~~~~~~~~~~~~~~~~~~~~~~~*/
/*                    Functions                                            */
/*~~~~~~~~~~~~~~~~~~~~~~~~~~~~~~~~~~~~~~~~~~~~~~~~~~~~~~~~~~~~~~~~~~~~~~~~~~~~~*/
void TransitionOne(void)

{
...
}


/*~~~~~~~~~~~~~~~~~~~~~~~~~~~~~~~~~~~~~~~~~~~~~~~~~~~~~~~~~~~~~~~~~~~~~~~~~~~~~*/
int TransitionTwo(void)

{
int TestValue;

...
 return(TestValue);
}
```

```
/*~~~~~~~~~~~~~~~~~~~~~~~~~~~~~~~~~~~~~~~~~~~~~~~~~~~~~~~~~~~~~~~~~~~~~~~~~~~*/
void TransitionThree()

{
 Synchronized      = FALSE;
 ChannelOne        = FALSE;
 ChannelTwo        = FALSE;
 For (LocalIndex = 0;
      LocalIndex < MAX_HISTORY;
      LocalIndex ++)
       PreviousPlaces[LocalIndex];
 ChannelOnePulseWidth   = 0;
 ChannelTwoPulseWidth   = 0;
 ChannelThreePulseWidth = 0;
 TemporaryPulseWidth    = 0;
 CounterBeginValue      = 0;
 CounterEndValue        = 0;
 TimerOneAddress        = null;
 TimerInterruptMaskRegisterTwo     = TMSK2;
 TimerControlRegisterTwoAddress    = TCTL2HIGH;      /* big endian    */
 TimerOneAddress                   = TIC1HIGH;
 *TimerInterruptMaskRegisterTwo = TMSK_TWO_INIT;     /* prescale 1    */
                                                     /* no ints       */
}

/*~~~~~~~~~~~~~~~~~~~~~~~~~~~~~~~~~~~~~~~~~~~~~~~~~~~~~~~~~~~~~~~~~~~~~~~~~~~*/
void TransitionFour(void)

{
...
...
}

/*~~~~~~~~~~~~~~~~~~~~~~~~~~~~~~~~~~~~~~~~~~~~~~~~~~~~~~~~~~~~~~~~~~~~~~~~~~~*/
int TransitionSix(void)

{
 ChannelOnePulseWidth = TemporaryPulseWidth;
 ChannelOne = TRUE;
}

/*~~~~~~~~~~~~~~~~~~~~~~~~~~~~~~~~~~~~~~~~~~~~~~~~~~~~~~~~~~~~~~~~~~~~~~~~~~~*/
void TransitionFifteen(void)

{

 TemporaryPulseWidth = ReadPulseWidth(CounterBeginValue);
 if (TemporaryPulseWidth < ONE_MILLISECOND)
     RealTimeInterrupt = FALSE;
 if ( (TemporaryPulseWidth >= ONE_MILLISECOND) &
      (TemporaryPulseWidth < TWO_MILLISECOND) )
     RealTimeInterrupt = FALSE;
 if (TemporaryPulseWidth >= TWO_MILLISECOND)
     {
      RealTimeInterrupt = FALSE;
      Synchronized = TRUE;
     }
}


/*~~~~~~~~~~~~~~~~~~~~~~~~~~~~~~~~~~~~~~~~~~~~~~~~~~~~~~~~~~~~~~~~~~~~~~~~~~~*/
unsigned int ReadAndSwapEndianBytes(unsigned int *HardwareAddress)

/* MC68HC11 requires read of Hight Byte first--verify that this         */
/* compiles to implement that sequence                                  */

 {
```

```c
unsigned short int LocalCounterValue;
unsigned short int LowByte;
unsigned short int HighByte;

 LocalCounterValue = *HardwareAddress;
 LowByte = (LocalCounterValue >> BYTE_LENGTH) & PASS_LOW_BYTE;
 HighByte = (LocalCounterValue << BYTE_LENGTH) & PASS_HIGH_BYTE;
 LocalCounterValue = HighByte | LowByte;
 return(LocalCounterValue);
}
/*~~~~~~~~~~~~~~~~~~~~~~~~~~~~~~~~~~~~~~~~~~~~~~~~~~~~~~~~~~~~~~~~~~~~~~~*/
unsigned int ReadPulseWidth(int CounterBeginValue)

{
unsigned int LocalPulseWidth;

 CounterEndValue = ReadAndSwapEndianBytes(TimerOneAddress);
 if (CounterEndValue >= CounterBeginValue)
        LocalPulseWidth = CounterEndValue - CounterBeginValue;
 else
    LocalPulseWidth = MAX_COUNTER_VALUE -
                      (CounterBeginValue - CounterEndValue);
 return(LocalPulseWidth);
}


/*~~~~~~~~~~~~~~~~~~~~~~~~~~~~~~~~~~~~~~~~~~~~~~~~~~~~~~~~~~~~~~~~~~~~~~~*/
int main(void)

{
 PassedSelfTest = FALSE;
 while(!PassedSelfTest)
  {TransitionOne();                       /* configure hardware        */
   PassedSelfTest = TransitionTwo();      /* Self-test                 */
  }
 TransitionThree();                       /* Initialize all variables  */
 for (;;)
  {
   if(Synchronized & !ChannelOne & !ChannelTwo)
       TransitionSix();
   else if(Synchronized & ChannelOne & !ChannelTwo)
       TransitionSeven();
   else if(Synchronized & ChannelOne & ChannelTwo)
       TransitionSeventeen();
   else if(!Synchronized & RealTimeInterrupt & RealTimeInterruptEnable)
       TransitionFive();
   else if(Synchronized & RealTimeInterrupt & RealTimeInterruptEnable)
       TransitionFifteen();

   else

       ;

  }                       /* end while                                 */
 return(0);               /* Normal return to operating system         */
}                         /* end main                                  */
```

APPENDIX III: LINKER

```
/*-------------------------------------------------------------------*/
/* File:  KJH.ld                                                     */
/* copied from c.ld and modified for Buffalo board, 2/10/90          */
/*                                                                   */
/*     vers=1.0                                                      */
/*                                                                   */
/*     KJH.ld - command file for C for 68HC11                        */
/*                                                                   */
/*     Configuration is a Motorola MC68HC11EVB evaluation board.     */
/*     This is designed for programs that are used WITH the          */
/*     BUFFALO monitor.  The executable code and initialized data will */
/*        placed in an 8Kx8 RAM at 0xC000.                           */
/*                                                                   */
/*     Modification history:                                         */
/*       KJH:  included modified printf and ofmt 2/14/90             */
/*       KJH:  KJH files are now assumed to be in c:\introl\kjh, 3/15/91*/
/*                                                                   */
/*                                                                   */
/*-------------------------------------------------------------------*/

set H11RAM = 0x00;                      /* page containing 68HC11 base page*/
set H11REG = 0x01;                      /* page containing 68HC11 registers*/

set H11VECSIZE = 42;                    /* size of 68HC11 vectors       */
set H11RAMREG = (H11RAM<<4)|H11REG;     /*mask for setting H11INIT in start.s*/
set H11REGORG = (H11REG<<12)            /* origin of 68HC11 registers   */
set H11VECORG = 0x10000-H11VECSIZE;     /* origin of 68HC11 vectors     */

section .base bss origin 0;           /* uninitialized base page storage */
section .text origin 0xC000;          /* exec. code put in 8K RAM of EVB */
section .bss origin endof (.text) bss comms;/* foll. by uninit storage */
section .data origin endof (.bss);         /* followed by data          */
section .const origin endof(.data);        /* followed by constants     */
section .strings origin endof(.const);     /* followed by strings       */
section .init origin endof(.strings);      /* data to be copied to RAM  */
section .heap bss origin endof (.init) minsize 128; /* followed by heap */

/*-------------------------------------------------------------------*/
/* The following line can be used to copy data from ROM into RAM     */
/* substitute in place of the section .data line above where the     */
/* underscore is replaced with the address of the ROM where the data */
/* is to be stored                                                   */
/*                                                                   */
/* section .data origin _____ copiedfrom .init = .data;             */
/*-------------------------------------------------------------------*/


/*-------------------------------------------------------------------*/
/* Sections particular to the registers and vectors in the EVB.      */
/* Used by the assembler to locate the registers and interrupt vectors */
/* at their proper location in memory since I/O is memory mapped.    */
/*-------------------------------------------------------------------*/

section .H11REGS bss origin H11REGORG;
section .H11VEC data origin H11VECORG maxsize H11VECSIZE;


/*-------------------------------------------------------------------*/
/* Checks for sections which exceed the limits of RAM on the EVB     */
/* If the optional RAM is installed, then modify the limits accordingly */
/*-------------------------------------------------------------------*/

check endof(.text)  > 0xdfff fatal "Code area too large";
check endof(.heap)  > 0xdfff fatal "Program too large for RAM";
check sizeof(.base) > 256 fatal "Base page too large";
check endof(.bss)   > 0xDFFF fatal "Uninitialized variable space too large";
```

```
/*------------------------------------------------------------------*/
/* Initializes some values which are used by the initialization code to */
/* initialize various areas of RAM                                   */
/*------------------------------------------------------------------*/

set _ramstart = startof(.bss);
set _ramend = endof(.bss);
set _heapstart = startof(.heap);
set _heapend = endof(.heap);
set _stackstart = endof(.heap);
set _stackend = 0xe000;          /* 0xdfff is the end of the RAM         */
set _initstart = startof(.init);
set _initend = endof(.init);
set _datastart = startof(.data);
set _dataend = endof(.data);

check _stackend - _stackstart < 64 fatal "Stack too small";


/*------------------------------------------------------------------*/
/*                                                                    */
/* These object files replace functions in the library that are unique */
/* to the Buffalo monitor version                                     */
/*                                                                    */
/* kjhstart.o, kjhiob.o and kjhbuff.o are recompiled versions of      */
/* (kjh)start.s, (kjh)iob.s, and (kjh)buffalo.s which allows using the */
/* Buffalo monitor routines for I/O                                   */
/*                                                                    */
/* printf.o has been compiled to use the BUFFALO I/O Calls and should */
/* not be included unless necessary since it takes up quite a bit of  */
/* space.                                                             */
/*                                                                    */
/* ofmt.o has been compiled with the #define FLOATS  and LONGS omitted */
/* to reduced size of code.   New name:  SHRTOFMT.O                   */
/*                                                                    */
/* ifmt.o has been compiled with the #define FLOATS  and LONGS omitted */
/* to reduced size of code.   New name:  SHRTIFMT.O                   */
/*                                                                    */
/*------------------------------------------------------------------*/

'c:\introl\kjh\kjhstart.o'    /Startup routine                          */
'c:\introl\kjh\kjhiob.o'         /* buffalo input output calls          */
'c:\introl\kjh\kjhbuff.o'
'c:\introl\kjh\shrtofmt.o'               /* KJH: shorter version of printf */
'c:\introl\kjh\shrtifmt.o'               /* KJH: shorter version of scanf  */

readline;              /* read the command line                        */

/*------------------------------------------------------------------*/
/* inclusion of the standard libraries to resolve external references */
/*                                                                    */
/*                                                                    */
/*------------------------------------------------------------------*/

-lc                      /* use the C library                           */
-lcio                    /* C i/o library                               */
-lm                      /* math library                                */
-lgen                    /* general library                             */
```

APPENDIX IV: IRTRACK

```
/*~~~~~~~~~~~~~~~~~~~~~~~~~~~~~~~~~~~~~~~~~~~~~~~~~~~~~~~~~~~~~~~~~~~~~~~~~*/
/* Partial C coding IR Tracker                                         */
/* for MC68HC11                                                        */
/*                                                                     */
/*~~~~~~~~~~~~~~~~~~~~~~~~~~~~~~~~~~~~~~~~~~~~~~~~~~~~~~~~~~~~~~~~~~~~~~~~~*/

/*~~~~~~~~~~~~~~~~~~~~~~~~~~~~~~~~~~~~~~~~~~~~~~~~~~~~~~~~~~~~~~~~~~~~~~~~~*/
/*                          Defines                                    */
/*~~~~~~~~~~~~~~~~~~~~~~~~~~~~~~~~~~~~~~~~~~~~~~~~~~~~~~~~~~~~~~~~~~~~~~~~~*/
#ifndef COMMENTS
#define COMMENTS                 /* if defined, prints comments        */
#endif

#ifndef FALSE
#define FALSE 0
#endif

#define TRUE !FALSE

#define ZERO_OFFSET        0x7F     /* 2.5 volt zero on A/D             */
#define DEGREES_PER_RADIAN 57.29577
#define COUTPUT            0xFF     /* All C Ports set to output        */
#define SPI_ENABLE         0x80     /* SPI enable                       */
#define SPI_ENABLE_INT     0xC0     /* Enable SPI with Interrupts       */
#define START_OF_MESSAGE   0xFF     /* First Byte Send in all Radar Msgs */
#define MODE_INDEX         1        /* location of mode in Radar Msg    */
#define TIMER_OVERFLOW_VECTOR  4         /* INTROL Specific 6811 vector  */
#define SPI_VECTOR         1            /* INTROL Specific 6811 vector   */
/*~~~~~~~~~~~~~~~~~~~~~~~~~~~~~~~~~~~~~~~~~~~~~~~~~~~~~~~~~~~~~~~~~~~~~~~~~*/
/*                        Hardware Addresses                           */
/*~~~~~~~~~~~~~~~~~~~~~~~~~~~~~~~~~~~~~~~~~~~~~~~~~~~~~~~~~~~~~~~~~~~~~~~~~*/
#define DELTAX_ADDR        0x1032        /*A/D Result -Delta X          */
#define DELTAY_ADDR        0x1033        /*A/D Result -Delta Y          */
#define IRLOCK_ADDR        0x1034        /*A/D Result -IRLock           */
#define OPTIONWD           0x90          /* Choose start up opts.       */
#define ADCTLWD            0x10          /* A/D Control word            */
#define CCF                0x80          /* Conversion complete         */
#define H11DDRC            0x1007        /* Port C Data Direction Reg   */
#define H11ADCTL           0x1030        /* A/D Control Register        */
#define ADR1               0x1031        /* A/D Result Register 1       */
#define ADR2               0x1032        /* A/D Result Register 2       */
#define ADR3               0x1033        /* A/D Result Register 3       */
#define ADR4               0x1034        /* A/D Result Register 4       */
#define SPCR               0x1028        /* SPI Control Register        */
#define SPSR               0x1029        /* SPI Status Register         */
#define SPDR               0x102A        /* SPI Data Register           */
#define H11OPTION          0x1039        /* Option Register             */
#define MOUNT_STATE_ADDRESS    0x6000    /* Mount Flags                 */
#define MOUNT_CONTROL_ADDRESS  0X6080    /* Mount D/A for control       */

/*~~~~~~~~~~~~~~~~~~~~~~~~~~~~~~~~~~~~~~~~~~~~~~~~~~~~~~~~~~~~~~~~~~~~~~~~~*/
/*                      Structure Definitions                          */
/*~~~~~~~~~~~~~~~~~~~~~~~~~~~~~~~~~~~~~~~~~~~~~~~~~~~~~~~~~~~~~~~~~~~~~~~~~*/
struct MountStateType
  {
  int Azimuth;
  int Elevation;
  int AzDot;
  int ElDot;
  long int StartTime;
  long int EndTime;
  };

struct MountControlType
  {
  unsigned short int Azimuth;
  unsigned short int Elevation;
  };
```

```
struct VideoTrackerStateType
   {
   short int DeltaX;
   short int DeltaY;
   short int VideoTrackerLock;
   long int Time;
   };


/*~~~~~~~~~~~~~~~~~~~~~~~~~~~~~~~~~~~~~~~~~~~~~~~~~~~~~~~~~~~~~~~~~~~~~~~~~~~~~~~~*/
/*               Uninitialized Global Variables                               */
/*~~~~~~~~~~~~~~~~~~~~~~~~~~~~~~~~~~~~~~~~~~~~~~~~~~~~~~~~~~~~~~~~~~~~~~~~~~~~~~~~*/
struct   MountStateType          CurrentMount;
struct   MountStateType          PreviousMount;
struct   MountStateType          PredictedMount;
struct   MountControlType        MountControl;
struct   VideoTrackerStateType   CurrentVideoTracker;
struct   VideoTrackerStateType   PreviousVideoTracker;
struct   VideoTrackerStateType   PredictedVideoTracker;
char     IncomingPositionMessage[];      /* msg assembled here              */
char     PositionMessage[];           /* double buffered message input      */

/*~~~~~~~~~~~~~~~~~~~~~~~~~~~~~~~~~~~~~~~~~~~~~~~~~~~~~~~~~~~~~~~~~~~~~~~~~~~~~~~~*/
/*               Uninitialized Global Variables                               */
/*~~~~~~~~~~~~~~~~~~~~~~~~~~~~~~~~~~~~~~~~~~~~~~~~~~~~~~~~~~~~~~~~~~~~~~~~~~~~~~~~*/
int PM              = FALSE;    /* Position Mode Semaphore          */
int TM              = FALSE;    /* Track Mode Semaphore             */
int IE              = FALSE;    /* Interrupt Enable Semaphore       */
int SRMsg           = FALSE;    /* Surv Radar Message Semaphore     */
int SRMsgRx         = FALSE;    /* Surv Radar Message Received Sem. */
int PST             = FALSE;    /* Passed Self-Test Semaphore       */
int EOM             = TRUE;     /* End of Message Semaphore         */
int RETIE           = FALSE;    /* Real time interrupt enable       */
int SOMReceived     = FALSE;    /* Start of Message received semaphore*/
int MessageLength   = 0;        /* expected length of radar message */
int MessageIndex    = 0;        /* number of characters actually rxd */
long int RealTime   = 0;        /* 32-bit real time clock           */

/*~~~~~~~~~~~~~~~~~~~~~~~~~~~~~~~~~~~~~~~~~~~~~~~~~~~~~~~~~~~~~~~~~~~~~~~~~~~~~~~~*/
/*                            Hardware Pointer Variables                      */
/*~~~~~~~~~~~~~~~~~~~~~~~~~~~~~~~~~~~~~~~~~~~~~~~~~~~~~~~~~~~~~~~~~~~~~~~~~~~~~~~~*/
unsigned char            *SPIControlAddress;
unsigned char            *SPIStatusAddress;
unsigned char            *SPIDataAddress;
unsigned char            *OptionRegisterAddress;
unsigned char            *DDRCAddress;
unsigned char            *ADControlAddress;
unsigned char            *AzDotAddress;
unsigned char            *ElDotAddress;
unsigned char            *IRLockAddress;
unsigned char            *DeltaXAddress;
unsigned char            *DeltaYAddress;
struct MountStateType    *MountStateAddress;
struct MountControlType  *MountControlAddress;

/*~~~~~~~~~~~~~~~~~~~~~~~~~~~~~~~~~~~~~~~~~~~~~~~~~~~~~~~~~~~~~~~~~~~~~~~~~~~~~~~~*/
/*                            Function Prototypes                             */
/*~~~~~~~~~~~~~~~~~~~~~~~~~~~~~~~~~~~~~~~~~~~~~~~~~~~~~~~~~~~~~~~~~~~~~~~~~~~~~~~~*/
      int TransitionOne(void);
      int TransitionTwo(void);
      int TransitionThree(void);
      void TransitionFour(void);
      void TransitionFive(void);
      void TransitionSix(void);
      void ComputeControl(struct MountStateType *PreviousMount,
                          struct MountStateType *CurrentMount,
                          struct MountControlType *MountControl);
      int CheckSum();
      void SPIInterruptServiceRoutine(void);
      int TimerISR(void);
```

```
/*~~~~~~~~~~~~~~~~~~~~~~~~~~~~~~~~~~~~~~~~~~~~~~~~~~~~~~~~~~~~~~~~~~~~~~~~~~~~~*/
/*                          Functions                                       */
/*~~~~~~~~~~~~~~~~~~~~~~~~~~~~~~~~~~~~~~~~~~~~~~~~~~~~~~~~~~~~~~~~~~~~~~~~~~~~~*/
int TransitionOne(void)

#ifdef COMMENTS
/* T1: Initializes Digital I/O, Analog I/O, and Serial comms               */
/*     returns TRUE if all values initialized correctly                    */
#endif

{
/* initialize pointers  */
 SPIControlAddress      =           (unsigned char *)SPCR;
 SPIStatusAddress       =           (unsigned char *)SPSR;
 SPIDataAddress         =           (unsigned char *)SPDR;
 OptionRegisterAddress  =           (unsigned char *)H11OPTION;
 DDRCAddress            =           (unsigned char *)H11DDRC;
 ADControlAddress       =           (unsigned char *)H11ADCTL;
 MountStateAddress      =           (unsigned char *)MOUNT_STATE_ADDRESS;
 MountControlAddress    =           (unsigned char *)MOUNT_CONTROL_ADDRESS;
 DeltaXAddress          =           (unsigned char *)DELTAX_ADDR;
 DeltaYAddress          =           (unsigned char *)DELTAY_ADDR;
 IRLockAddress          =           (unsigned char *)IRLOCK_ADDR;
/* initialize registers */
 *OptionRegisterAddress =           OPTIONWD;        /* Power up A/D        */
 *DDRCAddress           =           COUTPUT;         /* Set Port C for Output*/
 *SPIControlAddress     =           SPI_ENABLE;      /* Init ser periph intfc*/
/* Insure no erroneous command exists on output ports                      */
 (*MountControlAddress).Azimuth = 0;
 (*MountControlAddress).Elevation = 0;
 if ((*OptionRegisterAddress  == OPTIONWD) &&
     (*DDRCAddress == COUTPUT) &&
     ( *SPIControlAddress      == SPI_ENABLE))
   return(TRUE);
 else
   return(FALSE);
}

/*~~~~~~~~~~~~~~~~~~~~~~~~~~~~~~~~~~~~~~~~~~~~~~~~~~~~~~~~~~~~~~~~~~~~~~~~~~~~~*/
int TransitionTwo(void)

#ifdef COMMENTS
/* T2:  conducts self test and returns TRUE if tests are passed.           */
#endif

{
int TestValue;
 ...
 return(TestValue);
}

/*~~~~~~~~~~~~~~~~~~~~~~~~~~~~~~~~~~~~~~~~~~~~~~~~~~~~~~~~~~~~~~~~~~~~~~~~~~~~~*/
int TransitionThree(void)

#ifdef COMMENTS
/* T3:  Initializes all variables, tokens and clears interrupts            */
/*      tests each value and returns TRUE if all are set correctly         */
#endif

{
unsigned int TempChar;

 MountStateAddress = (unsigned char *)MOUNT_STATE_ADDRESS;
 MountControlAddress = (unsigned char *)MOUNT_CONTROL_ADDRESS;
 PM             = FALSE;
 TM             = FALSE;
 SRMsg          = FALSE;
 SRMsgRx        = FALSE;
 MessageLength  = 0;
 MessageIndex   = 0;
```

```
ATTACH(TimerISR,TIMER_OVERFLOW_VECTOR);
ATTACH(SPIInterruptServiceRoutine,SPI_VECTCR);
TempChar        = *SPIDataAddress;                    /* Clear SPI buffer       */
*SPIControlAddress   =          SPI_ENABLE_INT; /* Init ser periph int  */
RETIE          = TRUE;
if (PM || TM || !RETIE || SRMsg || SRMsgRx || !PST)
   {
   printf("Semaphores not initialized correctly, failure in T3.\n");
   return(FALSE);
   }
else
   return(TRUE);
}


/*~~~~~~~~~~~~~~~~~~~~~~~~~~~~~~~~~~~~~~~~~~~~~~~~~~~~~~~~~~~~~~~~~~~~~~~~~~*/
void TransitionFour(void)

#ifdef COMMENTS
/* T4:   Sets the MODE semaphore to Position Mode or Track Mode as a        */
/*       result of the message received from the Surveillance Radar         */
/*       First Character after length indicates mode (index = 1)            */
/*       if RadarMessage[1] = 0, then Position Mode                         */
/*       if RadarMessage[1] = 1, then Track Mode                           */
#endif

{
 if (PositionMessage[MODE_INDEX] == 0)
   {
   PM = TRUE;
   TM = FALSE;
   }
 else if (PositionMessage[MODE_INDEX] == 1)
   {
   PM = FALSE;
   TM = TRUE;
   }
}


/*~~~~~~~~~~~~~~~~~~~~~~~~~~~~~~~~~~~~~~~~~~~~~~~~~~~~~~~~~~~~~~~~~~~~~~~~~~*/
void TransitionFive(void)

#ifdef COMMENTS
/* T5:   Reads and stores Azimuth angle of mount in T51                     */
/*       Reads and stores Elevation angle of mount   T51                    */
/*       Reads and stores Azimuth rate of mount      T52/T53                */
/*       Reads and stores Elevation rate of mount    T54/T55                */
/*       Reads and stores the current time           T51/T55                */
/*       Computes the Azimuth and Elevation control signals T55            */
/*       Outputs the Azimuth and Elevation Control signals   T55           */
/*                                                                          */
/* Azimuth and Elevation are 16 bits mapped 0 = 000 degrees                */
/*                                FFFF  = 359 degrees  .                    */
/* AzDot and ElDot are corrected for sign.                                 */
/* assuming 2.5 volts ( 0x7F ) is the zero value                           */
#endif

{
/* T51 */
 CurrentMount.Azimuth    = (*MountStateAddress).Azimuth;
 CurrentMount.Elevation = (*MountStateAddress).Elevation;
 CurrentMount.StartTime = RealTime;
 *ADControlAddress = ADCTLWD;          /* initialize A/D converters    */

/* T52 */
 while (!(*ADControlAddress & CCF));     /*Wait for conversion complete  */
```

```
/* T53 */
 CurrentMount.AzDot = (int)*AzDotAddress - ZERO_OFFSET;  /* read        */
 *ADControlAddress = ADCTLWD;               /* initialize A/D converters  */

/* T54 */
 while (!(*ADControlAddress & CCF));        /*Wait for conversion complete */

/* T55 */
 CurrentMount.ElDot = (int)*ElDotAddress - ZERO_OFFSET;  /* read        */
 CurrentMount.EndTime = RealTime;
 ComputeControl(&PreviousMount,&CurrentMount,&MountControl);
 (*MountControlAddress).Azimuth = MountControl.Azimuth;
 (*MountControlAddress).Elevation = MountControl.Elevation;
}

/*~~~~~~~~~~~~~~~~~~~~~~~~~~~~~~~~~~~~~~~~~~~~~~~~~~~~~~~~~~~~~~~~~~~~~~~~~~~~*/
 void TransitionSix(void)

#ifdef COMMENTS
/* Outputs zero control signals to both the azimuth and elevation       */
/* actuators.                                                           */
#endif

{
 (*MountControlAddress).Azimuth = 0;
 (*MountControlAddress).Elevation = 0;
}
/*~~~~~~~~~~~~~~~~~~~~~~~~~~~~~~~~~~~~~~~~~~~~~~~~~~~~~~~~~~~~~~~~~~~~~~~~~~~~*/
 void ComputeControl(struct MountStateType *PreviousMount,
                     struct MountStateType *CurrentMount,
                     struct MountControlType *MountControl)

#ifdef COMMENTS
/* Computes the values of the control values for the azimuth and        */
/* elevation axes using the current and previous positions and          */
/* velocities.                                                          */
/*                                                                      */
/* The control values are returned through the pointer to the mount     */
/* control variable.  They are not passed in this function to the       */
/* physical output device.                                              */
#endif

{
 ...
}

/*~~~~~~~~~~~~~~~~~~~~~~~~~~~~~~~~~~~~~~~~~~~~~~~~~~~~~~~~~~~~~~~~~~~~~~~~~~~~*/
 int CheckSum(...)

{
 ...
}

/*~~~~~~~~~~~~~~~~~~~~~~~~~~~~~~~~~~~~~~~~~~~~~~~~~~~~~~~~~~~~~~~~~~~~~~~~~~~~*/
 void SPIInterruptServiceRoutine(void)

#ifdef COMMENTS
/* Reads character from SPI data register and transfers it to buffer.   */
/* MessageIndex always points to last received character.               */
/* First byte received indicates number of characters to follow.        */
/* Does not contain code to check whether first charcter received is     */
/* from middle of message.                                              */
/* Last received byte is msg checksum not including length of message.  */
/* MessageLength does not include SOM or MessageLength characters        */
#endif

{
int LocalIndex;
```

```
if (EOM)                              /* if message complete, start another  */
  {
  MessageLength = 0;
  MessageIndex  = 0;
  EOM = FALSE;
  return;
  }
if ((MessageLength == 0) &&      /* Is it the first character of new msg?*/
    (*SPIDataAddress == START_OF_MESSAGE))
  {
  SOMReceived = TRUE;
  return;
  }
else if ((MessageLength == 0) && SOM_Received) /* Read msg length       */
  {
  MessageLength = *SPIDataAddress;
  return;
  }
else if (MessageLength <> 0)              /* Enter character into buffer  */
  {
  MessageIndex ++;
  IncomingPositionMessage[MessageIndex] = *SPIDataAddress;
  }
  if ((MessageIndex == MessageLength) && CheckSum())
    {
    for (LocalIndex = 1;
         LocalIndex < MessageLength;
         LocalIndex++)
         {
         PositionMessage[LocalIndex] = IncomingPositionMessage[LocalIndex];
         }
    SRMsgRx = TRUE;
    EOM = TRUE;
    }
}
/*~~~~~~~~~~~~~~~~~~~~~~~~~~~~~~~~~~~~~~~~~~~~~~~~~~~~~~~~~~~~~~~~~~~~~~~~~~~~*/
int TimerISR(void)

#ifdef COMMENTS
/* Keeps track of real time in global long int variable RealTime          */
#endif

{
...
}

/*~~~~~~~~~~~~~~~~~~~~~~~~~~~~~~~~~~~~~~~~~~~~~~~~~~~~~~~~~~~~~~~~~~~~~~~~~~~~*/
int main(void)

{
  TransitionOne();                        /* configure hardware          */
  while(!PST)
    PST = TransitionTwo();                /* Self-test                   */
  TransitionThree();                      /* Initialize all variables    */
  while (TRUE)                            /* Executive                   */
    {
    if(IntEn && SRMsgRx)
      TransitionFour();                   /* Set Mode Semaphore          */
    else if(PM && !SRMsgRx)
      TransitionFive();                   /* Move to position            */
    else if(PM && SRMsgRx)
      TransitionSix();                    /* Zero control                */
    else if(TM && !SRMsgRx && IRTrkLock)
      TransitionSeven();                  /* Video tracking              */
    else if(
      (TM && !SRMsgRx && !IRTrkLock) ||
      (TM && SRMsgRx))
      TransitionEight();                  /* Lost track lock             */
```

```
    else if(SRMsg && IntEn)
      TransitionNine();                    /* Read Character                    */
    else
    ...
            ;
    }                              /* end while                                 */
    return(0);                     /* Normal return to operating system         */
}                                  /* end main                                  */
```

Index

A/D converter, 12, 41, 57, 99, 121, 154, 155, 165, 183, 190, 196, 206, 233, 247, 256, 257, 262, 269, 279, 281, 284, 288, 453, 456, 460
Absolute Addressing, 317
Ada, 328
Address bus, 12, 230, 240, 269, 288, 361, 362, 368
Address space, 15, 16, 32, 55, 57, 74, 163, 211, 231, 233, 236, 240, 251, 252, 311, 314, 315, 317, 328, 334, 336, 338, 359, 395, 396
Addressing modes, 45, 47, 48, 71, 72, 212, 226, 232, 244, 245, 252, 255, 261, 272, 273, 304, 316, 318, 320, 331, 396, 397, 407
ALU, 4-6, 16, 18, 20, 26, 29, 175, 182, 185, 199, 266, 268, 310, 359, 375
Am29000, 140, 372, 454
Am29050, 303, 372, 372, 375, 381, 385, 401, 404, 405, 453, 454
AMD (Advanced Micro Devices), 454
Arc, 94, 96, 101
Architecture, 18, 21, 27-29, 35, 37, 41, 43, 47, 77, 88, 118, 140, 176, 184, 192, 198-200, 205, 209, 210, 232, 235, 237, 239, 251, 265, 270, 272, 288, 289, 304, 305, 313, 316, 318, 321, 325, 326, 328-330, 332, 336, 338, 339, 343, 355, 372, 375, 385, 396, 399
Arithmetic control register, 325
Arithmetic logic unit, 4, 5, 266, 375
ASCII, 45, 125, 131, 299
ASM-960, 339, 343
Assembler, 127, 145, 151-153, 158, 160, 162, 163, 169, 191, 231, 273, 317, 325, 339, 455, 466
Assembly language, 8, 48, 70, 71, 79, 127, 128, 130-132, 136, 137, 141-143, 145, 146, 147, 149, 150, 152, 153, 155, 156, 158-171, 231, 236, 265, 339, 455, 460, 463
Asynchronous, 8-10, 12, 28, 29, 31, 32, 35, 41, 78, 86, 141, 209, 210, 228, 236, 246, 256, 279, 289, 350, 385
Atomic instructions, 324
Autodecrement, 52, 318

Autoincrement, 52, 272, 318

Bandwidth, 10, 52, 116, 312, 392, 435
Barrel-shifter, 6
BASIC, 14, 15, 18, 29, 37, 45, 56, 57, 65, 71, 72, 118, 140-142, 155, 183, 195, 197, 232, 257, 262, 272, 304, 312, 322, 336, 397, 399, 400, 409, 454
Big endian, 16355, 471
BIOS, 15
Bootstrap, 14, 15, 36, 37, 231, 235, 239, 240
Branch instructions, 194, 245, 325, 326, 407
Branch prediction, 325, 326, 342
Branch prediction flag, 325, 326
Branch target cache, 373, 401, 404, 454
Breakpoint, 34, 67, 359
Burst mode, 435, 439
Bus, 12, 16, 18, 20, 21, 27, 28, 33, 41, 50, 53, 55-57, 62, 88, 174, 176, 184, 190, 192, 196, 200, 208, 211, 229, 230, 236, 240, 251, 252, 266, 268, 269, 282, 285, 286, 288, 289, 305, 307, 310, 312, 313, 324, 329, 331, 339, 342, 344, 347, 350, 353, 361, 362, 368, 369, 372, 392, 396-399, 405, 409, 417, 418, 419, 430, 432, 435, 437, 439, 441-443, 445, 446
Bus controller, 268, 305, 307, 313
Byte, 6, 9, 10, 35, 51, 60, 64, 72, 118, 166, 167, 184, 187, 190, 192, 194, 195, 199, 200, 211, 212, 226, 227, 229, 233, 239, 240, 251, 252, 260, 268, 270, 272, 273, 275, 285, 296, 307, 311, 313, 321, 322, 324, 329, 330, 331, 336, 344, 353, 355, 362, 370, 375, 395-397, 407, 418, 419, 454, 460, 469, 471, 472, 473, 478

C, 5, 6, 12, 60, 67, 94, 124, 127, 128, 131, 132, 136, 137, 140, 142, 142, 143, 143, 143, 145, 145, 146, 147, 147-149, 149-156, 159-171, 226, 228, 231, 235, 236, 240, 246, 276, 296, 324, 328, 339, 339, 342, 343, 385, 395, 443, 456, 459-461, 463, 465, 467, 469, 473, 474, 476
C-960 compiler, 339, 342

Cache, 15, 16, 28, 139, 304, 305, 307, 307, 308, 311-313, 329, 342, 343, 353, 359, 373, 398, 400, 401, 404, 407, 435, 454
Call instructions, 326, 331
Call mechanism, 331, 332
Carry flag, 185, 276
CDC 6600, 46
Central processing unit (CPU), 5, 16
Channel, 10, 13-15, 26, 41, 57, 61, 107, 108, 112-116, 156, 171, 183, 190, 194, 196, 231, 234, 247, 249, 256, 262, 279, 285, 289, 290, 312, 315, 343, 347, 392, 398-400, 405, 419, 430, 435, 437, 439, 453, 470
CHMOS, 265, 305, 344, 442
Clock, 10-12, 18, 20, 21, 24, 27, 29, 36, 41, 80, 88, 101, 109, 121, 138, 144, 183, 187, 190-192, 196, 206, 208, 210, 228, 231, 233, 234, 236, 237, 240, 246-248, 255, 256, 262, 289, 293, 302, 307, 310-312, 329, 342, 344, 350, 399, 435, 444, 470, 475
Clock cycle, 36, 289, 307, 310-312, 344, 435, 444
CMOS, 182, 183, 194, 207, 231, 235, 236, 288, 372, 392
Colored Petri nets (CPN), 92
Command File, 128, 131, 141, 143, 153, 163-165, 169, 465
Compiler, 48, 49, 84, 127, 131, 136, 140-143, 145, 150-154, 158, 160, 162-164, 166, 168, 169, 231, 262, 305, 339, 342, 347
Condition Code, 26, 212, 230, 232, 240, 325, 326, 381, 406, 407
Conditional branch, 226, 325, 326
Control, 3-6, 8, 12, 15, 16, 18, 20, 21, 21, 24, 28, 29, 31, 32, 34, 37, 45, 46, 51-53, 55, 60, 61, 65-69, 71, 72, 74, 78, 78-80, 84, 85, 89, 92, 93, 98, 99, 103, 104, 106-109, 112, 114, 116, 118, 121, 124, 125, 128, 130, 131, 137, 137-139, 141, 150, 152, 154, 157-159, 166-168, 175, 176, 182, 187, 190, 192, 195, 196, 198, 200,, 206, 208, 209, 211, 228, 230-232, 234-236, 239-241, 244-248, 252, 256, 257, 266, 269, 273, 276, 282, 285, 289, 304, 307, 308, 311, 313, 314, 318, 319, 325-328, 335, 336, 339, 344, 347, 353, 358, 359, 361, 368, 369, 385, 392, 395, 396, 398, 399, 404, 407, 409, 417-419, 430, 432, 439, 445, 449, 454, 455, 457, 461, 464, 469, 470, 474-479
Control signal generator (CSG), 21
Control-store, 21, 24
Controller, 4, 16, 18, 20, 33, 37, 41, 42, 52, 60, 61, 68, 74, 77-80, 84-88, 101, 106, 108, 109, 114, 125, 137-139, 141, 143, 147, 155, 170, 265, 266, 268, 290, 305, 307, 311-313, 315, 318, 335, 343, 347, 353, 355, 359, 361, 362, 369, 370, 392, 395, 398-400, 405, 409, 417-419, 435, 439, 444, 449, 453

COP, 236, 239, 248, 432
Core architecture, 304, 336
CPU, 5, 12, 13, 16, 27-29, 31-33, 35, 44, 52, 55, 61, 62, 176, 191, 192, 196, 197, 207, 210, 236, 246, 252, 261, 265, 266, 273, 288, 289, 297, 304, 305, 309, 310, 312, 313, 316, 318, 343, 344, 347, 353, 355, 359, 371, 372, 392, 395, 399, 404, 405, 417-419, 439, 442-444, 446, 453, 454
CPU32, 289
Cross-assembler, 127, 145, 160, 169, 191, 231
Cross-compiler, 127, 145, 150, 160, 162-164, 166, 169, 231
Cyclone Microsystems, 347

D/A converter, 13
Data bus, 18, 28, 56, 62, 211, 240, 252, 282, 288, 313, 339, 344, 372, 392, 397, 405, 417-419, 445, 446
Data cache, 305, 353, 359, 398, 404
Data controllers, 139
Data path, 174, 175, 307, 430, 435
Data types, 44, 51-53, 55, 74, 142, 270, 321, 322, 324, 355, 375
Debugging, 34, 147, 156-158, 165, 282, 326, 339, 342, 343, 347, 371, 395, 435
Decoder, 21, 199, 318, 344, 368, 417-419, 437, 442, 443
Destination, 24, 46, 72, 241, 309, 310, 312, 318, 321, 324, 326, 376, 385, 398
Development tools, 191, 269, 339, 343
Digital signal processing, 3, 62, 266, 385, 392, 399
Direct addressing, 48, 195, 211, 212, 227, 232, 245, 273
Direct memory access (DMA), 16, 33, 88, 312
Directed graph, 81, 93
Directive, 145, 150, 152, 154, 158, 159, 161, 163, 168
Displacement, 317-320, 325, 331, 397
DMA, 16, 33, 37, 42, 88, 249, 284, 290, 305, 312, 313, 315, 335, 343, 347, 353, 369, 392, 395, 398-400, 405, 409, 419, 435, 437, 439, 453
DMA controller, 16, 33, 88, 290, 312, 315, 335, 343, 347, 392, 395, 398-400, 405, 409, 419, 435, 453
Double-precision, 304, 310, 375, 392, 395
Doubleword, 270, 397
DRAM (dynamic RAM), 353, 361
DRAM controller, 361, 362, 369, 370, 405, 417, 418, 444, 453
Dynamic RAM, 15, 16, 353, 361

Effective address, 5, 16, 44, 48, 49, 53, 212, 311, 318, 320
Embedded controller, 265, 318, 353, 355, 405
80188, 339
8088, 339

8096, 8, 288
80960, 303, 303-305, 310, 314, 316, 318,
  321, 336, 338, 339, 342-344, 347, 353, 430
80960CA, 303, 304, 304, 305, 307, 310-313,
  315, 316, 318, 322, 324-334, 336, 342-344,
  347, 353, 401, 404-409, 417, 419, 441,
  443, 444, 453, 454
80960KA, 304, 336, 339
80960KB, 304, 336, 339, 343, 453
80960MC, 304, 336, 338, 454
80960SA, 338, 353
80960SB, 338, 339
8096BH, 269
80C196KB, 265, 276, 286, 288, 297
80C196KC, 265, 286, 291, 453
83C196KC, 269, 453
87C196KC, 269, 453
EPROM, 14, 14, 29, 36, 37, 41, 148, 149,
  192, 231, 232, 235-237, 249, 268-270, 285-
  287, 293, 344, 347, 350, 353, 418, 439,
  441-443, 453
Event controllers, 136, 139, 142, 143, 169
Exception, 182, 232, 262, 269, 313, 331,
  358, 359

Facsimile (FAX), 9
Facsimile accelerator module (FAM), 62
Fault, 35, 326, 327, 336, 407
FIFO (first-in, first-out), 344
Finite-state machine (FSM), 78
Firmware, 15, 1
Flag, 31, 67, 98, 106, 110, 114, 121, 154,
  159, 184, 185, 190, 248, 251, 256, 275,
  276, 296, 325-327
Flag-driven transition, 114
Floating-point, 45, 145, 305, 310, 318, 336,
  339, 342, 374, 375, 385, 392, 395, 397,
  405, 454
FORTH, 60, 141, 141, 142
FORTRAN, 128, 142
Frame pointer, 313, 330
Framing error, 229, 246
FSM, 78-81, 83, 84, 86, 92, 94-96, 103
Full-duplex, 246, 256
Function, 3, 18, 33, 35-37, 42, 44, 45, 61,
  78-81, 83-86, 94, 98, 108, 131, 141, 143,
  145, 146, 149, 150, 154, 156, 157, 159,
  160, 161, 166, 167, 171, 232, 248, 251,
  252, 261, 266, 269, 273, 289, 313, 335,
  338, 339, 392, 430, 432, 437, 460, 470,
  475, 478

Global registers, 275, 310, 313, 318, 319,
  331, 376, 381
Graph, 79, 81, 93, 101

Half-duplex, 256
Handshaking, 28, 29, 32, 84-86, 289
Hard real-time, 139, 140, 145
HCMOS, 236, 289
Heap, 164, 465, 466
High-level language (HLL), 128

Hitachi, 13, 37, 53, 60, 208, 1
Horizontal microcoding, 24
Host, 131, 139, 191, 343, 347, 371, 430,
  432, 439

I/O, 3, 4, 8, 9, 11-13, 27-29, 31, 32, 35-37,
  45, 51-53, 56, 57, 60-65, 74, 98, 118, 120,
  128, 140, 142, 143, 147, 158, 164, 168,
  175, 176, 183, 185-187, 190-192, 194, 195,
  198, 200, 205-212, 226, 230-232, 235, 236,
  237, 239, 240, 245-247, 249, 252, 257,
  269, 276, 284, 289, 312, 342, 347, 353,
  372, 396, 398, 405, 419, 430, 435, 439,
  441, 443, 448-450, 453, 466, 467, 475
I/O port, 32, 53, 62, 63, 98, 142, 143, 168,
  183, 187, 200, 210, 211, 247, 249, 312,
  448
IBM 360, 46, 51, 52
IBM 370, 52
ICE, 37
IEEE, 152, 304, 336, 344, 347, 375
Immediate addressing, 48, 212
In-circuit emulator (ICE), 37
Indexed addressing, 48, 212, 252, 397
Indirect addressing, 52, 195, 273, 321
Initialization, 10, 11, 149, 154, 157, 184,
  307, 312, 314, 316, 336, 339, 430, 432,
  466
Input, 4, 6, 8, 9, 11-13, 15, 29, 31, 32, 34,
  52, 53, 61, 70, 81, 86, 93-97, 101, 108,
  109, 113, 116, 125, 137, 146, 147, 169,
  171, 175, 176, 182, 191, 194, 196, 197,
  205, 206, 208, 209, 211, 227, 228, 230,
  233, 234, 235, 237, 245-248, 252, 255,
  256, 262, 268, 276, 281, 282, 289, 293,
  296, 311, 312, 334, 335, 355, 361, 370,
  398, 405, 419, 442, 463, 467, 469, 475
Instruction cache, 304, 305, 307, 308, 312,
  343, 353, 359, 398, 435
Instruction format, 27, 44, 56, 385, 398
Instruction register (IR), 16, 374
Instruction scheduler, 305, 307, 308
Instruction set, 3, 18, 21, 31, 43-45, 45, 47,
  49, 51-53, 61, 71, 74, 140, 182, 183, 187,
  192, 195, 198, 200, 200, 205, 209, 226,
  231, 232, 236, 244, 252, 265, 276, 290,
  304, 310, 322, 324, 342, 355
Integer, 6, 8, 29, 45, 46, 52, 65, 145, 146,
  154, 161, 167, 168, 236, 262, 270, 304,
  305, 308, 310, 311, 318, 321, 322, 324,
  325, 326, 374, 375, 385, 392, 397, 406,
  407, 453, 454
Integer execution unit, 305, 310, 311
Intel, 8, 9, 13, 27, 29, 29, 32, 34, 47, 61,
  139, 143, 155, 174, 209, 212, 249, 249,
  252, 265, 265, 266, 269, 272, 303, 304,
  304, 338, 342-344, 347, 409, 417, 418,
  442-444, 448,, 455, 463
Interrupt, 12, 13, 31-34, 36, 41, 68, 86-89,
  91, 92, 98, 99, 101, 103, 106, 110, 112-
  114, 121, 124, 126, 137-140, 164, 165,
  166, 167, 171, 184, 185, 187, 190, 191,

194, 196-198, 200, 205-208, 211, 212, 226,
    228, 230, 232-236, 240, 241, 245-249, 252,
    255-257, 260, 266, 268, 269, 275, 276,
    282, 284, 285, 289, 296, 305, 307, 311-
    313, 315, 328, 332-336, 343, 375, 385,
    392, 395, 396, 399, 404, 405, 407, 408,
    409, 435, 437, 439, 445, 446, 450, 453,
    466, 475
Interrupt controller, 266, 305, 311, 335,
    343, 405, 453
Interrupt handling routine, 407
Interrupt service routine (ISR), 32, 86, 88,
    99, 137
Interrupt vector, 234, 284, 332-335, 408
Interrupt-driven transition, 113
INTROL Corporation, 127, 1
ISR, 32, 86-89, 91, 92, 99, 137, 140, 165-
    167, 230, 245, 246, 395

Kermit, 131, 131, 151
Kernel, 131, 139, 140, 343, 347, 359
Kernel mode, 359

Limits.h, 148
Line, 9-11, 16, 29, 32, 33, 69, 125, 143,
    145-147, 149-154, 159, 162-165, 175, 176,
    211, 227, 229, 256, 276, 307, 307, 342,
    344, 353, 368, 398, 419, 435, 439, 443,
    446, 448, 466, 467
Linkage, 313, 328, 330, 407
Linker, 48, 127, 128, 131, 135, 143, 153,
    160, 163-165, 169, 339, 465
Little endian, 355
Load, 23, 31, 34, 41, 51, 63, 131, 138, 140,
    152, 182, 191, 205, 207, 211, 212, 230,
    239, 266, 304, 308-311, 314, 317, 318,
    320-322, 324, 342, 355, 374, 397, 419,
    454, 459
Local calls, 331, 408
Local registers, 275, 310, 311, 313, 318,
    328-330, 334, 376, 381
Locked, 140, 286, 307, 312
Long integer, 161, 167, 270, 322
LR33000, 303, 353, 355, 358, 359, 361, 362,
    369-371, 400, 405, 453, 454
LSI logic, 303, 353, 371, 405

M68000, 288, 289
M68300, 288, 290
M68HC11, 36, 114, 154, 155, 157, 1
Macro 18, 21, 29, 143-146, 150, 151, 153,
    162, 170, 171, 339
Macro assembler, 151, 153, 339
Main memory, 15, 16, 18, 28, 265, 266,
    268, 304, 307, 311, 312, 343, 372, 401,
    453, 454
Mapping ROM, 23, 42, 47
Mark, 9, 162, 210, 1
Marking, 9, 94, 101, 229
Maskable interrupts, 33, 89, 232, 236, 312
MC68000, 288

MC68008, 339
MC68010, 288, 290
MC68332, 265, 288, 288, 289, 289, 290, 296
MC68340, 290
MCS-96, 61, 63-65, 67, 68, 265, 266, 269,
    270, 272, 276, 279, 288
MCU, 13, 31, 35-37, 41, 43, 53, 60, 61, 65,
    68, 70, 72-74, 77-79, 86, 87, 91, 92, 96,
    98, 101, 103, 106-109, 113, 114, 116, 118,
    120, 125-128, 131, 137-140, 142, 143, 149,
    154-156, 158, 169-171, 173-176, 182, 183,
    187, 190-192, 196-198, 207, 209, 226, 228-
    230, 233, 237, 239, 246-249, 256, 262
Memory, 3, 4, 8, 9, 13-16, 18, 20, 21, 27-
    29, 31-34, 36, 37, 41, 45-48, 50, 52, 53,
    55-57, 60-63, 65, 67, 74, 85, 88, 110, 120,
    121, 128, 131, 132, 136, 139-142, 148,
    149, 153, 154, 160, 163, 164, 167, 174,
    176, 182, 184, 185, 187, 191, 192, 194,
    195, 197, 199, 205-208, 210-212, 227, 230,
    232, 236, 237, 239, 240, 244, 245, 249,
    251, 252, 255, 257, 260, 261, 265, 266,
    268, 269, 272, 273, 275, 278, 285, 286,
    289, 297, 302, 304, 305, 307-315, 317,
    318, 320, 322, 324, 328, 329, 331, 334,
    336, 342-344, 350, 353, 359, 362, 368-370,
    372, 373, 375, 395-398, 400, 401, 404,
    407, 408, 417, 419, 435, 437, 439, 441,
    443, 444, 446, 453, 454, 456, 459, 466
Memory address register (MAR), 18
Memory data register (MDR), 18
Memory-mapped I/O, 8, 32, 74, 120, 232,
    236, 237, 252, 257, 342, 396, 443
Memory-to-memory, 265
Microcoding, 24
MIPS (microprocessor without interlocked
    pipeline stages), 454
MIPS Computer Systems Co., 353
MIPS R2000, 353
Motorola, 7, 8, 29, 32, 36, 60, 61, 106, 118,
    127, 132, 138, 173, 174, 209, 209, 212,
    231, 232, 235, 236, 237, 239, 241, 249,
    251, 252, 255-257, 262, 265, 288, 288,
    296, 439, 455, 460, 463, 465
Multiprocessing, 324
Multiprocessor, 229

National Semiconductor, 8, 55, 60, 61, 175,
    198, 198, 206, 208, 262, 303, 385, 385
Negative flag, 275
Nonmaskable Interrupts, 33, 68, 230
NS32000, 55-57, 61-63, 65, 67, 68, 385,
    399, 400
NS32CG16, 399
NS32CG160, 399
NS32FX16, 8, 61, 399
NS32GX32, 399, 400
NS32GX320, 399, 400
NS32SF640, 303, 385, 400, 453, 454
Numerics architecture, 304

Object File, 131, 150-152, 162, 163
Offset, 48, 67, 125, 205, 227, 233, 261, 262, 317, 321, 336, 381, 460, 473, 477
OPCODE, 21, 23, 44, 46, 47, 50, 53, 57, 162, 276, 284, 318, 385, 397
Operand, 5-7, 44, 46-48, 65, 67, 72 75, 155, 156, 187, 195, 205, 212, 245 268, 272, 307, 309, 316-318, 322, 325, 331, 355, 376, 381, 385, 392, 397, 398
Operating system, 128, 131, 139, 147, 148, 163, 311, 328, 395, 472, 479
Operator, 5, 35, 51, 72, 88, 137, 430
ORG, 158, 163, 169, 456
Organization, 18, 27, 28, 376
Output, 4-6, 8-13, 15, 20, 24, 29, 31, 32, 36, 44, 52, 61, 69, 70, 79-81, 83, 85, 86, 93, 94, 97, 98, 101, 108, 112, 113, 114, 115, 125, 128, 137, 143, 144, 147 151-153, 162-164, 168, 175, 176, 182, 186, 196, 199, 200, 205, 206, 208, 209 211, 227, 228, 231, 235, 237, 246, 248 252, 256, 262, 268, 276, 278, 281, 282 289, 293, 302, 361, 417-419, 430, 435, 442, 445, 446, 456, 461, 467, 473, 476 478
Overflow, 31, 33, 65, 157, 205, 206, 226, 228, 234, 245, 247, 248, 251, 255 256, 275, 276, 296, 310, 326, 329, 395 407, 469, 473, 476
Overflow flag, 31, 275, 326
Overrun error, 256

Parallel, 3, 6, 8, 10, 11, 18, 28, 34, 35, 53, 56, 68, 70, 84, 92, 96, 125, 168, 194, 206, 207, 210, 235, 246, 304, 307, 308 310, 311, 347, 405, 441, 448, 449
PC, 3, 15, 16, 18, 20, 21, 46, 48, 49, 67-70, 127, 131, 144, 184, 199, 205, 207 208, 212, 227, 232, 252, 255, 257, 260 261, 266, 268, 318, 342, 343, 355, 395
PC relative addressing, 48, 49
PDP-11, 45, 47, 52, 142, 1
Peripheral, 3, 12, 32, 33, 37, 56, 166, 171, 173, 175, 190, 192, 195, 196, 206 207, 233, 236, 237, 246, 247, 249, 260 261, 262, 266, 269, 276, 284, 289, 312 314, 396, 449
Petri net (PN), 78
Phoenix Technologies Ltd., 347
Physical address, 15, 16, 48, 338, 375
Pin, 36, 118, 176, 187, 197, 198, 207, 230, 235, 246, 248, 251, 255, 260, 265 270, 273, 276, 285, 289, 312, 313, 331 334, 335, 344, 362, 370-372, 375, 392, 409, 419, 443
Pipeline, 27, 50, 52, 307-311, 353, 359, 374, 375, 392, 404, 405, 407, 435, 454
PLA, 175
Places, 93, 94, 96-98, 101, 103, 106, 109, 118, 160, 207, 439, 470
PMU, 12, 32
PN, 78, 92, 94, 96, 98, 103

Pocket Rocket, 371
Pointer, 12, 48, 142, 146, 153, 154, 160, 161, 166-168, 184, 185, 187, 192, 195, 212, 232, 233, 241, 244, 251, 252, 260, 261, 269, 273, 313, 318, 325, 329-331, 376, 381, 385, 395, 435, 470, 475, 478
Port, 9-13, 32, 34, 36, 53, 57, 62, 63, 92, 98, 125, 135, 142-144, 153, 154, 159, 168, 183, 187, 190, 196, 200, 206, 210, 211, 228-230, 235, 245-249, 252, 256, 279, 281, 282, 299, 305, 310, 312, 347, 374, 419, 435, 437, 448, 453, 456, 459-461, 474, 476
Prefetch queue, 268
Primary memory, 33
Priority, 33, 98, 103, 166, 235, 249, 257, 268, 284, 289, 307, 327, 328, 333, 335, 395, 399, 407, 408, 432, 437
Procedure, 6, 140, 141, 145, 299, 310, 313, 326, 328-336, 339, 342, 405, 408
Procedure linkage, 330
Process, 34, 37, 41, 68, 70, 74, 78, 78, 80, 84, 92, 98, 102-106, 110, 112, 113, 128, 130, 135-140, 145, 163, 165, 167, 170, 205, 234, 247, 288, 302, 314, 327, 328, 333, 336, 343, 347, 392, 405, 407, 430, 435, 439, 446
Process control, 68, 74, 78, 78, 327, 328, 336, 347, 407
Process control register, 327, 328, 407
Program counter (PC), 16, 46, 232, 266, 355
Programmable logic array (PLA), 175
Protected architecture, 304, 336
Pulse width modulation (PWM), 61, 276

Quadword, 317, 397

R2000, 353
R3000, 353
RAM, 14-16, 28, 29, 37, 57, 61, 65, 149, 158, 159, 163, 164, 175, 176, 183, 184, 191, 192, 195, 197-200, 205, 207, 208, 209, 230-233, 235, 237, 239, 240, 249, 251, 252, 255, 257, 260, 269, 273, 286, 287, 289, 293, 302, 305, 307, 310, 311, 329, 330, 334-336, 339, 343, 344, 347, 353, 361, 407, 417, 453, 456, 465, 466
Real-time output ports (RTOP), 12
Register file, 266, 269, 272, 273, 275, 304, 305, 310, 311, 359, 372, 374, 405, 407
Register-to-register, 265, 308, 405, 454
Relocatable, 131, 150, 153, 162, 163
Return instruction, 313, 320, 325, 331, 336
Return mechanism, 313, 326, 328, 331, 334
ROM, 14, 15, 21, 23, 24, 28, 29, 37, 41, 42, 47, 57, 60, 136, 137, 139, 141, 149, 158, 163, 174, 175, 183, 184, 187, 191, 192, 198, 199, 205, 207-209, 230-232, 235-237, 239,, 249, 252, 257, 268-270, 285, 287, 302, 305, 312, 339, 342, 344, 347, 453, 466

RTOP, 12

S-Record, 132
Scoreboarding, 309
Scoreboarding register, 309
Secondary memory, 33
Sections, 4, 6, 16, 35, 128, 136, 140, 151,
    152, 155, 156, 158, 163, 273, 339, 400,
    466
Segment, 135, 146, 208
Semaphore, 99, 101-103, 106, 109, 110,
    113, 114, 118, 126, 156, 166, 167, 475,
    477, 479
Semaphore-driven transition, 113
Serial, 8-11, 28, 35, 41, 64, 86, 91, 92, 98,
    116, 166, 171, 173, 183, 186, 187, 190,
    191, 195-198, 200, 206, 207, 209, 210,
    211, 227-231, 239, 240, 246, 247, 252,
    256, 257, 262, 279, 289, 297, 302, 342,
    344, 347, 350, 371, 439, 441, 448, 450,
    453, 475
Set, 3, 6, 12, 16, 18, 20, 21, 23, 31, 34, 36,
    43-45, 45, 47, 49, 51-53, 55, 61, 65, 68,
    71, 74, 79-81, 86, 94, 98, 102, 103, 109,
    113, 114, 121, 124-126, 139-142, 150, 151,
    156-159, 163, 166, 167, 175, 176, 182,
    183, 185, 187, 192, 195, 198, 200, 200,
    205, 208, 209, 212, 226, 228, 230-233,
    236, 237, 239, 240, 244-248, 251, 252,
    255, 256, 261, 265, 275, 276, 284, 290,
    304, 305, 307, 309, 310, 311, 313, 321,
    322, 324-328, 330, 333-335, 339, 342, 344,
    355, 373, 398, 401, 407, 417, 432, 446,
    456, 457, 461, 465, 466, 473, 476, 479
Short integer, 270, 322
Single-precision, 304, 392, 395
Slushware, 15, 37
Soft real-time, 139
Software, 4, 9, 32-34, 37, 40, 43, 64, 68,
    77, 77, 78, 79, 79, 80, 83, 86, 88, 89, 92,
    93, 96, 98-101, 103, 104-106, 109, 110,
    112, 114, 118, 121, 127, 128, 130, 132,
    133, 137, 139, 141-143, 147, 156, 158,
    169, 171, 183, 187, 190, 191, 194, 196,
    206, 208-211, 226, 228-237, 239, 245, 246,
    247-249, 255-257, 260, 262, 282, 302, 307,
    311, 326, 328, 332, 339, 342-344, 347,
    353, 359, 370, 372, 376, 379, 385, 398,
    399, 417, 437, 444, 446
Source, 4, 24, 27, 46, 69, 70, 85, 96-98,
    101, 106, 130-132, 143, 147, 149-152, 154,
    156, 160, 162, 163, 190, 205-207, 228,
    233, 241, 262, 309, 310, 312, 313, 318,
    319, 321, 324, 328, 334, 335, 342, 347,
    376, 381, 385, 398, 455, 460, 463
Source file, 132, 150, 152, 162
Space, 2, 9, 15, 16, 27, 32, 41, 55, 57, 74,
    140, 144, 158, 163, 164, 210, 211, 231,
    233, 236, 240, 249, 251, 252, 265, 268,
    269, 273, 303, 304, 311, 314-317, 328,
    329, 334, 336, 338, 359, 395, 396, 441,
    466

Stack, 33, 44, 47, 52, 68, 74, 138, 140, 141,
    143, 145, 158, 160, 161, 164, 166, 171,
    176, 184, 185, 187, 192, 195, 198, 199,
    200, 207, 212, 226, 232, 233, 244, 251,
    252, 255, 260, 261, 269, 273, 291, 293,
    296, 313, 328-332, 334, 336, 342, 376,
    381, 395, 407, 456, 458, 459, 466
Stack frame, 313, 328-331, 334, 407
Stack pointer, 160, 161, 184, 187, 192, 195,
    212, 232, 233, 244, 251, 252, 260, 261,
    269, 273, 313, 329, 376, 381, 395
State Table, 83, 85, 87, 121, 130
Static RAM, 184, 289, 311, 417
Store, 14, 15, 20, 21, 24, 41, 46, 51, 60, 62,
    63, 74, 113, 131, 140, 141, 159, 208, 209,
    273, 293, 304, 310, 311, 318, 320, 321,
    322, 326, 342, 355, 362, 373, 374, 395,
    397, 406, 439, 454, 457-459, 464, 470
Superscalar, 37, 303, 304, 305, 318,
    353, 385, 435, 453
Superscalar processor, 304, 305, 318
Supervisor call, 331
Supervisor mode, 296, 314, 315, 327, 328,
    395, 407
Symbol table, 153, 160, 162
Synchronous, 8, 10, 11, 28, 29, 31, 35, 41,
    80, 166, 187, 196, 198, 206, 207, 210,
    228, 236, 246, 247, 256, 279, 289, 385
System calls, 147, 331, 332, 342, 408
System controllers, 139
System procedure table, 331

Tadpole Technology Inc., 347
Target, 3, 115, 116, 118, 125, 127, 128,
    131, 132, 143, 145, 147, 148, 150-153,
    164, 169, 231, 325, 331, 339, 342, 343,
    347, 373, 401, 404, 454
Timer, 13, 31, 35, 61, 69, 71, 74, 109, 113-
    115, 138, 167, 169, 176, 183, 187, 191,
    195-197, 205, 207-209, 211, 227, 228, 233-
    236, 239, 247, 248, 252, 255-257, 262,
    276, 278, 287, 289, 293, 343, 353, 385,
    392, 395, 399, 405, 441, 445, 446, 448,
    449, 469, 470, 473, 476
TMS1000, 174, 175, 176, 182, 184, 185, 198
Token, 92, 94-98, 101, 103-106, 109, 110,
    112-114, 118, 121, 145, 150
Transition, 9, 79-81, 83, 86, 94-97, 101-103,
    105, 106, 109, 112-114, 118, 121, 124,
    125, 156, 157, 166, 170, 198, 206, 230,
    248, 255, 282
Trap, 67, 110, 262, 276, 282, 284, 381, 385,
    395

UART, 10, 86, 87, 92, 249, 252, 256, 289
UNIX, 142, 147, 342, 343, 347
User mode, 2, 315, 327, 328, 359, 381

VAX, 52, 318, 343
Vector, 8, 52, 62, 65, 94, 155, 191, 234,
    245, 282, 284, 312, 332-336, 385, 395,
    399, 408, 456, 457, 461, 462, 473, 476

Vertical microcoding, 24
Vertical register windowing, 269
Virtual address, 15, 32, 55, 57, 60, 314, 336
VME, 347, 350, 430, 432, 435, 437, 439
VMS, 343
VxWorks OS, 343

Wait-State, 29, 370
Watchdog timer, 35, 61, 236, 239, 248, 257, 262, 441, 445, 446

Word, 52, 53, 55, 60, 64, 67, 72, 77, 83, 105, 108, 131, 158, 159, 173, 185, 251, 266, 269, 270, 272, 273, 275, 291, 293, 296, 307, 310-313, 317, 320, 322, 323, 329, 344, 355, 361, 375, 396, 397, 407, 417, 418, 455, 457, 461, 464, 474

Zero flag, 275

ABOUT THE AUTHORS

Kenneth J. Hintz is an associate professor of electrical and computer engineering at George Mason University, Fairfax, Virginia. He has worked extensively in the application of microprocessors and microcontrollers to the real-time analysis of electronic warfare signals, as well as the control of tracking mounts and other government R&D controls systems. His current research is in the areas of genetic algorithms, infrared image analysis, and neural network controllers. He earned his Ph.D. in electrical engineering at the University of Virginia in 1981 and is a senior member of IEEE and a member of SPIE and the European Neural Network Society.

Daniel Tabak is a professor of electrical and computer engineering at George Mason University, Fairfax, Virginia. He has conducted research and published extensively in the areas of computer-based applications, process control, computer architecture, multiprocessors, and microprocessors. He is the author of *Advanced Microprocessors* (McGraw-Hill, 1991), and an associate editor of the journals *Automatica, Engineering Applications of AI, Journal of Microcomputer Applications,* and *Microprocessors and Microsystems*. Dr. Tabak is a senior member of the IEEE, a member of ACM, and director of Euromicro, representing the United States.